Research on Intelligent Manufacturing

Research on Intelligent Manufacturing (RIM) publishes the latest developments and applications of research in intelligent manufacturing—rapidly, informally and in high quality. It combines theory and practice to analyse related cases in fields including but not limited to:
Intelligent design theory and technologies

Intelligent manufacturing equipment and technologies
Intelligent sensing and control technologies
Intelligent manufacturing systems and services

This book series aims to address hot technological spots and solve challenging problems in the field of intelligent manufacturing. It brings together scientists and engineers working in all related branches from both East and West, under the support of national strategies like Industry 4.0 and Made in China 2025. With its wide coverage in all related branches, such as Industrial Internet of Things (IoT), Cloud Computing, 3D Printing and Virtual Reality Technology, we hope this book series can provide the researchers with a scientific platform to exchange and share the latest findings, ideas, and advances, and to chart the frontiers of intelligent manufacturing.

The series' scope includes monographs, professional books and graduate textbooks, edited volumes, and reference works intended to support education in related areas at the graduate and post-graduate levels.

More information about this series at https://link.springer.com/bookseries/15516

Jiandong Tian

All Weather Robot Vision

華中科技大學出版社
Huazhong University of Science and Technology Press

Springer

Jiandong Tian
State Key Laboratory of Robotics,
Shenyang Institute of Automation
Chinese Academy of Sciences
Shenyang, Liaoning, China

ISSN 2523-3386 ISSN 2523-3394 (electronic)
Research on Intelligent Manufacturing
ISBN 978-981-16-6428-1 ISBN 978-981-16-6429-8 (eBook)
https://doi.org/10.1007/978-981-16-6429-8

Jointly published with Huazhong University of Science and Technology Press
The print edition is not for sale in China Mainland. Customers from China Mainland please order the print book from: Huazhong University of Science and Technology Press.

This Springer imprint is published by the registered company Springer Nature Singapore Pte Ltd.
The registered company address is: 152 Beach Road, #21-01/04 Gateway East, Singapore 189721, Singapore

Contents

Chapter 1
Introduction

Abstract In outdoor applications, there is no escape from complex illumination such as shadows, highlight, or dynamic lighting during a day and bad weather conditions with poor visibility such as rain, snow, or fog. These wide ranges of illumination and weather conditions usually change the pixel value distribution of images, which affects the basis of robot vision algorithms and influences some basic capabilities of a robot such as vSLAM, autonomous navigation, visual servoing, and environmental understanding. In this chapter, we first analyze how complex illumination and bad weather conditions affect images from the perspective of atmospheric physics and imaging optics. We then summarize and review the basic principles of color matching function and scene reproduction. We also summarize the key technologies for all-weather robot vision including lights, scattering transmission, surface reflection, and digital camera imaging modeling. Finally, we outline the main contents and structure of this book.

1.1 Robot Vision Overview

Robot vision is a system that provides visual perception for robots. It is widely used in sea, land, and sky robots, as shown in Fig. 1.1. Robot vision systems usually use cameras as tools and images as mediums. They can extract useful information for robots. This enables the robots to achieve localization and navigation, path planning, obstacle avoidance, environmental understanding, target detection, and so forth. Similar to the importance of the eyes to humans, robot vision plays a significant role in the composition of the robot system.

The analysis and processing of outdoor scene images have always been one of the research focuses of robot vision and its related tasks such as image segmentation, feature extraction, target recognition, target tracking, and scene understanding. After decades of development, robot vision technologies have attained remarkable achievements. With the development of deep learning in recent years, some vision tasks such as detection, recognition, scene understanding, tracking, navigation, and visual servo show great improvements. However, the problems it faces cannot be ignored. One of the important issues is how to improve the stability and reliability of a robot vision

J. Tian, *All Weather Robot Vision*, Research on Intelligent Manufacturing,
https://doi.org/10.1007/978-981-16-6429-8_1

Fig. 1.1 Robot vision is widely used in various robots

system in complex illumination and weather conditions. Pixel information, such as color and brightness of an image and various image features extracted from it, (such as angle point, edge, gradient, and texture) is the basis of robot vision. Changes in lighting and weather alter the color and brightness information of image pixels, thereby fundamentally reducing the performance and robustness of vision algorithms. For example, when there are shadows, reflections, lighting changes, or in scattering weather conditions, existing vision algorithms still lack robustness and adaptability. Robot vision systems usually do not work in lighting environments that are manually controllable like the machine vision, which can obtain clear images through good lighting conditions and overcome problems such as shadows and reflections. Robot vision needs to face dynamic and unstructured natural environment. This requires image preprocessing technologies under various complex natural illumination and weather conditions. These are also the main concerns of this book. The solution to these problems will promote the development of robot vision and has abroad application prospects.

1.2 Robot Vision in Complex Illumination and Bad Weather

The outdoor light varies significantly with different weather and time. As shown in Fig. 1.2, the appearance of visible imaging highly depends on the illumination conditions. For example, the light is close to white at noon on a sunny day and golden yellow at sunrise or sunset. Some illumination phenomena, such as shadow and highlight, also have close relation to lighting conditions. While these complex illumination variations bringing us a colorful world and various beautiful images,

Fig. 1.2 Beautiful images under different illumination and weather conditions

Fig. 1.3 Photographs taken from the same scene at different illumination and weather conditions look very different from each other

they will greatly affect the appearance of objects and bring many negative effects to the research and application of computer vision. It degrades the image quality and contrast, blurs the objects and scenes, and then affects the performance and robustness of some vision algorithms, such as feature extraction, object recognition, and tracking [4, 9]. Adini et al. [1] used experiments to show that the difference between the same face under different illumination is sometimes greater than that between different faces under the same illumination. Figure 1.3 also shows that images taken for the same scene under different illumination and weather conditions vary a lot. For these reasons, image illumination modeling and illumination processing are of great practical significance and have attracted a lot of attention in recent years.

According to how the illumination changes the image, the illumination variation on an image can be classified as global and local illumination variation. A majority of methods for global illumination variation processing are designed for single light source. Compared with global illumination variations, the local illumination variations on an image (shadow and highlight reflection) are more challenging and have much greater influence in real applications. For local illumination processing, in previous methods, a lot of efforts have been made to design some handcrafted

(a) (b)

Fig. 1.4 Decreased image qualities under dynamic weather conditions like rain, snow, and fog

illumination-invariant features, or some learning-based methods are applied to reduce the bad effect of illumination variation on robot vision tasks. The main problem with these methods may be that they are purely based on image data, without considering the inherent physically imaging mechanism. However, as physical phenomenons observed in most natural scenes, the image illumination variations are determined by the complex interactions of illumination, geometry, and reflectance. In this book, the research on illumination processing, starting with outdoor lighting modeling, spectral reflection calculation, and camera imaging simulation, puts forward a new outdoor light source spectral calculation method and image illumination processing algorithms.

For robot vision systems, it is also required for them to work normally in bad weather. Bad weather is generally classified into two categories: static bad weather (e.g., fog) and dynamic bad weather (e.g.,rain and snow). Static bad weather is usually caused by small floating particles in the air, which mainly reduce the sharpness of the image, as shown in Fig. 1.4a. The reduction of image sharpness under dynamic bad weather is usually caused by large moving particles, which make the image partly obscured, as shown in Fig. 1.4b. Therefore, the restoration of the image under static bad weather conditions is mainly to improve the clarity of the image, while the restoration of the image under dynamic bad weather conditions is mainly to complement the part that is occluded by the particles.

With the rapid development of deep learning technology in recent years, task-oriented researches in vision such as visual servoing, target recognition, tracking, and image understanding have made remarkable progress. However, it should be noted that these vision algorithms are developed for images of good quality and stable illumination conditions (as shown in Fig. 1.5c). In fact, robots often need to work in complex, dynamic,and unstructured environments. Under complex outdoor natural lighting and severe weather conditions, image quality may vary greatly (as shown in Fig. 1.5b). In this case, robot vision systems tend to perform poorly, lacking robustness and adaptability to the environment. Image is the basis of vision algorithms, and changes in lighting and weather alter the pixel value distribution of the image (as shown in Fig. 1.5b, c).

The author believes that illumination and weather variations, as natural phenomenons, are related to some physical factors such as light source, reflection, transmission, and imaging. If we only consider the problem from the perspective of image data, it is not easy to grasp the essence of the problem. Therefore, this book studies and solves the illumination and meteorological problems in robot vision from

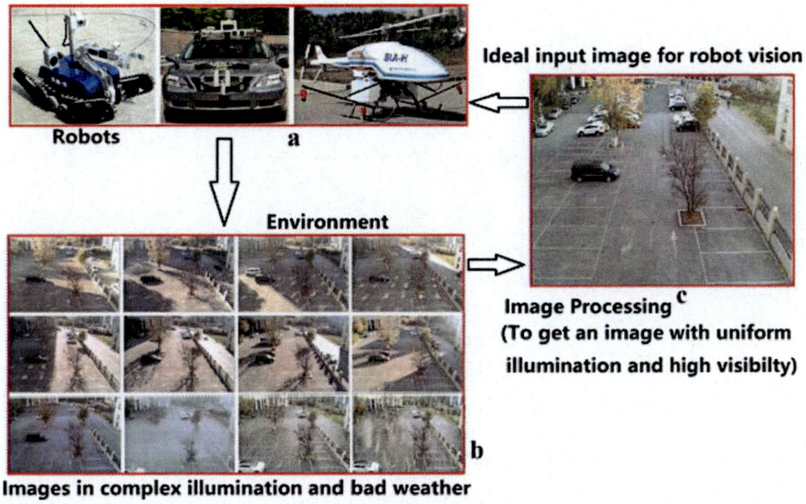

Fig. 1.5 Image processing of complex illumination and bad weather is of great significance for robot vision. As shown in figure c, the ideal inputs of robot vision are images with uniform lighting and good visibility. But robots often face complex illumination and bad weather in the environment. At this time, the images captured by vision systems are very different from the ideal input images, as shown in figure b. These differences seriously affect the performance of robot vision systems. Therefore, image preprocessing for this problem is of great significance to robot vision

the perspective of atmospheric physics and optical imaging. The main models and algorithms in this book are based on the theories of physical optics and have clear physical explanations, and the image features will also have physical meaning, which are the main characteristics and contributions of this book.

1.3 Color Matching Functions and Scene Reproduction

This book analyzes how complex illumination and bad weather conditions affect images from the perspective of atmospheric physics and imaging optics and then develops corresponding image processing algorithms, which involve scene reproduction calculation and imaging element analysis. This section firstly summarizes and reviews the basic principles of color matching function and scene reproduction.

Color is the human brain's perception of the formation of light entering the human eye. It is not only a physical quantity and involves visual physiology and visual psychology, which is more complicated. To measure and calculate color, researchers have done a lot of work. In color science, the most famous is the Grassmann color addition law and the International Commission on Illumination (CIE) color matching experiment. They provide a theoretical basis for modern color science and lay down a color science system based on three stimulus values. The color addition law of Grassmann can be expressed as follows: If $A \equiv B$ and $A1 \equiv B1$, then $A + A1 \equiv B + B1$, where $\equiv$ means that the visual effects are equal.

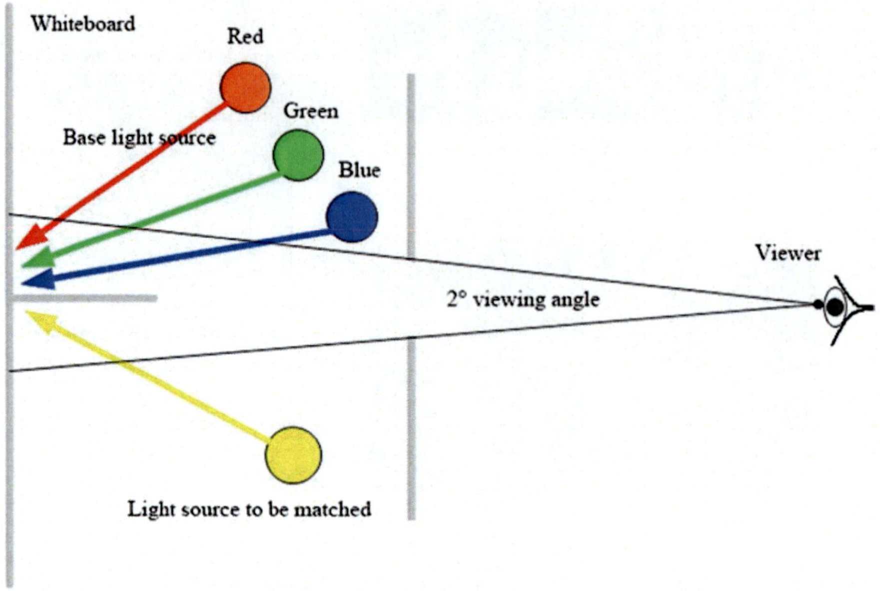

Fig. 1.6 Color matching experiment

The color matching experiment uses a mixture of three primary colors of light to achieve the same visual perception of the color of the matched light. As shown in Fig. 1.6, the leftmost side is a white screen. The three primary colors of red, green, and blue are illuminated on the upper half of the white screen. The color light is illuminated on the lower half of the white screen. The viewing angle of the human eye is approximately 2°. If (R), (G), and (B) are used to represent the unit quantities of the three primary colors of red, green, and blue (wavelengths are 700 nm, 546.1 nm, and 435.8 nm, respectively) that produce mixed colors, R, G, and B represent the required number of units of the three primary colors of red, green, and blue (also called tristimulus value) of the colors to be matched. If the mixed light has the same color perception as the C units light (C) that is to be matched, then the following formula holds:

$$C\,(C) \equiv R\,(R) + G\,(G) + B\,(B) \tag{1.1}$$

The result of the color matching experiment is the color matching functions (CMF), as shown in Fig. 1.7. The color matching functions represent the request of the intensity of primary colors to match the color of the monochromatic light at the wavelength indicated by the horizontal scale. A negative value of the matching function indicates that in the case where the chromaticity of the matched color is saturated, where cannot be matched with the three primary colors, and it is necessary to add color at the matched color.

It should be noted that the color matching experiment and the color matching function depend on the selection of the three primary colors. There are no other

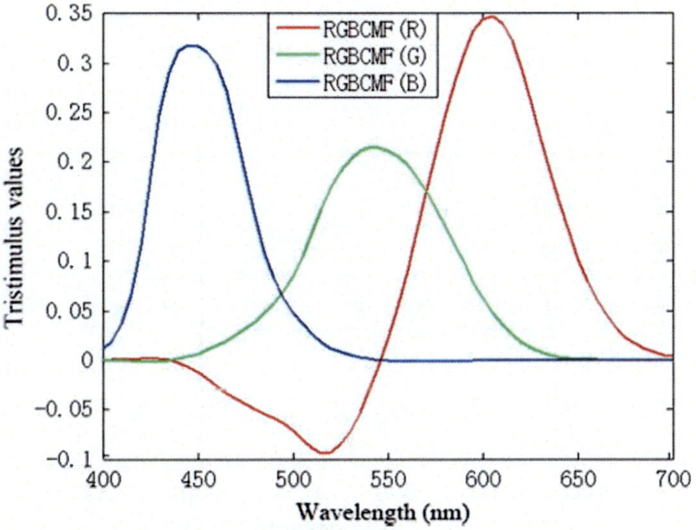

Fig. 1.7 CIE 1931 RGB color matching functions

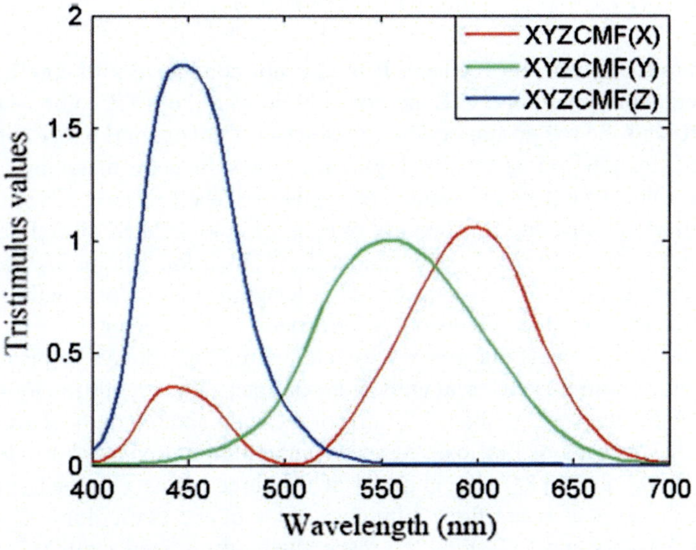

Fig. 1.8 CIE 1931 XYZ color matching functions

rules for selecting the three primary colors except that each primary color cannot be generated by mixing the other two primary colors. Different primary colors will result in different matching functions, which will result in different color systems. Just for RGB color systems, there are CIE RGB, Adobe RGB, sRGB, and Apple RGB, etc. More importantly, because the RGB color matching functions have negative values

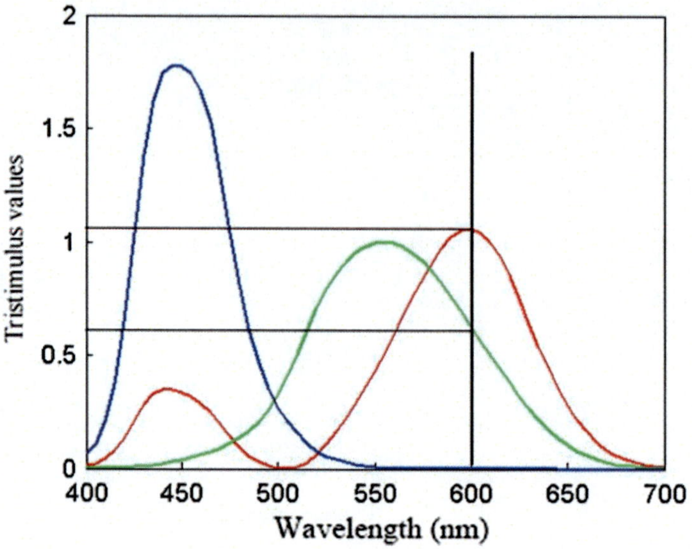

Fig. 1.9 Light at any wavelength can be matched by different units of three primary lights

and cannot be physically realized, it is also not conducive to theoretical analysis and calculation. Therefore, CIE theoretically derived the XYZ color system, where (X) represents the red primary color, (Y) represents the green primary color, and (Z) represents the blue primary color. Figure 1.8 shows the color matching functions of the XYZ color system, whose values are all positive. The CIE 1931 XYZ color system has a milestone meaning for the development of color-related disciplines, and it is widely used. It is a hub for the conversion of various color systems and various color management systems. It is also the basis for designing color reproduction equipment such as camera, scanner, television, color printer, and projector.

Figure 1.9 represents that for any wavelength of light, it can be matched by the three primary colors of different units. For example, a light at 600 nm looks the same as the combined effect of about 1.1 units of red light and 0.6 units of green light.

So, what does it look like to give a spectrum of light sources as shown in Fig. 1.10? Or how many units $[X]$, $[Y]$, and $[Z]$ of the three primary color lights look the same visual effect? If the three stimulus values of the two colors $(C1)$ and $(C2)$ are $[X1\ Y1\ Z1]$ and $[X2\ Y2\ Z2]$, respectively, the mixed color (C) is satisfied $(C) \equiv (C1) + (C2)$. By the color addition law of Grassmann, there are as follows:

$$\begin{aligned}(C) &\equiv (C1) + (C2) \\ &\equiv [X1\,(X) + Y1\,(Y) + Z1\,(Z)] + [X2\,(X) + Y2\,(Y) + Z2\,(Z)] \\ &\equiv (X1 + X2)\,(X) + (Y1 + Y2)\,(Y) + (Z1 + Z2)\,(Z)\end{aligned} \tag{1.2}$$

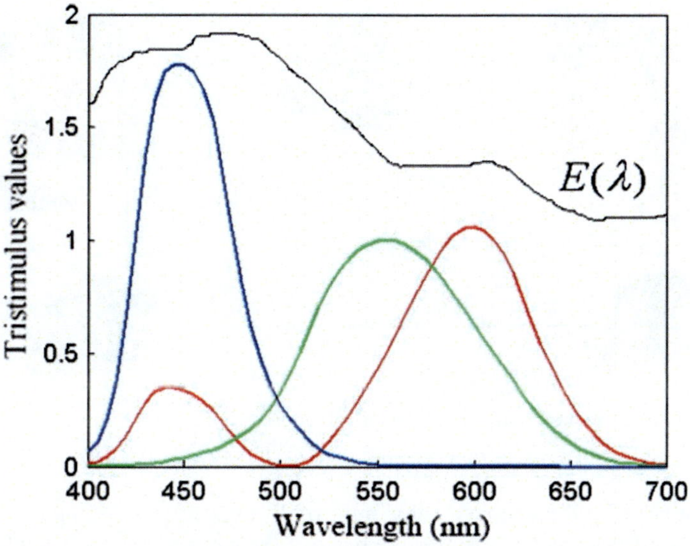

Fig. 1.10 Three stimuli for visible light bands

It shows that the tristimulus values of the color (C) are $X = X1 + X2; Y = Y1 + Y2;$ $Z = Z1 + Z2$. The tristimulus value of the mixed color is the sum of the corresponding tristimulus values of the constituent colors. Obviously, the above principle can be generalized to the mixing of multiple color light. For n kinds of color light, the tristimulus value of the mixed color light is as follows:

$$\begin{cases} X = \sum_{i=1}^{n} X_i \\ Y = \sum_{i=1}^{n} Y_i \\ Z = \sum_{i=1}^{n} Z_i \end{cases} \tag{1.3}$$

Therefore, for the spectrum of any light source, there are as follows:

$$\begin{cases} X = \int_{400}^{700} E(\lambda)\overline{x}(\lambda)\mathrm{d}\lambda \\ Y = \int_{400}^{700} E(\lambda)\overline{y}(\lambda)\mathrm{d}\lambda \\ Z = \int_{400}^{700} E(\lambda)\overline{z}(\lambda)\mathrm{d}\lambda \end{cases} \tag{1.4}$$

where $\overline{x}(\lambda)$, $\overline{y}(\lambda)$, and $\overline{z}(\lambda)$ represent the CIE XYZ color matching functions. So we have got $E(\lambda) \equiv X[X] + Y[Y] + Z[Z]$, which means that the color $E(\lambda)$ can

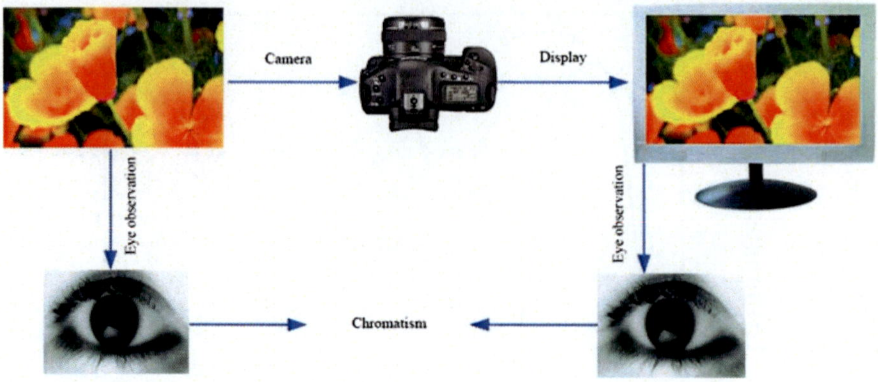

Fig. 1.11 Overall frame diagram of scene reproduction

be reproduced by the primary color light of X units of $[X]$, Y units of $[Y]$, and Z units of $[Z]$. Equation (1.4) is also the basis of the camera design. The photoelectric conversion function of CCD is to convert the spectrum $E(\lambda)$ into tristimulus values (i.e., the three-color channel values) using Eq. (1.4). If the display's electron gun can accurately emit X units of $[X]$, Y units of $[Y]$, and Z units of $[Z]$, then the color of light we see on the monitor is the same as the effect of looking $E(\lambda)$ directly. In fact, color scanners, televisions, projectors, and other color reproduction equipment are also based on this principle.

Figure 1.11 shows the overall frame diagram of scene reproduction. The so-called scene reproduction is to use a camera or other photosensitive equipment to record the scene light intensity (i.e., forming an image) and then use a display device to regenerate the image data into a light signal. The ultimate goal is to achieve consistency between the scene viewed on the display and the scene viewed directly in the physical world. The CIE 1931 XYZ color space is a space commonly used to describe human color characteristics. As mentioned in the previous section, if the camera's spectral response functions (SRF) are consistent with XYZ CMF, the three primary colors of the display are the same as the three primary colors of XYZ, the display is consistent with the white field of XYZ space, and then an error-free color reproduction system can be implemented. However, the pictures recorded by the camera are generally viewed through a display device. Unfortunately, since the three primary colors of the XYZ system are derived from theory and cannot be physically realized, they cannot be used as the primary colors of the electron gun of the display. At present, the sRGB color system is a standard space generally accepted in the display manufacturing industry. The sRGB color space is a standard RGB color space developed by HP and Microsoft in 1997 for monitors, printers, and the Internet.

This standard is generally supported by industry vendors. Therefore, it is necessary to convert XYZ to sRGB in the real system via Formula (1.5).

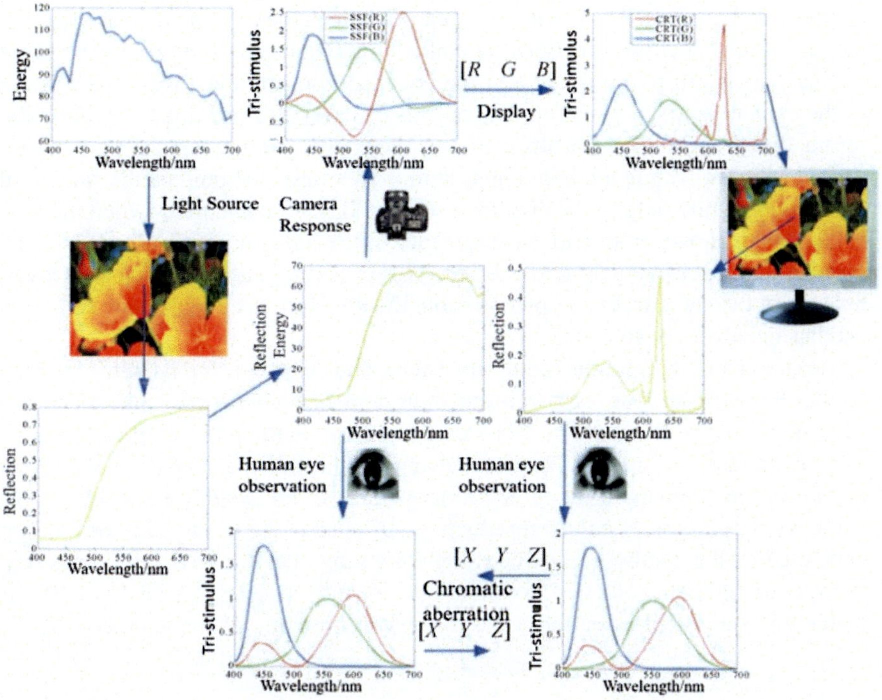

Fig. 1.12 Principle of scene reproduction

$$\begin{bmatrix} R_{\text{sRGB}} \\ G_{\text{sRGB}} \\ B_{\text{sRGB}} \end{bmatrix} = \begin{bmatrix} 3.2406 & -1.5372 & -0.4986 \\ -0.9689 & 1.8758 & 0.0415 \\ 0.0557 & -0.2040 & 1.5070 \end{bmatrix} \begin{bmatrix} X \\ Y \\ Z \end{bmatrix} \tag{1.5}$$

The sRGB color matching functions can be obtained by Formula (1.5). For any light, if the camera's spectral response functions are strictly consistent with the sRGB color matching functions, then this color system (including camera and display) can perfectly reproduce the color of the scene. Regarding the brightness of the scene, due to the nonlinear characteristics of the response of the display and the human eye, the display hardware and image coding need to be separately gamma corrected.

The physical imaging process of outdoor images is that the light reflected by an object entering a camera is weighted by the camera response curves and then photoelectrically converted by the CCD integration. The camera then performs gamma correction and sRGB space conversion to finally form an image. It is worth noting here that the response curves (combined results of Bayer filters and camera color correction matrices) of most cameras are consistent with the sRGB color matching functions (Engelhardt 1993 and Haneishi 1995). Although due to manufacturing reasons, the spectral response functions of the camera may be significantly different from the sRGB or XYZ color matching functions, camera manufacturers generally correct the error caused by the inconsistency between the camera response functions and the color matching functions by embedding a correction matrix in the camera circuit. It is worth pointing out that the conclusion of the ideal response functions of the camera is given based on the analysis of CIE color matching experiments. It means that the white field of the primary color lights and the test light are required to be consistent. While the imaging system is shooting, it should be the same as the background white field of the display (usually CIE D65 white field) when the image is displayed. However, due to the variability of the light source in practice, such an ideal situation is obviously not satisfied. This problem is usually solved by a white balance in the camera. In computer vision, this problem is called "color constancy" and has been widely studied.

As shown in Fig. 1.12, the light source shines on an object, and its light is reflected into the human eye to form a tristimulus value, and the brain perceives the scene information. The same light enters the camera to complete the photoelectric conversion to generate a scene image. The three channel values of the image pixels are used to control the intensity of the three primary colors of the display electron gun. This is the whole process of color reproduction. Of course, the scene light and the light generated by the display cannot be exactly the same, but due to the existence of the heterochromatic phenomenon, we do not perceive their difference (in fact, the color matching experiment essentially uses the heterochromatic phenomenon).

1.4 Overview of Key Technologies for All-Weather Robot Vision

1.4.1 Imaging Elements

Figure 1.13 describes the imaging elements of an outdoor scene. An image of the outdoor scene is mainly determined by four elements: light source (including direct sunlight, scattered skylight, and the sum of them: sunlight), light transmission, object reflection, and observer (or camera). The light from a light source is transmitted through the medium, and part of the light is reflected when it irradiates to the surface of the object. After this part of the light is reflected by the surface, it is transmitted through the medium into the imaging system, and the image is finally obtained through the photoelectric conversion and processing of the camera. It can be seen from the above process that the imaging light source, the object surface reflection, the medium transmission, and the camera are the key factors affecting the target imaging. As shown in the figure, the complex illumination in the image is closely related to these weather and imaging elements. The scattering of light in the atmosphere produces the skylight, which forms the shadow character of the ground; the scattering of light in the air forms haze and in the water forms turbidity; the non-Lambertian reflection of light forms highlight. Therefore, in addition to the spectral

Fig. 1.13 Outdoor imaging elements

response characteristics of the cameras, camera imaging modeling also needs to know the environmental illumination, object reflection, and medium transmission characteristics.

1.4.1.1 Lights

Light source plays an important role in the process of imaging. In physics, light is an electromagnetic wave that includes visible and invisible light (ultraviolet, infrared, X-ray, etc.). In general, the wavelength range that the human eye can feel is between 390 and 720 nm. Therefore, the main research in robot vision and color science focuses on visible light. Many robot vision algorithms benefit from the prior knowledge of spectral irradiance of light sources, such as object reflection spectral curves recovery [26], color constancy [7], image rendering [6], augmented reality [24], and high dynamic range imaging [2].

In our real life, there are many kinds of light sources. On one hand, sunlight is the main light source in outdoor scenes, with the skylight also playing an important role. On the other hand, light sources are mainly artificial in indoor scenes, such as fluorescent lamps, halogen tungsten lamps, and LED light sources. For the convenience of research and calculation, CIE specifies a set of standard illuminants and standard light sources to represent different light sources in practical application, including four outdoor illuminants A, B, C, and D. CIE standard illuminant A is intended to represent tungsten–filament lighting. Its relative spectral power distribution is that of a Planckian radiator at a temperature of approximately 2856 K. Standard illuminant B uses direct daylight when the color temperature reaches 4874k to represent the middle daylight. Standard illuminant C represents the light when the color temperature reaches 6774k and usually represents the average sunlight. Standard illuminant D represents daylight other than standard illuminants, such as D50. For indoor light sources, CIE uses F2, F7, and F11 to represent fluorescent lamps. The color

Table 1.1 CIE standard illuminants

Illuminants	Color temperature (Kelvin)	Typical light source
A	2855.6	Tungsten filament lighting
B	4874	Direct sunlight at noon
C	6744	Average daylight
D50	6500	Horizontal daylight
D65	6504	Midday daylight
D75	7500	Cloudy daylight
F2	4100	Cold white light
F7	6500	Broadband fluorescent lamp
F11	4000	Narrow band fluorescent lamp

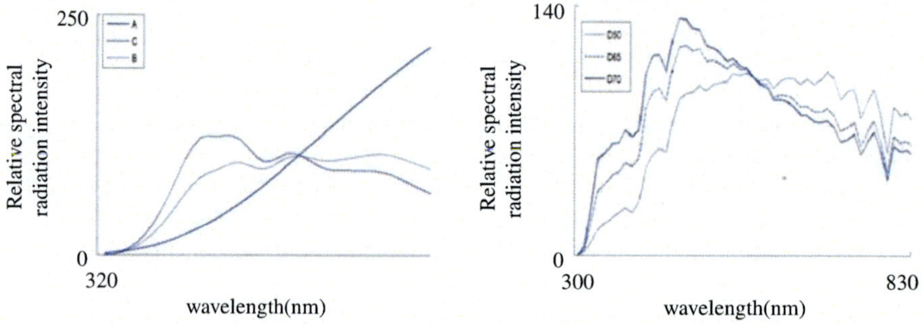

Fig. 1.14 Spectral power distributions of outdoor light sources

temperatures and representative light sources of the above standard illuminants are shown in Table 1.1, and the relative spectral radiance distributions are shown in Figs. 1.14 and 1.15, respectively.

The spectral radiation of F series standard illuminants is strong in several bands, comparing to their weak radiation in other bands. They have a low color rendering index because the spectrum of their spectral radiation distributions is limited.

Outdoor light sources often change with time and weather, such as clear sky and cloudy days. The image of the same target also varies with the changes of the light source. Same objects will show different colors under different light sources, as shown in Fig. 1.16. Therefore, it is important to calculate the spectral power distribution (SPD) of outdoor light sources for computer vision research. At present, in computer vision, the theory of blackbody radiation is often used to approximate or the CIE D65 spectral irradiance is directly used to replace the outdoor light source. However, these are very different from the actual SPD of the outdoor light sources. In an outdoor environment, the sun is the main light source, so the SPD calculation of direct sunlight and scattered skylight has become important.

The light emitted by the sun passes through the Earth's atmosphere and will be affected by atmospheric absorption, reflection, and scattering. This results in the

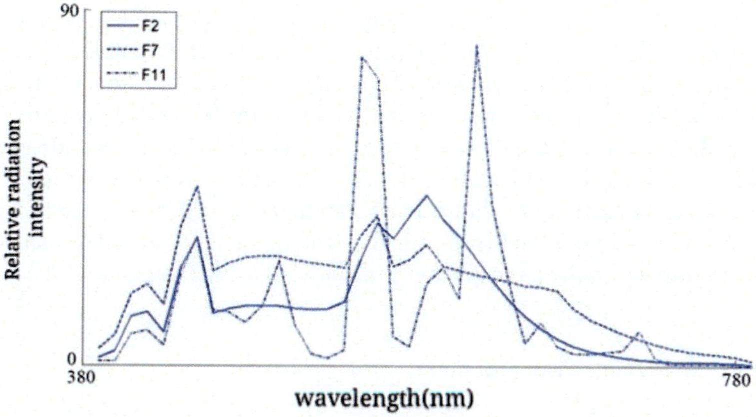

Fig. 1.15 Spectral power distributions of indoor light sources

Fig. 1.16 Images under different illumination

spectral irradiance to the ground changing with time (zenith angle) and atmospheric factors, which further leads to the complexity and diversity of outdoor scene images. In Chap. 2 of this book, we will describe a simple and effective method to calculate the SPD of direct sunlight and scattered skylight in the visible spectrum. In this method, the accuracy of the spectrum near the peaks of the color matching functions is specially considered, which aims to reduce the deviation when replacing the real light for imaging calculation with our calculation results. From the literature we have reviewed, this method is the first physical calculation method based on analyzing the light transmission in the atmosphere for robot vision and imaging.

1.4.1.2 Scattering Transmission

Visible light scattering refers to the phenomenon that part of the visible light deviates from the initial direction due to the inhomogeneity of the propagation medium. The light deviating from the original direction is called scattering light. The ratio of the scattering particle radius r to the incident light wavelength λ determines the scattering degree α:

$$\alpha = 2\pi r/\lambda \tag{1.6}$$

From Formula (1.6), it can be concluded that the scattering degree varies greatly with the size of scattering particles. Therefore, researchers often use α as the criterion to distinguish and classify the scattering degree, that is, when α is far less than 1 ($r < \lambda$), it belongs to Rayleigh scattering; when $1 < \alpha < 50(r \approx \lambda)$, it belongs to Mie scattering; when $\alpha > 50(r \gg \lambda)$, it belongs to geometrical optical scattering. The scattering involved in this book is Rayleigh scattering and Mie scattering. The shadow in an outdoor image is mainly caused by Rayleigh scattering, and the fog and underwater turbidity in images are caused by Mie scattering.

The law of Rayleigh scattering was proposed by the British physicist Lord Rayleigh in 1871. This scattering is mainly caused by atoms and molecules in the atmosphere, such as nitrogen, carbon dioxide, ozone, and oxygen molecules. The scattering light intensity of the scattering particle varies according to directions. The light shares the same degree of Rayleigh scattering in the forward direction and reverse direction, while having the lowest degree (Nave 2016) in the direction perpendicular to the incident light, which is shown in Fig. 1.17. Mie scattering refers to the scattering that occurs when the diameter of particles in the atmosphere is equal to the wavelength of radiation. This scattering is mainly caused by particles in the atmosphere, such as smoke, dust, small water droplets, and aerosols. The scattering intensity of the Mie scattering is inversely proportional to the quadratic of the wavelength, and the scattering is stronger in the front of the light than that in the back. When the r is large, the relationship between the scattering light intensity and the wavelength is no longer obvious. When the particle scale is close to or larger than the incident light wavelength, the scattering light intensity is asymmetric in all directions [10]. As shown in Fig. 1.17, most of the incident light is scattered along the forward direction.

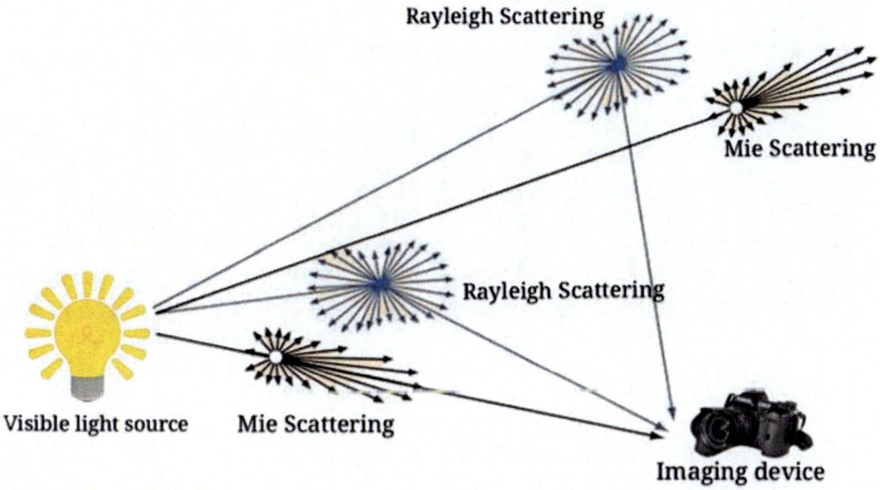

Fig. 1.17 Spatial distribution of Rayleigh scattering and Mie scattering

When the atmosphere is clear, there are fewer large particles in the air, and molecular Rayleigh scattering primarily happens. The blue light with shorter wavelength, however, is scattered and diffused in the sky, and the sky will exhibit a high saturated blue. At this time, ground shadows are obvious, and their edges are clear. In cloudy conditions, the number of large particles in the air increases, and the sky gradually appears gray white. At this time, the contrast between the skylight and the sunlight decreases, and the ground shadow also becomes slight or even disappears. In hazy weather, water vapor and aerosol particles with relatively large sizes are suspended in the air, and Mie scattering mainly occurs. Visible light is almost equally scattered, making the sky white or gray. The atmospheric scattering model, which is proposed based on Mie scattering theory, is used to describe the physical characteristics of light transmission in hazy weather. It is the main basis to understand the degradation mechanism of hazy weather images and restore the clear image of the scene. In clean water, there are only a few suspended particles with very small radius. At this time, Rayleigh scattering, which is dominated by molecular scattering, plays a major role, making the water dark blue. Figure 1.18 is a clear sky and a clean water image dominated by Rayleigh scattering. There are many inorganic and organic suspended substances in turbid water, such as offshore, whose particle size is much larger than the wavelength of visible light. At this time, the law of Mie scattering should be followed. Figure 1.19 includes an image in dense fog weather and an image in turbid waters that are dominated by the Mie scattering.

Fig. 1.18 Clear sky and clean water images that are dominated by the Rayleigh scattering

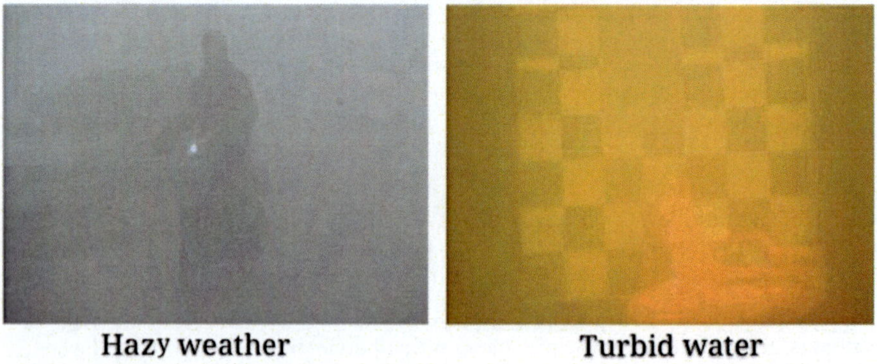

Fig. 1.19 Dense fog and turbid water images that are dominated by the Mie scattering

1.4.1.3 Surface Reflection

When the light from a light source hits the surface of an object, the light interacts with the object. Due to the selective absorption of light by objects, one part of the wavelength band is absorbed, and the other part is reflected. The reason why objects have different colors is that they selectively absorb certain wavelengths of incident light. For example, the reason why flowers show red is that when white light hits their surfaces, they absorb the blue and green light and reflect the red light. Therefore, they appear red as reflected light enters the human eye. Another example is that when a green light illuminates a red surface, all the energy of the green light is absorbed by the red surface, thus all the light, of course, is absorbed, and no light is reflected into the human eye. So the surface of the object that the human eye sees in this case appears black. The color of an object's surface, therefore, is different when it is irradiated by light sources with different spectra. From the above process, it can be seen that the process of color formation of objects is that when the incident light reaches the surface of an object, the object only absorbs a part of the light with specific wavelengths, and the remaining light that is not absorbed presents the color we see. The color characteristics of opaque objects are mainly represented by their spectral reflectance ratio. The spectral reflectance of an object is usually expressed by the ratio of the reflected spectral radiation to the incident spectral irradiance, that is

$$S(\lambda) = \frac{E_r(\lambda)}{E_i(\lambda)} \tag{1.7}$$

where $S(\lambda)$ represents the spectral reflectance ratio, $E_i(\lambda)$ represents the spectral incidence, $E_r(\lambda)$ represents the spectral reflectance, and λ represents the wavelength. It can be seen from Formula (1.7) that this spectral reflectance ratio of an object can be expressed as a function of wavelength. In addition to the influence of wavelength, the angle of incidence and angle of reflection also affect the spectral reflection ratio of the object. A more obvious example is highlighted on surfaces. The highlight phenomenon is due to the large angle of reflection, which reaches the mirror reflection. Therefore, the angle of incidence and angle of reflection also have an important influence on the spectral reflectance ratio. The characteristics of reflection generally need to be described by the bidirectional reflection model, which will be described in detail in Chap. 6 of this book.

The reflection spectrum includes the essential properties of the material characteristics of objects, which is defined as the "fingerprint" of objects, due to its robustness to incident light changes. Different objects will absorb and reflect different wavelengths. When the light is reflected into an imaging device, the image of the object will present different colors. Therefore, the study of the spectral reflectance of objects will help us to analyze the imaging characteristics of the objects. In the second chapter of this book, we will describe the wavelength-sensitive function-controlled reflectance reconstruction method from a single color image. This method can obtain more accurate reconstruction at wavelengths where sensors have high sensitivities, and therefore, it can also obtain better imaging calculation or scene reproduction results.

1.4.1.4 Digital Camera Imaging Modeling

Figure 1.20 shows the general diagram of camera imaging. Light enters a camera through its lens after the object reflection. Because CCD/CMOS is only a photoelectric conversion device that does not have the color separating ability, it usually needs to cover the color filter array (CFA) in front of CCD/CMOS to allow different colors to penetrate. The most commonly used color filter array is the Bayer filter. The data generated after CCD sensing is called raw data. At present, imaging Formula (1.8) is widely used in camera modeling (such as [7, 11]).

$$F_H^{RAW} = \int_{400}^{700} E(\lambda)S(\lambda)Q_H(\lambda)\mathrm{d}\lambda;\ H = R, G, B \tag{1.8}$$

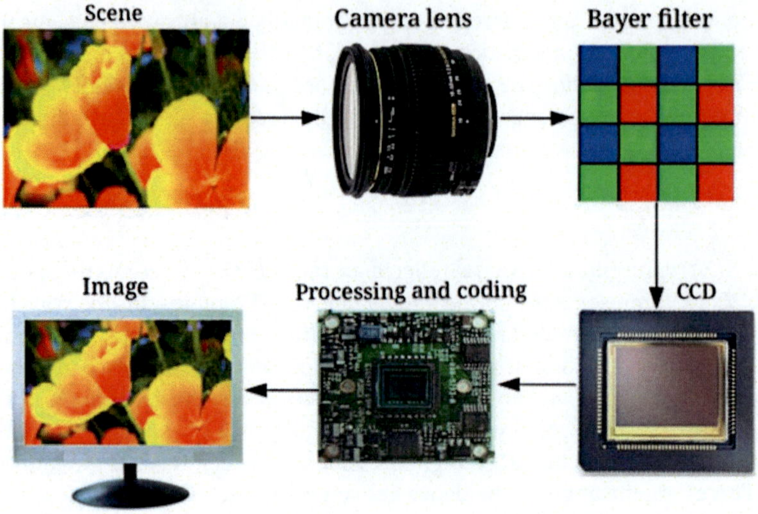

Fig. 1.20 Block diagram for a typical camera imaging process

where F represents image pixel value, H represents one of the three color channels, $E(\lambda)$ represents spectral irradiance, and $S(\lambda)$ represents object spectral reflectance. Formula (1.8) only considers the sensor response functions of a camera, which is more suitable for modeling the raw format. The camera response functions are determined by the spectral response function of CCD, CFA filter, and infrared cutoff filter. Many professional industrial cameras usually give the sensor spectral response function in their product manuals. For off-the-shelf cameras, the methods in literature [3, 23] can be used to calibrate the sensor spectral response functions of the camera. After the photoelectric conversion of cameras, the original data should be processed by software in DSP, including white balance, gamma correction, camera characterization, color correction, etc. Relevant details will be described in the third chapter.

1.4.2 Image Processing of Complex Illumination

The images of a scene tend to change with the variation of its illumination and the outdoor illumination changes with time. Illumination processing for images can be divided into global and local illumination processing. Among them, the global one is mainly for the problem of color constancy, and the local one is for the problem of highlight and shadow. At present, in the illumination processing and relevant research, the color constancy under a single light source has been solved well. The so-called color constancy refers to dealing with the problem of color deviation of an object under different illumination, for example, white paper is red under red light and is blue under blue light. The recent cameras have been able to make it adaptive to illumination variation through "white balance", but the cameras cannot

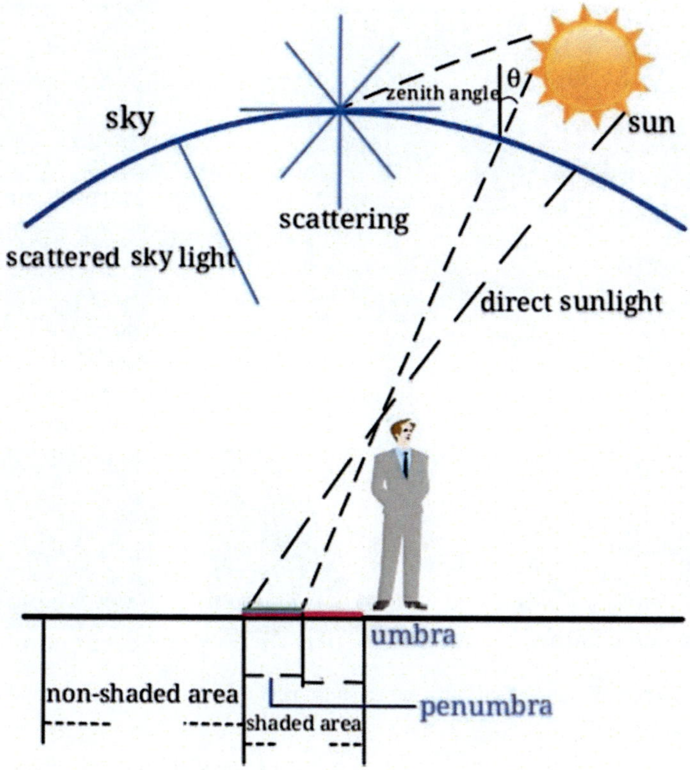

Fig. 1.21 Physical explanation of shadow formation

do anything about the local illumination changes including shadow and highlight. Local illumination has a great influence on the robustness of robot vision algorithms, which is still a difficult problem to be solved in this field. The appearance of shadows is equivalent to an additional light source appeared in the scene—as shown in Fig. 1.21. A shadow contains the umbra and the penumbra. The light source in the umbra is the scattered light from the sky and that in the penumbra is the scattered skylight and a little direct light from the sun. The light source in the non-shaded area is the scattered skylight plus the direct sunlight. Therefore, it is necessary to model and calculate outdoor light sources (sunlight and skylight) to analyze their spectral distribution characteristics. Combining the spectral power distribution characteristics of skylight and sunlight and the formation process of images, we can analyze the physical imaging characteristics of shadows, such as the relationship between the inside and outside of shadow regions, the relationship between the color channels of a shadow region, etc. These characteristics will be the basis for the establishment of our shadow detection, verification, and removal algorithms. The calculation of the spectral characteristics of the outdoor light source will bring great convenience to the analysis of the illumination variation in outdoor scenes. However, the illumination

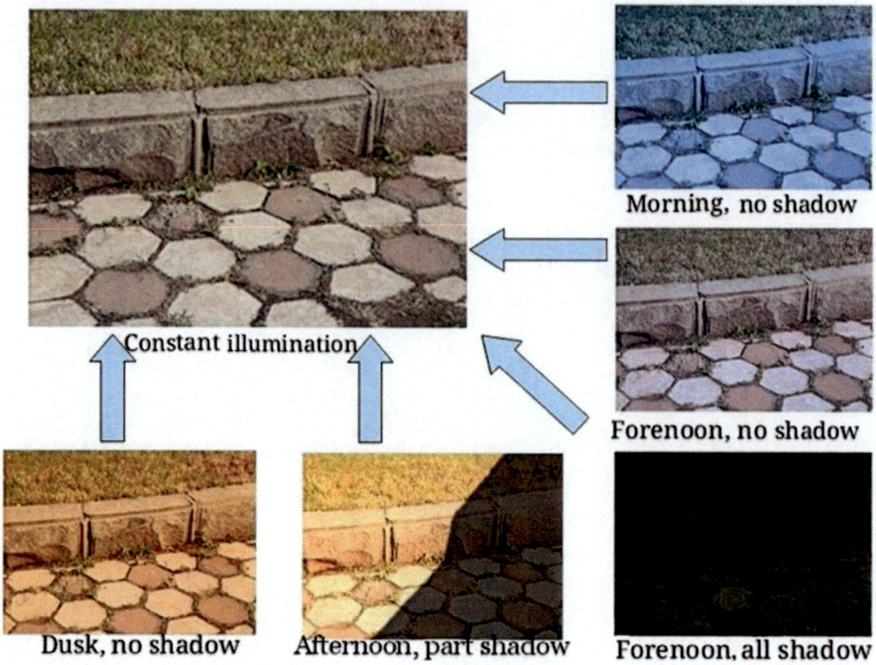

Fig. 1.22 Expected illumination constancy results under different illumination conditions

phenomenon in the image is also related to the reflection spectrum of objects and the response characteristics of cameras. The study of illumination processing in this monograph starts from three aspects as follows: outdoor illumination modeling, spectral reflection calculation, and camera response analysis. The new spectral calculation method of the outdoor light source and some image illumination processing algorithms are described. These algorithms mainly include shadow detection and removal, along with illumination-reflectance separation and relighting.

The goal of light processing is to achieve image illumination constancy by proposing the models that are suitable for computer vision calculation, such as computable illumination model, imaging modeling method, intrinsic-image decomposition, and relighting algorithms. The models are expected to have illumination constancy for both local shadows and global illumination variation, as shown in Fig. 1.22.

1.4.3 Image Processing of Bad Weather

In outdoor environment, the negative effects of weather on robot vision are mainly haze, rain, and snow. The corresponding research contents mainly include image/video fog, rain, and snow removal.

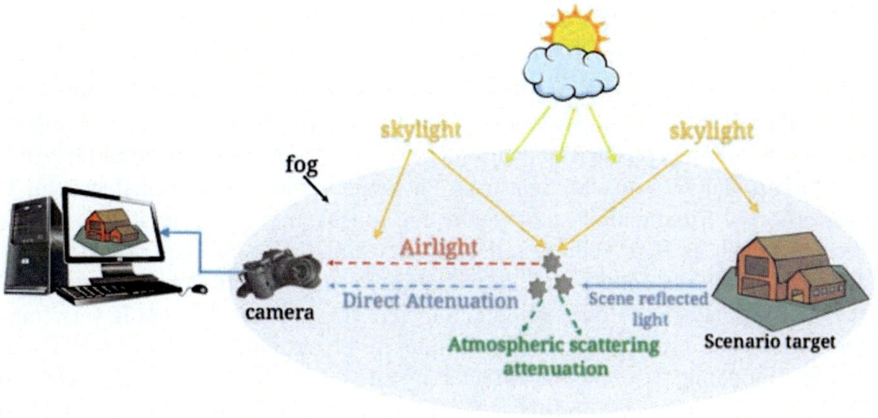

Fig. 1.23 Schematic diagram of fog formation

Light is scattered by particles when transmitted in foggy weather or muddy underwater media, which causes the light entering the camera to be attenuated, at the same time some stray light outside the light path enters the camera. The image is blurred, the saturation is reduced, and the contrast is declined. Figure 1.23 is a schematic diagram of the imaging model of fog. After the sunlight being scattered and refracted through the atmosphere, it forms the skylight together with the direct sunlight at low altitude. Components of skylight scattered through aerosol particles into the camera field of view from the airlight. The skylight reflected by the scene objects forms the reflected light of the scene, and the components scattered into the camera's field of view through the aerosol particles form the direct attenuation. Finally, a fog image is formed by the combination of the airlight and the direct attenuation component. Image defogging is a difficult problem, and the core of which is the calculation of transmission of optical media. In Chaps. 8 and 9, the depth estimation algorithms based on region optimization and light field image are described, respectively.

Compared with the research of image defogging, there is less research on image desnow and derain, especially the research of image desnow. The light transmission characteristics in rainy and snowy weather change obviously, which includes the scattering of raindrops and snowflakes as well as the occlusion problem. Raindrops and snowflakes have both sparsity and density, with dynamic characteristics. Traditional snow and rain removal algorithms usually assume that raindrops and snowflakes are sparse in the scene, so these algorithms can detect raindrops or snowflakes based on their brightness changes, falling direction, and shape. Although these methods are effective for light rain and snow and static scenes, they are difficult to deal with heavy rain and snow, high dynamic scenes, and camera motion. In the cases of heavy rain and snow, raindrops and snowflakes show sparse and dense characteristics simultaneously. The scene is not only occluded by sparse rain and snow, but also blurred by dense rain and snow. In this book, the composition of a rainy or snowy scene is divided into sparse rain and snow, dense rain and snow, background, and foreground.

1.5 The Main Contents and Structure of This Book

From the perspective of atmospheric physics and imaging optics, this book studies image illumination, weather modeling, and the processing methods. The image of a real scene is formed by the illumination, the transmission of light in different weather conditions, and the reflected light from the target surface to the visual system. The expression of illumination and weather on images are closely related to imaging elements such as light source, reflection, transmission, and camera, as shown in Fig. 1.24. Since the complexity and variability of natural lighting and weather, the parametric analysis of illumination and meteorological environment is very important, which can greatly reduce the complexity of the illumination processing and weather modeling. An image is affected by environmental lighting parameters, but these parameters are disappeared after imaging. Through the theoretical modeling and analysis for imaging elements, image features are correlated with environmental parameters, and then we can build a connection bridge between the lighting environments and image data. The overall structure of the book is divided into four parts: imaging elements modeling, processing of complex illumination on images, processing of bad weather on images, and application examples.

The main contents of this book are summarized and reorganized from the academic papers published by the author. The work of imaging elements modeling is described in Chap. 2 about outdoor spectral radiography calculation [19] and spectral reflection recovery from a single image [17], in Chap. 9 about underwater scattering transmission modeling [21], and in Chap. 3 about camera response function estimation [25]. Since color cameras are only sensitive to visible light, by thoroughly analyzing the calculation of spectral irradiance, we find that ozone, water vapor, nitrogen, carbon dioxide, and other mixed air absorption, as well as surface multiple reflections in traditional meteorological methods have very little effect on visible light, so these calculations can be ignored. This greatly reduces the computational complexity and the number of parameters and thus makes it possible to be applied in imaging calculation. In our calculation method, the outdoor illumination is modeled by two parameters: zenith angle and aerosol, in which zenith angle is the key parameter for complex illumination analysis such as shadow and highlight, and aerosol is the key parameter for bad weather analysis such as rain, snow, and fog.

On the basis of imaging elements modeling, the work of this book in the field of image processing for complex illumination includes shadow tricolor attenuation model [18], shadow linear model [16], and shadow detection algorithm [20] in Chap. 4, intrinsic images [12] and illumination decomposition [8] in Chap. 5, and shadows and highlight removal in Chap. 6 [13, 15]. The linear model of shadow plays the role of connecting the preceding and the following. We proposed the shadow linear model in [16] and proved that the parameters of the model do not depend on the spectral reflectance curves of objects, but only depend on the solar angle. This model reveals the intrinsic relationship between images and outdoor light, which is equivalent to providing additional information for image shadow analysis. Based on this, we developed shadow detection, removal, intrinsic-image deriving, and illumination decomposition algorithms.

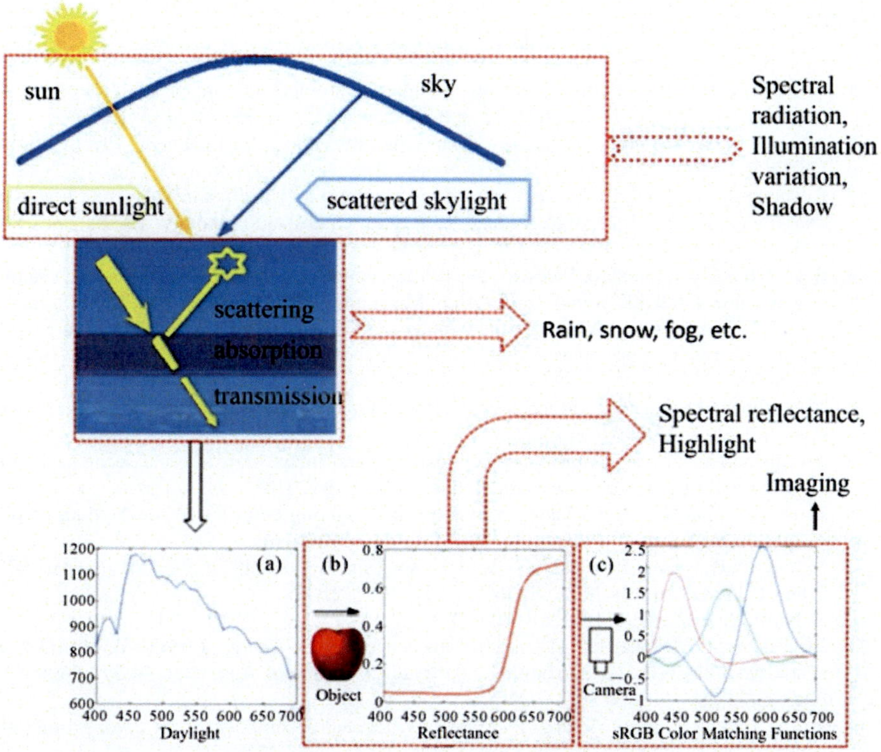

Fig. 1.24 Complex illumination and bad weather in images are closely related to imaging elements

The work of this book on processing for bad weather in images includes rain and snow removal algorithms in Chap. 7 [14, 22], fog and scattering removal algorithms in Chaps. 8 and 9 [5, 21]. The image scattering model has a great relationship with light conditions. In hazy night or muddy water, the light source is generally an active one, which belongs to the near-field illumination. While in hazy weather during a day, its light source is sunlight and belongs to the infinite parallel illumination. For near-field scattering, we described the depth estimation and scattering removal algorithm based on light field images. For far-field scattering, we described the defogging algorithm based on region optimization. For rain and snow videos, we supposed that the background of a scene is stationary. Even if the background of the scene may have motion, it can also be aligned through the existing registration algorithms. For a rainy/snowy scene, the background has a low-rank characteristic, and thus the background can be constructed by low-rank expression. In addition, it can also be considered that moving objects' morphological changes are relatively slow. When the moving objects in different frames are matched together, they also have low-rank characteristics. Similarly, the matrix decomposition method can be used to separate the foreground from rain and snow.

References

1. Y. Adini, Y. Moses, S. Ullman, Face recognition: the problem of compensating for changes in illumination direction. IEEE Trans. Pattern Anal. Mach. Intell. **19**, 721–732 (1997)
2. A.O. Akyuz, E. Reinhard, Color appearance in high-dynamic-range imaging. J. Electron. Imaging **15**(3), 033001 (2006)
3. K. Barnard, B.V. Funt, Camera calibration for color research, in *Human Vision and Electronic Imaging IV*, volume 3644 (International Society for Optics and Photonics, 1999), pages 576–585
4. R. Cucchiara, C. Grana, M. Piccardi, A. Prati, Detecting moving objects, ghosts, and shadows in video streams. IEEE Trans. Pattern Anal. Mach. Intell. **25**(10), 1337–1342 (2003)
5. T. Cui, J. Tian, E. Wang, Y. Tang, Single image dehazing by latent region-segmentation based transmission estimation and weighted l 1-norm regularisation. IET Image Proc. **11**(2), 145–154 (2016)
6. T. Gierlinger, D. Danch, A. Stork, Rendering techniques for mixed reality. J. Real-Time Image Proc. **5**(2), 109–120 (2010)
7. A. Gijsenij, T. Gevers, Color constancy using natural image statistics and scene semantics. IEEE Trans. Pattern Anal. Mach. Intell. **33**(4), 687–698 (2010)
8. Z. Han, J. Tian, L. Qu, Y. Tang, A new intrinsic-lighting color space for daytime outdoor images. IEEE Trans. Image Process. **26**(2), 1031–1039 (2016)
9. S. Nadimi, B. Bhanu, Physical models for moving shadow and object detection in video. IEEE Trans. Pattern Anal. Mach. Intell. **26**(8), 1079–1087 (2004)
10. R. Nave. hyperphysics: Hysteresis (2016)
11. J.L. Nieves, E.M. Valero, S.M. Nascimento, J. Hernández-Andrés, J. Romero, Multispectral synthesis of daylight using a commercial digital CCD camera. Appl. Opt. **44**(27), 5696–5703 (2005)
12. L. Qu, J. Tian, Z. Han, Y. Tang, Pixel-wise orthogonal decomposition for color illumination invariant and shadow-free image. Opt. Express **23**(3), 2220–2239 (2015)
13. L. Qu, J. Tian, S. He, Y. Tang, R.W. Lau, Deshadownet: a multi-context embedding deep network for shadow removal, in *Proceedings of the IEEE Conference on Computer Vision and Pattern Recognition, 2017*, pages 4067–4075
14. W. Ren, J. Tian, Z. Han, A. Chan, Y. Tang, Video desnowing and deraining based on matrix decomposition, in *Proceedings of IEEE CVPR* (2017a), pages 4210–4219
15. W. Ren, J. Tian, Y. Tang, Specular reflection separation with color-lines constraint. IEEE Trans. Image Process. **26**(5), 2327–2337 (2017)
16. J. Tian, Y. Tang, Linearity of each channel pixel values from a surface in and out of shadows and its applications, in *CVPR 2011* (IEEE, 2011), pages 985–992
17. J. Tian, Y. Tang, Wavelength-sensitive-function controlled reflectance reconstruction. Opt. Lett. **38**(15), 2818–2820 (2013)
18. J. Tian, J. Sun, Y. Tang, Tricolor attenuation model for shadow detection. IEEE Trans. Image Process. **18**(10), 2355–2363 (2009)
19. J. Tian, Z. Duan, W. Ren, Z. Han, Y. Tang, Simple and effective calculations about spectral power distributions of outdoor light sources for computer vision. Opt. Express **24**(7), 7266–7286 (2016)
20. J. Tian, X. Qi, L. Qu, Y. Tang, New spectrum ratio properties and features for shadow detection. Pattern Recogn. **51**, 85–96 (2016)
21. J. Tian, Z. Murez, T. Cui, Z. Zhang, D. Kriegman, R. Ramamoorthi, Depth and image restoration from light field in a scattering medium, in *Proceedings of the IEEE International Conference on Computer Vision* (2017), pages 2401–2410
22. J. Tian, Z. Han, W. Ren, X. Chen, Y. Tang, Snowflake removal for videos via global and local low-rank decomposition. IEEE TMM **20**(10), 2659–2669 (2018)
23. M.J. Vrhel, H.J. Trussell, Color device calibration: a mathematical formulation. IEEE Trans. Image Process. **8**(12), 1796–1806 (1999)

24. J. Wither, S. DiVerdi, T. Höllerer, Annotation in outdoor augmented reality. Comput. Graph. **33**(6), 679–689 (2009)
25. D. Wu, J. Tian, B. Li, Y. Wang, Y. Tang, Recovering sensor spectral sensitivity from raw data. J. Electron. Imaging **22**(2), 023032 (2013)
26. K.-J. Yoon, E. Prados, P. Sturm, Joint estimation of shape and reflectance using multiple images with known illumination conditions. Int. J. Comput. Vision **86**(2–3), 192 (2010)

Chapter 2
Spectral Power Distributions and Reflectance Calculations for Robot Vision

Abstract The spectral power distributions (SPD) of outdoor light sources are not constant over time and atmospheric conditionas, which causes the appearance variation of a scene and common natural illumination phenomena, such as twilight, shadow, and haze/fog. Calculating the SPD of outdoor light sources at different time (or zenith angles) and under different atmo spheric conditions is of interest to physically based vision. In this chapter, for robot vision and its applications, we propose a feasible, simple, and effective SPD calculating method based on analyzing the transmittance functions of absorption and scattering along the path of solar radiation through the atmosphere in the visible spectrum. Our model has fewer parameters and is accurate enough to be directly applied in robot vision. It can be applied in robot vision tasks including spectral inverse calculation, lighting conversion, and shadowed image processing. The experimental results of the applications demonstrate that our calculation methods have practical values in robot vision. It establishes a bridge between image and physical environmental information, e.g., time, location, and weather conditions. Spectral reflectance is defined as the "fingerprint" of an object and is illumination invariant. It has many applications in color reproduction, imaging, robot vision, and computer graphics. In previous reflectance reconstruction methods, spectral reflectance is treated equally over the whole wavelength. However, human eyes or sensors in an imaging device usually have different weights over different wavelengths. In this chapter, we propose a method to reconstruct reflectance considering wavelength-sensitive function (WSF) that is constructed from sensor sensitive functions (or color matching functions). Our main idea is to achieve more accurate reconstruction at wavelengths where sensors have high sensitivities. This more accurate reconstruction can achieve better imaging or color reproduction performance. In our method, we generate a matrix through the Hadamard product of the reflectance matrix and the WSF matrix. We then obtain reconstructed reflectance by applying the singular value decomposition on the generated matrix. The experimental results show that our method can reduce 47% mean-square error and 55% Lab error compared with the classical PCA method.

J. Tian, *All Weather Robot Vision*, Research on Intelligent Manufacturing,
https://doi.org/10.1007/978-981-16-6429-8_2

2.1 Introduction

Solar irradiance is changed by atmospheric transmittance effects including absorption, reflecting, and scattering, which causes the spectral power distribution (SPD) of the light that ultimately reaches the Earth's surface to vary with time and air conditions. As shown in Fig. 2.1, atmospheric transmittance effects lead to many natural things related to robot vision applications such as the variation of sun and sky appearance, twilight, shadow, haze/fog, and cloud. Therefore, in robot vision, a SPD calculation method of outdoor light sources is useful to deal with the problems caused by different time (or zenith angles) and weather conditions.

In the robot vision community, the researchers usually employ Planck's blackbody radiation law to approximately calculate the SPDs of outdoor light sources. Finlayson et al. [8] and Tian et al. [28] applied Planck's blackbody radiation law to approximately estimate the SPDs of daylight and skylight for deriving shadow-invariant images and for detecting shadows, respectively. Planck's blackbody radiation law is usually applied to calculate the SPDs of outdoor light sources. It is well known that extraterrestrial solar irradiance can be approximated by the blackbody radiation at 5777 K. However, from Fig. 2.2, we can see that the extraterrestrial solar irradiance roughly follows the blackbody radiation at 5777 K in the near infrared spectrum while it does not follow blackbody radiation law well enough in the visible region. Figure 2.3 shows the chromaticity of the blackbody radiation and that of sunlight, skylight, and daylight. We find that the chromaticity of daylight is very near to Planckian locus, and so it can be accurately approximated by the blackbody radiation. In contrast, the chromaticity of sunlight and skylight deviates from that calculated by blackbody radiation noticeably, which indicates that noticeable errors occur when blackbody radiation is employed to model the SPDs of sunlight and skylight.

In meteorology, there are SPD calculation methods based on analyzing the transmittance functions of absorption and scattering along the path of solar radiation through the atmosphere. The latest versions of two representative SPD calculation models with wavelength resolution are SMARTS2 [9] and MODTRAN [5]. The SPDs calculated by these two methods can approximate real measured ones well. However, they may be too complex to be used in robot vision, because they contain

Fig. 2.1 Outdoor appearances vary with time and air conditions

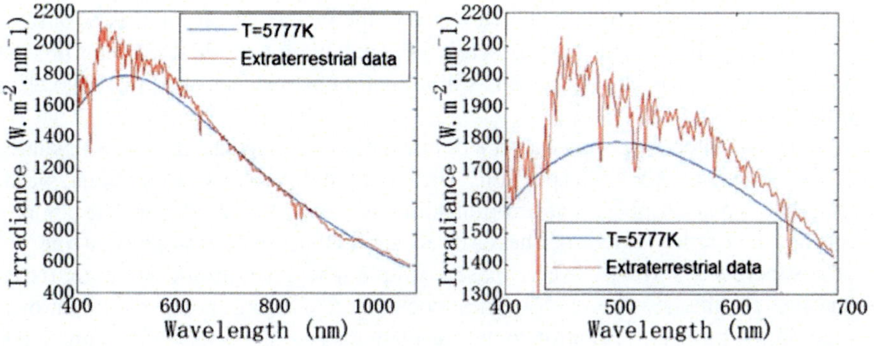

Fig. 2.2 Using blackbody radiation to approximate the true extraterrestrial data. The right image is the close-up view of the left one. Reprinted with permission from Tian et al., "Simple and Effective Calculations about Spectral Power Distributions of Outdoor Light Sources for Computer Vision," Optics Express, Vol. 24, Issue 7, pp. 7266–7286, 2016. © The Optical Society

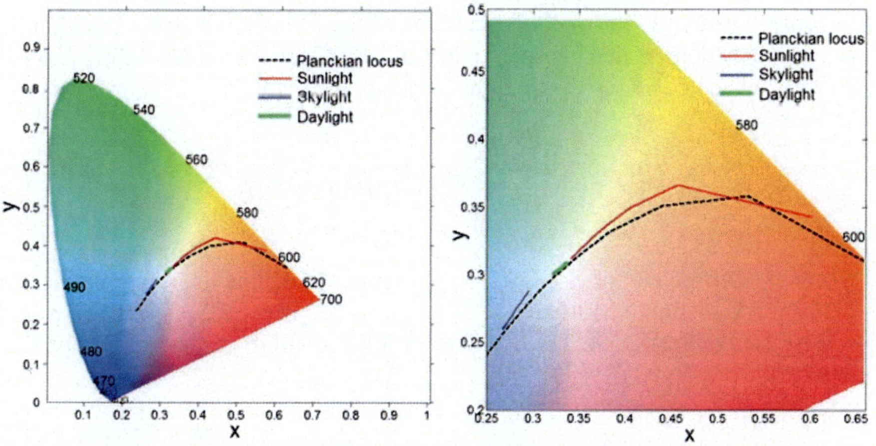

Fig. 2.3 CIE Chromaticity of sunlight, skylight, and daylight is compared with Planckian locus formed by color temperatures from 1000k to 500,000k. Reprinted with permission from Tian et al., "Simple and Effective Calculations about Spectral Power Distributions of Outdoor Light Sources for Computer Vision," Optics Express, Vol. 24, Issue 7, pp. 7266–7286, 2016. © The Optical Society

many parameters to be determined and unfortunately most of these meteorological parameters are not readily available for robot vision applications. Bird and Riordan [6] proposed a simpler method to calculate the SPD of outdoor light sources. Though Bird's method is much simpler than those in [5, 9], it is still difficult to be used in robot vision. In general, the existing methods are not developed for robot vision. They consider the full spectrum and involve many factors and parameters that have no effects on the visible spectrum. Besides, in all existing methods, the characteristics of human vision are not taken into account. Therefore, they may be more suitable for the research on meteorology than robot vision. Because most robot vision tasks concentrate on analyzing images captured in the visible spectrum, it is possible for us to develop the simpler and effective SPD calculation method that can be easily applied in robot vision tasks, such as illumination processing, shadow removal, and image relighting.

In this chapter, we propose a very simple method to calculate the SPDs of sunlight and skylight in the visible spectrum. Different from previous calculations, we first calculate the atmosphere's absorption, since some light emitted from the sun never reaches the Earth's surface neither as direct sunlight nor as diffuse skylight due to the atmosphere's absorption. We propose a simple new absorption calculation method and provide the new "total absorption coefficient." During the development of our method, we pay more attention to the wavelengths near the peaks of the human color matching functions (CMFs), to guarantee the accuracy of color reproduction when using our calculated SPDs to substitute the real ones.

Because our SPDs calculations are for imaging and robot vision, we develop our method based on image formation theories. Figure 2.4 shows the image formation procedure in outdoor environments. Since both our eyes and conventional cameras are only sensitive to the visible spectrum, all the calculations and analysis in this paper will be done within the spectrum 400–700. Given illumination SPD $E(\lambda)$ and object's reflectance $S(\lambda)$, the tristimulus values are,

$$F_H = \int_{400}^{700} E(\lambda)S(\lambda)Q_H(\lambda)\mathrm{d}\lambda \tag{2.1}$$

where $Q_H(\lambda)$ denotes the XYZ or sRGB color matching functions in three channels. In this work, tristimulus values in XYZ will be used in Sect. 2.2.3 and those in sRGB will be used in Sect. 2.2.4. The detail about XYZ to sRGB conversion can be found in [7].

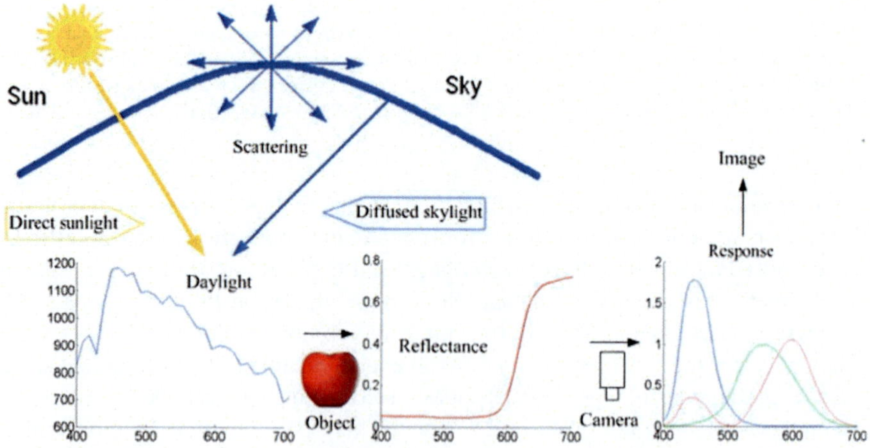

Fig. 2.4 Illustration of the image formation procedure

2.2 Outdoor Spectral Power Distributions Calculations for Robot Vision

2.2.1 Absorption of Extraterrestrial Irradiance Passing Through the Atmosphere

When the extraterrestrial irradiance passes through the atmosphere, it is attenuated by absorption, reflection, and scattering processes, and thereby its spectral composition is changed considerably. Some solar radiation will never reaches the Earth's surface neither as direct sunlight nor as diffuse skylight due to the atmosphere's absorption by ozone, nitrogen dioxide, mixed gases, and water vapor. We denote $E_{o\lambda}$ as extraterrestrial irradiance at mean Earth–sun distance for wavelength λ. After absorption, the irradiance becomes,

$$E_{\text{in}} = E_{o\lambda} \cdot T_{o\lambda} \cdot T_{N\lambda} \cdot T_{w\lambda} \cdot T_{u\lambda} \tag{2.2}$$

where $T_{o\lambda}$, $T_{N\lambda}$, $T_{w\lambda}$, and $T_{u\lambda}$ denote the transmittance functions for molecular ozone, nitrogen dioxide, molecular water vapor, and mixed gases, respectively. Extraterrestrial irradiance also varies with the sun–Earth distance which is determined by the Day Number of a year. Since this factor is independent of wavelength, it can be taken outside the integration and can be considered as exposure time in Eq. (1). Therefore, this factor can be neglected for robot vision application. $T_{o\lambda}$ and $T_{N\lambda}$ can be represented by,

$$T_{o\lambda} = \exp(-0.35 a_{o\lambda} m) \text{ [15]} \tag{2.3}$$

$$T_{N\lambda} = \exp(-0.0016 a_{n\lambda} m) \text{ [9, 25]} \tag{2.4}$$

where $a_{o\lambda}$ and $a_{n\lambda}$ are the absorption coefficients of molecular ozone and nitrogen dioxide, respectively. In [6, 15], $T_{w\lambda}$ and $T_{u\lambda}$ are given by,

$$T_{w\lambda} = \exp[-0.2385 a_{w\lambda} W m / (1 + 20.07 a_{w\lambda} W m)^{0.45}] \tag{2.5}$$

$$T_{u\lambda} = \exp[-1.41 a_{u\lambda} m / (1 + 118.93 a_{u\lambda} m)^{0.45}] \tag{2.6}$$

where $a_{u\lambda}$ and $a_{w\lambda}$ are the absorption coefficients of molecular water vapor and mixed gases, respectively. W is water vapor amount, and its typical value is 1.6. m is air mass that is calculated by,

$$m = \sec(\theta) \tag{2.7}$$

θ denotes zenith angle in this paper. Unlike that $T_{o\lambda}$ and $T_{N\lambda}$ take noticeable effects in the visible spectrum, $T_{u\lambda}$ mainly takes effect on wavelengths longer than 1000 nm and in the visible spectrum takes effect in 687–691 nm [22]; $T_{w\lambda}$ mainly takes effect on wavelengths that are longer than 800 nm and in the visible spectrum takes effect in 690–700 nm [14]. $T_{u\lambda}$ and $T_{w\lambda}$ have little effect in the visible spectrum but their computations are more complex than those of $T_{o\lambda}$ and $T_{N\lambda}$. More importantly, $T_{u\lambda}$ and $T_{w\lambda}$ have different forms with $T_{o\lambda}$ and $T_{N\lambda}$, which brings difficulties for merging these four terms. Therefore, we here want to seek simple and effective expressions for $T_{u\lambda}$ and $T_{w\lambda}$. Equations (2.5) and (2.6) have similar formulas, thus we first merge $T_{u\lambda}$ into $T_{w\lambda}$, i.e.,

$$T_{w\lambda}(a_{w\lambda} + \varepsilon a_{u\lambda}|m) = T_{w\lambda}(a_{w\lambda}|m) \cdot T_{u\lambda}(a_{u\lambda}|m) \tag{2.8}$$

where ε can be determined by,

$$\varepsilon = \arg\min \sum_m \|T_{w\lambda} \cdot T_{u\lambda} - T_{w\lambda}(a_{w\lambda} + \varepsilon a_{u\lambda})\|_2 \tag{2.9}$$

The result is $\varepsilon = 4.3$. We have,

$$T_{w\lambda}(a_{w\lambda} + 4.3a_{u\lambda}) \approx T_{w\lambda} \cdot T_{u\lambda} \tag{2.10}$$

The two attenuations caused by water vapor and mixed gases are merged into a single attenuation coefficient. We define the new term as $T'_{w\lambda}$,

$$T'_{w\lambda} = \exp[-0.2385a'_{w\lambda}Wm/(1 + 20.07a'_{w\lambda}Wm)^{0.45}] \tag{2.11}$$

where $a'_{w\lambda} = a_{w\lambda} + 4.3a_{u\lambda}$.

For different m, we find that data counterparts of $[a'_{w\lambda}, \log(T'_{w\lambda})]$ satisfy a power function. So in the following we write $T'_{w\lambda}$ as the form of $T_{o\lambda}$ and $T_{N\lambda}$. We define a new transmittance function of water vapor and uniform mixture gas as,

$$T^{\text{new}}_{w\lambda} = \exp[p(a'_{w\lambda})^q m] \tag{2.12}$$

with,

$$p, q = \arg\min \sum_m \left\| p(a'_{w\lambda})^q m - \log(T'_{w\lambda}) \right\|_2 \tag{2.13}$$

The results of Eq. (2.13) are $p = -0.055$ and $q = 0.56$. We apply the simple exhaustive search method to solve the optimization problem in Eqs. (2.9) and (2.13). The new attenuation expression of water vapor and mixed gases is approximated as a decreasing exponential function of the combined coefficient.

$$T^{\text{new}}_{w\lambda} = \exp[-0.055(a'_{w\lambda})^{0.56} m] \tag{2.14}$$

Then since all the absorption factors can be expressed by a decreasing exponential function, we rewrite Eq. (2.2) as,

$$E_{\text{in}} = E_{o\lambda} \cdot T_\lambda \tag{2.15}$$

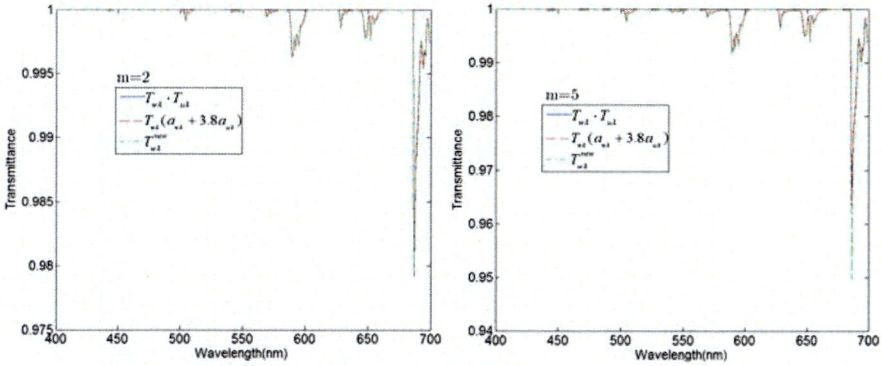

Fig. 2.5 New attenuation and transmittance expressions of water vapor and uniformly mixed gases produce similar results as the original expressions. Reprinted with permission from Tian et al., "Simple and Effective Calculations about Spectral Power Distributions of Outdoor Light Sources for Computer Vision," Optics Express, Vol. 24, Issue 7, pp. 7266–7286, 2016. © The Optical Society

where,

$$T_\lambda = \exp(-\tau'(\lambda) \cdot m) \tag{2.16}$$

The results in Fig. 2.5 show that our new simple expression produces similar results compared with the combined effects of the original two terms in the visible spectrum. We name the proposed T_λ as "total absorption transmittance function" and $\tau'(\lambda)$ as "total absorption coefficient" that is represented by,

$$\tau'(\lambda) = 0.35a_{o\lambda} + 0.0016a_{n\lambda} + 0.055{a'_{w\lambda}}^{0.56} \tag{2.17}$$

The total absorption coefficient can be found in the Appendix (Table 4) and is plotted in Fig. 2.6, in which $a_{o\lambda}, a_{n\lambda}, a_{u\lambda}$, and $a_{w\lambda}$ are derived from [9] with $a_{o\lambda} = A_{o\lambda}$, $a_{n\lambda} = A_{n\lambda}, a_{u\lambda} = 100A_{g\lambda}$, and $a_{w\lambda} = 100A_{w\lambda}$. The SPD of extraterrestrial irradiance and an example of its result after absorption at $m = 2$ is shown in Fig. 2.7.

Using our proposed total absorption coefficient and transmittance function, the absorption calculation of ozone, nitrogen dioxide, water vapor, and mixed gases can be considered all together. Equation (2.16) is much simpler to use.

2.2.2 Computing Direct Sunlight

The direct irradiance at Earth's surface normal to the direction of the sun can be calculated by,

$$E_{d\lambda} = E_{in} \cdot T_{r\lambda} \cdot T_{a\lambda} \tag{2.18}$$

where E_{in} is the SPD of extraterrestrial irradiance after absorption, and $T_{r\lambda}$, $T_{a\lambda}$ denote the transmittance functions of Rayleigh scattering and Aerosol scatting, respectively. The transmittance function for the Rayleigh scattering in [10] is adopted in this paper.

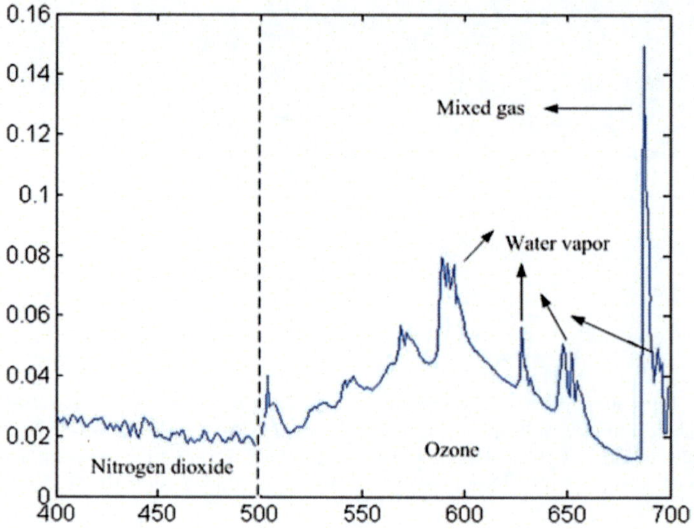

Fig. 2.6 Total absorption coefficient. Reprinted with permission from Tian et al., "Simple and Effective Calculations about Spectral Power Distributions of Outdoor Light Sources for Computer Vision," Optics Express, Vol. 24, Issue 7, pp. 7266–7286, 2016. © The Optical Society

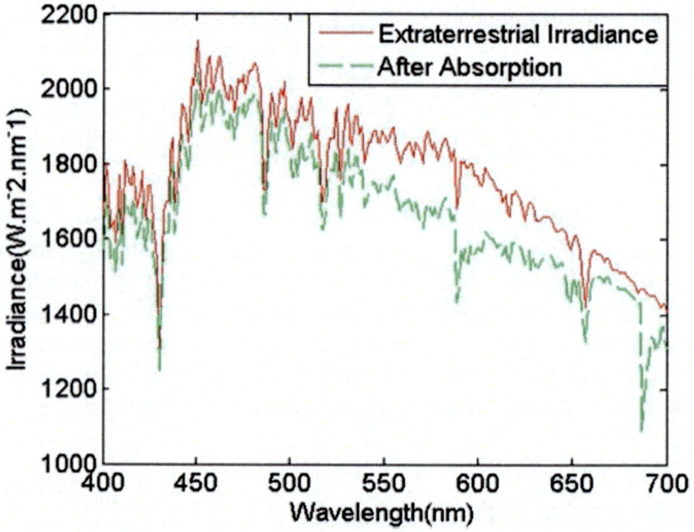

Fig. 2.7 SPD of extraterrestrial irradiance reported by World Radiation Center and that after absorption at air mass equals to 2. Reprinted with permission from Tian et al., "Simple and Effective Calculations about Spectral Power Distributions of Outdoor Light Sources for Computer Vision," Optics Express, Vol. 24, Issue 7, pp. 7266–7286, 2016. © The Optical Society

$$T_{r\lambda} = \exp(-0.008735\lambda^{-4.08} \cdot m) \tag{2.19}$$

The transmittance function for the aerosol scattering in [10] is adopted in this paper.

$$T_{a\lambda} = \exp(-\beta\lambda^{-\alpha} \cdot m) \tag{2.20}$$

where α is wavelength exponent and the value of 1.3 is commonly employed. The parameter β is the cleanliness index (turbidity) which varies from 0.0 to 0.4. For clear weather, turbid weather, and very turbid weather, its value is about 0.1, 0.2, and 0.3, respectively [10].

2.2.3 Computing Diffuse Skylight

The calculations of diffuse skylight in other works are complicated. Here we propose a very simple method. From Eqs. (2.19) and (2.20), we know that $T_{r\lambda}$ and $T_{a\lambda}$ describe the direct transmittance functions of Rayleigh scattering and Aerosol scattering, respectively. Thus, (1-$T_{r\lambda} \cdot T_{a\lambda}$) actually describes the sum of scattered light that can reach the Earth's surface and that is lost in the atmosphere. So the key point becomes how to determine the proportion of the scattered light that can reach the ground. Therefore, the diffuse skylight on the horizontal surface can be calculated by,

$$E_{s\lambda} = E_{in} \cdot \cos(\theta) \cdot (1 - T_{r\lambda} \cdot T_{a\lambda}) \cdot \kappa \tag{2.21}$$

where κ accounts for the proportion of the scattered light that can reach the ground.

As shown in Fig. 2.8, κ should be zenith-dependent. The longer light passes through the atmosphere, the more light will be lost, i.e., less scattered light can reach the ground. We determine κ by fitting CIE ΔE_{Lab} between the calculated SPDs by Eq. (2.21) and those from the true measurements by a SOC710 hyperspectral imager. In detail, the ideal white reflectance, $S(\lambda) = 1$, is applied in calculation. For the calculated SPD by our method, the XYZ tristimulus values are,

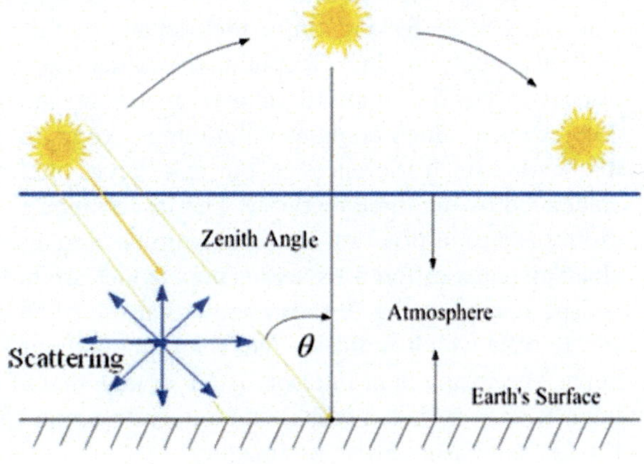

Fig. 2.8 Schematic diagram of the solar radiation onto the Earth surface. Reprinted with permission from Tian et al., "Simple and Effective Calculations about Spectral Power Distributions of Outdoor Light Sources for Computer Vision," Optics Express, Vol. 24, Issue 7, pp. 7266–7286, 2016. © The Optical Society

Table 2.1 Proportion of the scattered light that can reach the ground with different zenith angles

Angle	20	30	40	50	60	70	80
κ	0.65	0.65	0.62	0.62	0.60	0.58	0.52

$$F_H = \int_{400}^{700} E_{s\lambda} S(\lambda) Q_H(\lambda) \mathrm{d}\lambda. \quad (2.22)$$

For the measured SPD, the XYZ tristimulus values are,

$$F_H^{'} = \int_{400}^{700} E_{s\lambda}^{'} S(\lambda) Q_H(\lambda) \mathrm{d}\lambda. \quad (2.23)$$

The XYZ tristimulus values are converted to CIELAB color space and ΔE_{Lab} is applied to evaluate the error between $E_{s\lambda}$ and $E_{s\lambda}^{'}$.

$$\Delta E_{\mathrm{Lab}} = \sqrt{\Delta L^2 + \Delta a^2 + \Delta b^2} \quad (2.24)$$

κ can be determined by,

$$\kappa = \arg\min \Delta E_{\mathrm{Lab}} \quad (2.25)$$

The reason that we apply XYZ values and $\Delta E_{\mathrm{Lab}}$to determine κ is to keep higher accuracy near the peaks of CMFs. The values of κ with different zenith angles are tabulated in Table 2.1.

2.2.4 *Experiments and Comparisons*

Figure 2.9 shows the comparisons of the calculated SPDs by our method with those by SMARTS2 method [9] and Bird's method [6] in clear weather, i.e., $\beta = 0.1$. The results generally denote that there are no significant discrepancies between the calculated SPDs by our simple method and by the other two complex meteorology ones, especially for direct sunlight. To evaluate the influences of the discrepancies on imagery, we simulate sRGB values (see the 2nd and 4th columns in Fig. 2.9) of Xrite colorchecker which contains 24 common colors in our daily life. From the results, it is hard to find differences by our naked eyes on the simulated images under SPDs calculated by the three methods. The last column in Fig. 2.9 shows the quantitative discrepancies of pixel values of the simulated images. For each color patch, in three channels, max differences versus max values are plotted as percent errors.

The results in Fig. 2.9 also show that the SPDs do not vary significantly from 20° to 40°, which indicates that the outdoor light sources are quite stable during noon. The variation of intensity is larger than that of spectrum. Because variation of intensity can be taken outside the integration and can be put into exposure time in Eq. (2.1), it has no effect on imagery.

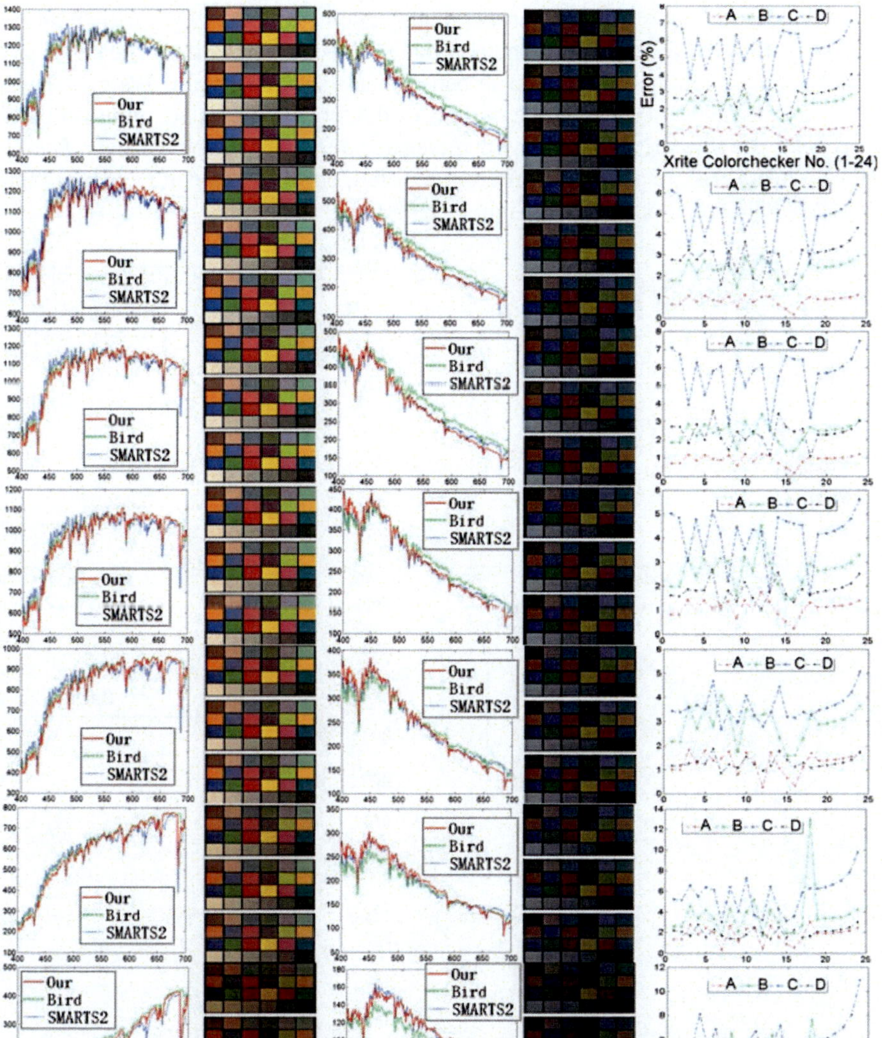

Fig. 2.9 Comparison of the calculated SPDs by our method and by Bird's method [6] and SMARTS2 method [9] at different zenith angles for $\beta = 0.1$. From top to bottom, the zenith angle equals to 20, 30,....,80°, respectively. The x-coordinate denotes wavelength (nm) and the y-coordinate denotes irradiance power ($\mathrm{W} \cdot \mathrm{m}^{-2} \cdot \mathrm{nm}^{-1}$). The 2nd column shows simulated Xrite colorchecker appearances under sunlight and the 4th column is that under skylight. Upper: Our; Middle: Bird's; Lower: SMARTS2. The last column is pixel value differences (percent). A: Direct sunlight compared with Bird's method. B: Direct sunlight compared with SMARTS2 method. C: Diffuse skylight compared with Bird's method. D: Diffuse skylight compared with SMARTS2 method. Reprinted with permission from Tian et al., "Simple and Effective Calculations about Spectral Power Distributions of Outdoor Light Sources for Computer Vision," Optics Express, Vol. 24, Issue 7, pp. 7266–7286, 2016. © The Optical Society

Figure 2.10 shows the comparison of the calculated SPDs by our method with those by SMARTS2 and Bird's method under different turbidities. The consistency of the three methods on sunlight is better than that on skylight. We can see that for very clear weather $\beta = 0$, the outputs of the three methods are quite similar for skylight. The similarity degrades with increasing β. For very turbid weather, $\beta = 0.3$, the three outputs are obviously different. Overall, our results are more close to those of SMARTS2 method in the skylight calculation.

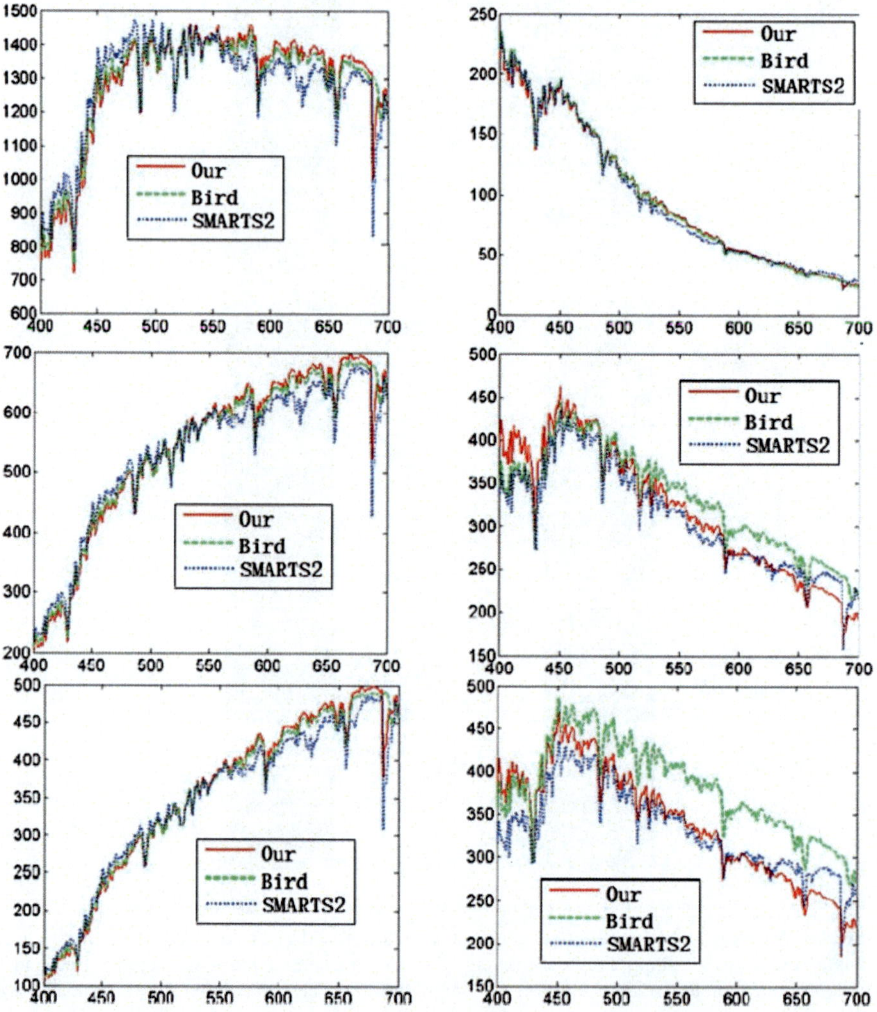

Fig. 2.10 Comparisons of the calculated SPDs by our method and those by Bird's method [6] and SMARTS2 method [9] under different turbidities. From top to bottom are $\beta = 0$, $\beta = 0.2$, and $\beta = 0.3$, respectively. Reprinted with permission from Tian et al., "Simple and Effective Calculations about Spectral Power Distributions of Outdoor Light Sources for Computer Vision," Optics Express, Vol. 24, Issue 7, pp. 7266–7286, 2016. © The Optical Society

The comparison on the formula expressions of our method with those of SMARTS2 and Bird's method is listed in Table 2.2. From the table, we can see that our expressions are much simpler than those in the other two methods. More importantly, in the other two methods, there are many parameters that may be hard to be determined in robot vision.

Based on Perez sky model [21, 23] proposed an improved sky model. It suggests an analytical approximation of the five distribution coefficients in Perez sky model to be a linear function of a single parameter, turbidity, by fitting the Perez formulas to the reference images. The Preetham model [23] also only requires the zenith angle and turbidity as varying parameters. Using Judd's characteristic vectors, this model can produce radiance spectral information of all points on the skyphere. Details about recovering radiance spectral information from sky chromaticity values based on Judd's characteristic vectors can be found in Appendix 5 in [23]. After integration of the incoming light from all the directions in the hemisphere, SPD of skylight irradiance can be calculated. Figure 2.11 shows the comparison between our method, SMARTS2, Bird's method, and Preetham method. We can find that, compared with our method, Preetham method is not accurate enough to match the two meteorology methods. This is caused by that some errors may arise when recovering full spectra from sky XYZ values. It only uses three principle basis. Another disadvantage of using the Preetham sky model to recover SPD is that it mainly recovers skylight SPD while it is difficult to recover SPD of direct sunlight and daylight.

We first demonstrate estimating illumination, zenith angle, turbidity, and camera spectral sensitivities from a single 24 color checker image. Rewriting $E(\lambda)$ in Eq. (2.1), we have,

$$F_H = \int_{400}^{700} E(\theta, T, \lambda) S(\lambda) \mathrm{Q}_H(\lambda) d\lambda \ , H = R, G, \ B \tag{2.26}$$

Jiang's et al. [11] show that camera spectral sensitivities (CSS) can be decomposed as $\mathrm{Q}_H = \sigma_\mathrm{H} \cdot \mathrm{C}_\mathrm{H}$, where $\sigma_H = \left[\sigma_{H,1}, \sigma_{H,2}\right]$; $C_H = \left[c_{H,1}, c_{H,2}\right]^T$ is the eigenvector matrix that is precomputed by a dataset including different CSS. Given a 24 color checker image and the reflectance spectrum as shown in Fig. 2.12, parameters σ_H, θ, T can be recovered by iteratively minimizing the following RMS, and then illumination SPDs and CSS can be recovered.

$$\sum_{n=1}^{24} \left\| F_H^n - \sum_{\lambda} E(\theta, T) \cdot S^n \cdot [\sigma_\mathrm{H} \cdot \mathrm{C}_\mathrm{H}] \right\| , H = R, G, B \tag{2.27}$$

Compared with Jiang's method that can recover CSS and one illuminations-SPD, as shown in Fig. 2.13, our method can recover CSS and SPDs of three lights, i.e., sunlight, daylight, and skylight simultaneously. More importantly, using our method, sun angle and turbidity can be recovered from the image while Jiang's method can only recover color temperature.

Table 2.2 Comparison of the formula expressions of our methods with those of SMARTS2 and Bird's method

	SMARTS2	Bird	Our
	$T_{u\lambda} = \exp[-1.41a_{u\lambda}m/(1+118.93a_{u\lambda}m)^{0.45}]$ $m_u = [\cos\theta + 4.57\theta^{0.07}(96-\theta)^{-1.7}]^{-1}$	$T_{u\lambda} = \exp[-(ma_{u\lambda}u_g)^{0.56}]$ $m = [\cos(\theta) + 0.15(93.885-\theta)^{-1.253}]^{-1}$	
	$T_{o\lambda} = \exp(-u_o a_{o\lambda} m_o)$ $m_o = [\cos\theta + 268\theta^{0.5}(115-\theta)^{-3.3}]^{-1}$	$T_{o\lambda} = \exp(-u_o a_{o\lambda} m_o)$ $m_o = (1+h_o/6370)/[\cos^2(\theta) + 2h_o/6370]^{0.5}$	
Absorption	$T_{N\lambda} = \exp(-u_n a_{n\lambda} m_n)$ $m_n = [\cos\theta + 602\theta^{0.5}(117-\theta)^{-3.5}]^{-1}$	Not mentioned	$T_\lambda = \exp(-\tau'(\lambda)\cdot m)$ $\tau'(\lambda) = 0.35a_{o\lambda} + 0.0016a_{n\lambda} + 0.055a_{w\lambda}^{\prime 0.56}$ $a'_{w\lambda} = a_{w\lambda} + 4.3a_{u\lambda}$ $m = \sec(\theta)$
	$T_{w\lambda} = \exp\{-[m_w w^{1.05} f_w{}^n B_w a_{w\lambda}]^c\}$ $f_w = [0.39 - 0.27\lambda + (0.46 + 0.24\lambda)P]$ $n = 0.88 + 0.025\lambda - 3.59\exp(-4.54\lambda)$ $c = 0.54 + 0.0033\lambda + 1.52\exp(-4.29\lambda)$ $B_w = h(m_w w)\exp(0.19 - 0.079m_w + 4.71E - 4m_w{}^2)$ $h(m_w w) = 0.62m_w w^{0.46} \text{if} a_{w\lambda} < 0.01$ $h(m_w w) = (0.53 + 0.25m_w w)^{0.45} otherwise$ $m_w = [\cos\theta + 3.11\theta^{0.1}(92-\theta)^{-1.38}]^{-1}$	$T_{w\lambda} = \exp[-0.2385a_{w\lambda}Wm/$ $(1 + 20.07a_{w\lambda}Wm)^{0.45}]$ $m = [\cos(\theta) + 0.15(93.885-\theta)^{-1.253}]^{-1}$ $W = 1.6$	

(continued)

Table 2.2 (continued)

	SMARTS2	Bird	Our
Diffuse irradiance	$I_{s\lambda} = I_{r\lambda} + I_{a\lambda} + I_{g\lambda}$ $I_{r\lambda} = \cos(\theta) \cdot E_{o\lambda} \cdot D \cdot T_{aa\lambda} \cdot T_{o\lambda} \cdot T_{n\lambda} \cdot$ $T_{w\lambda}(1 - T_{r\lambda}{}^{0.9})0.5F_{R2}$ $I_{a\lambda} = \cos(\theta) \cdot E_{o\lambda} \cdot D \cdot T_{aa\lambda} \cdot T_{o\lambda} \cdot T_{n\lambda} \cdot$ $T_{u\lambda} \cdot T_{w\lambda} \cdot T_{r\lambda}(1 - T_{as\lambda})F_a$ $I_{g\lambda} = (I_{d\lambda} \cos\theta + I_{r\lambda} + I_{a\lambda})r_{s\lambda}r_{g\lambda}/(1 - r_{s\lambda}r_{g\lambda})$ $F_{R2} = 1, \text{if} \tau_{R\lambda} < \tau_{Rm}$, and $F_{R2} = \exp\left[-\left(\frac{\tau_{R\lambda}-\tau_{Rm}}{\sigma_R}\right)^{0.72+\cos\theta}\right]$ $\sigma_R = 3.65 - 2.3\exp(-4\cos\theta)$ $\tau_{Rm} = 0.17[1 - \exp(-8\cos\theta)]$ $T_{aa\lambda} = \exp[-m_a(1 - \varpi_0)\tau_{a\lambda}]$ $\varpi_0 = 0.94 - 0.09\exp(1 - 3.38\lambda)$ $T_{as\lambda} = T_{a\lambda}/T_{aa\lambda}$ $F_{a1} = 1 - 0.5\exp[(a_{s0} + a_{s1}\cos\theta)\cos\theta]$ $a_{s0} = [1.5 + (0.16 + 0.41F_g)F_g]F_g$ $a_{s1} = [0.08 + (0.38 + 0.58F_g)F_g]F_g$ $F_g = \ln(1 - g)$ $F_{a2} = 1 \text{for} \tau_{as\lambda} \leq 2$, orelse $F_{a2} = \exp\left[-\left(\frac{\tau_{as\lambda}-2}{\sigma_a}\right)^{\xi}\right]$ $\varsigma = -0.5 + \exp[0.24(\cos\theta_0)^{-1.24}]$ $\cos\theta_0 = Max(0.05, \cos\theta)$ $\sigma_a = Max\{1, 3.5 - (4.53 - 0.82\tau_{as\lambda})\cos\theta +$ $(8.26 - 6.02\tau_{as\lambda})\cos^2\theta\}$	$I_{s\lambda} = I_{r\lambda} + I_{a\lambda} + I_{g\lambda}$ $I_{r\lambda} = \cos(\theta) \cdot E_{o\lambda} \cdot D \cdot T_{aa\lambda} \cdot T_{o\lambda} \cdot T_{w\lambda}(1 - T_{r\lambda}{}^{0.95})0.5$ $I_{a\lambda} = \cos(\theta) \cdot E_{o\lambda} \cdot D \cdot T_{aa\lambda} \cdot T_{o\lambda}$ $\cdot T_{w\lambda} \cdot T_{r\lambda}{}^{1.5}(1 - T_{as\lambda})F_s$ $I_{g\lambda} = (I_{d\lambda}\cos\theta + I_{r\lambda} + I_{a\lambda})r_{s\lambda}r_{g\lambda}/(1 - r_{s\lambda}r_{g\lambda})$ $T_{as\lambda} = \exp(-\omega_r \beta\lambda^{-\alpha} m)$ $T_{aa\lambda} = \exp(-(1 - \omega_r)\beta\lambda^{-\alpha} m)$ $\omega_r = 0.945\exp\{-0.095[\ln(\lambda/0.4)]^2\}$ $F_s = 1 - 0.5\exp[(\text{AFS} + \cos(\theta)\text{BFS})\cos(\theta)]$ $r_{s\lambda} = T'_{o\lambda} \cdot T'_{w\lambda} \cdot T'_{aa\lambda}[0.5(1 - T'_{r\lambda}) +$ $(1 - F'_s)T'_{r\lambda}(1 - T'_{as\lambda})]$ AFS = ALG[1.459 + ALG(0.1595 + ALG0.4129)] BFS = ALG[0.0783 + ALG(−0.3824 − ALG0.5874)] ALG = ln(1 − 0.65)	$I_{s\lambda} = E_{in} \cdot \cos(\theta) \cdot (1 - T_{r\lambda} \cdot T_{a\lambda}) \cdot \kappa$ $E_{in} = E_{o\lambda} \cdot T_{\lambda}$
Parameters	Earth–sun distance, pressure, ozone, nitrogen dioxide,mixed gases, water vapor, carbon dioxide, albedo, tilt albedo, zenith angle, turbidity	Earth–sun distance, pressure, ozone, water vapor, albedo, aerosol optical depth, zenith angle, turbidity	zenith angle, turbidity

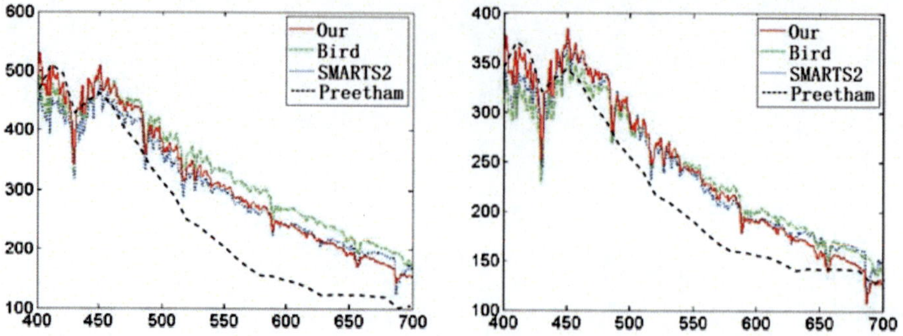

Fig. 2.11 Compared with Preetham model. Left: at 30° zenith angle. Right: at 60° zenith angle. Reprinted with permission from Tian et al., "Simple and Effective Calculations about Spectral Power Distributions of Outdoor Light Sources for Computer Vision," Optics Express, Vol. 24, Issue 7, pp. 7266–7286, 2016. © The Optical Society

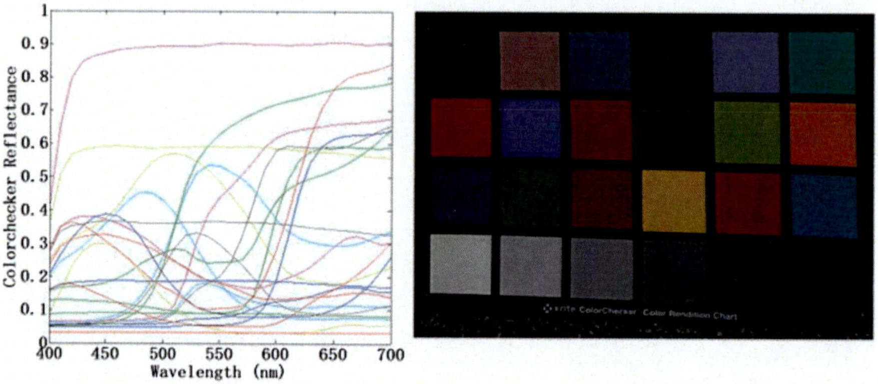

Fig. 2.12 A 24 color checker image captured by a Canon 5D Mark II camera under skylight and its corresponding spectral reflectance. Reprinted with permission from Tian et al., "Simple and Effective Calculations about Spectral Power Distributions of Outdoor Light Sources for Computer Vision," Optics Express, Vol. 24, Issue 7, pp. 7266–7286, 2016. © The Optical Society

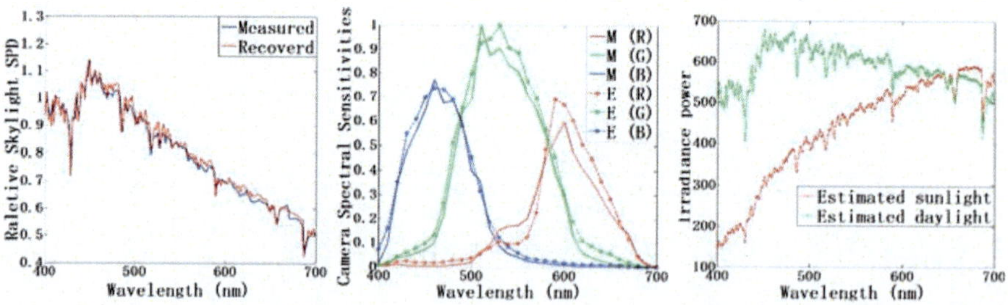

Fig. 2.13 Recovered skylight spectrum and the recovered CSS of Canon 5D Mark II. In the second figure, M and E are the abbreviations of "Measured" and "Estimated," respectively. Reprinted with permission from Tian et al., "Simple and Effective Calculations about Spectral Power Distributions of Outdoor Light Sources for Computer Vision," Optics Express, Vol. 24, Issue 7, pp. 7266–7286, 2016. © The Optical Society

Illumination is one of the important components of imaging. Understanding the properties of spectral distributions of outdoor light and their dynamical changes at different times and under different atmospheric conditions is of interest to robot vision. In this chapter, we proposed a simple method for computing SPDs of sunlight and skylight for a given zenith angle and turbidity. In the robot vision community, researchers usually apply the blackbody radiation or the eigenvectors of daylight [12] to approximate real SPDs. Compared with these two kinds of methods, our method has two advantages. First of all, SPDs calculated by our method can approximate those calculated by meteorology ones. The second is that our method can simultaneously calculate SPDs of daylight, sunlight, and skylight. The advantages especially the second one are important for our applications. These applications are difficult to accomplish by the blackbody or the eigenvector model since they cannot simultaneously calculate the corresponding SPDs of daylight, sunlight, and skylight.

Our method is much simpler than the existing atmospheric methods and shows its advantages in a number of robot vision applications. Different from other models that need many parameters such as ozone, nitrogen dioxide, mixed gases, and water vapor that have no close relation to robot vision and are not easily obtained in robot vision, our model only requires two parameters that are most related to physically based robot vision: sun angle (geometry) and turbidity (weather). Our model establishes a bridge between image and physical environmental information, e.g., time, location, and weather conditions. Since we focus on robot vision applications rather than atmosphere physics or meteorology, the basic calculations in atmosphere physics, e.g., transmission functions of scattering, are still based on previous atmospheric sciences literature.

2.3 Wavelength-Sensitive Function Controlled Reflectance Reconstruction

Spectral reflectance is a significant physical property of materials and is robust to illumination variation. It is useful in many applications. For example, in computer graphics, the increased dimensionality from tristimulus values to full spectral reflectance allows more accurate illumination computation [1, 4]. In image relighting, the reflectance spectra can be used to estimate the color appearance of a scene under various illuminants [18, 27]. In scene classification, spectral reflectance has been successfully applied as a classification and recognition descriptor that is robust to changes in illumination and geometry [24]. To summarize, more accurate reflectance reconstruction will promote the progress of these research topics.

A set of reflectance vectors $\mathbf{r}_i \in \boldsymbol{R}^n (i = 1 \cdots m)$ can be represented by $\mathbf{r}_i = \sum_{j=1}^{n} \alpha_j \mathbf{b}_j, m \geq n$, where $\mathbf{b}_j$ are known as basis function vectors. Spectral reflectance can be approximated by using only a few basis function vectors [17, 19], i.e., $\hat{\mathbf{r}}_i \approx \sum_{j=1}^{d} \alpha_j \mathbf{b}_j, d \ll n$. In practice, the number of basis functions d usually corresponds to the channel number of an imaging device. The existing methods use either a multispectral camera system ($d > 3$) [26] or a tricolor camera system ($d = 3$) [16] for reflectance reconstruction.

Let $\mathbf{R} = [\mathbf{r}_1\, \mathbf{r}_2 \cdot\cdot\cdot \mathbf{r}_m]^T$ and $\hat{\mathbf{R}} = [\hat{\mathbf{r}}_1\ \hat{\mathbf{r}}_2\ \cdots\ \hat{\mathbf{r}}_m]^T$ be two matrices that contain original and reconstructed reflectance vectors, respectively. The basis functions can be chosen by minimizing the following error function:

$$\min_{\mathbf{b}_j} \left\| \mathbf{R} - \hat{\mathbf{R}} \right\|_2^2 \qquad j = 1{,}2 \cdots d \tag{2.28}$$

The solution of Eq.(2.28) can be obtained by classical methods, such as singular value decomposition (SVD) or principal component analysis (PCA) [1, 2, 13, 20]. We first define a centered matrix as $\mathbf{X} = [\mathbf{r}_1-\bar{\mathbf{r}}\ \ \mathbf{r}_2-\bar{\mathbf{r}} \cdot\cdot\cdot \mathbf{r}_m-\bar{\mathbf{r}}]^T$, where $\bar{\mathbf{r}}$ denotes the mean reflectance vector, i.e., $\bar{\mathbf{r}} = \frac{1}{m}\sum_{i=1}^{m}\mathbf{r}_i$. The SVD expression of $\mathbf{X}$ can be written as $\mathbf{X} = \sum_{j=1}^{n}\sigma_j\mathbf{u}_j\mathbf{v}_j^T$, where σ_j, $\mathbf{u}_j$, and $\mathbf{v}_j$ denote the singular values, m-dimensional left singular vectors, and n-dimensional right singular vectors of $\mathbf{X}$, respectively. The basis functions $\mathbf{b}_j$ correspond to the column vectors $\mathbf{v}_j$, i.e., $\mathbf{b}_j = \mathbf{v}_j$ $(j = 1, 2, ...n)$. If we define the vector $\mathbf{h} = [1, 1, \cdots, 1]_m^T$, matrix $\mathbf{R}$ can be expressed by

$$\mathbf{R} = \sum_{j=1}^{n}\sigma_j\mathbf{u}_j\mathbf{v}_j^T + \mathbf{h}\otimes\bar{\mathbf{r}} \tag{2.29}$$

Here ⊗ denotes the tensor product of vectors. Similarly, the reconstructed spectral reflectance matrix by PCA (or SVD) method can be obtained by

$$\hat{\mathbf{R}}_0 = \sum_{j=1}^{d}\sigma_j\mathbf{u}_j\mathbf{v}_j^T + \mathbf{h}\otimes\bar{\mathbf{r}} \tag{2.30}$$

Figure 2.14 shows one example of the original and reconstructed reflectance obtained by the PCA method. We note that the entire spectral reflectance receives equal treatments along different wavelengths in the PCA-based reconstruction process. Therefore, the error distribution between the original and reconstructed

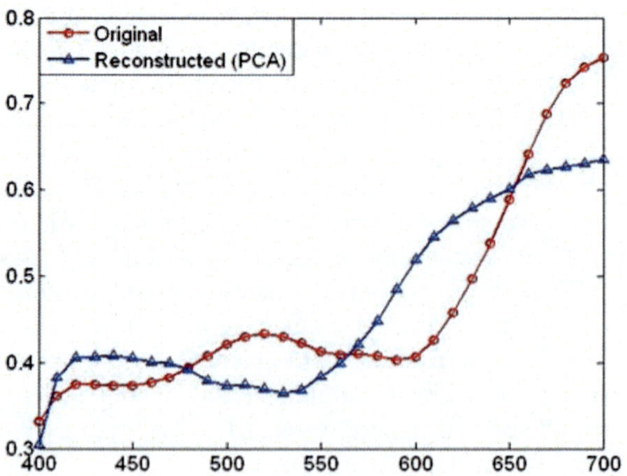

Fig. 2.14 One example of reflectance reconstructed by the PCA method. Reprinted with permission from Jiandong Tian and Yandong Tang, "Wavelength-sensitive-function controlled reflectance reconstruction," Optics Letters, Vol. 38, Issue 15, pp. 2818–2820, 2013. © The Optical Society

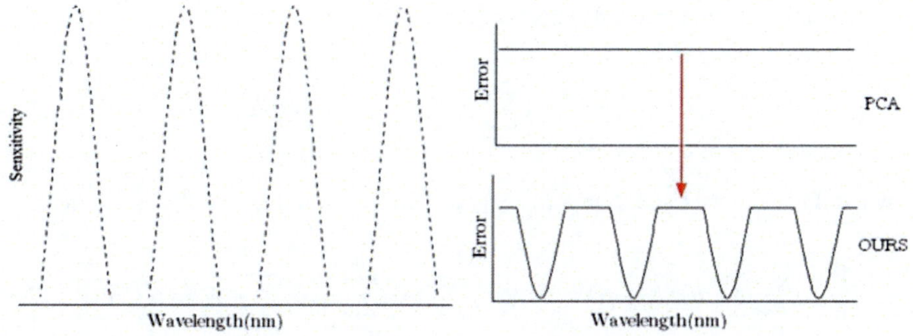

Fig. 2.15 Error distribution of PCA and our method. Left, illustration of sensor spectral sensitivities; upper right, error distribution of PCA method; lower right, expected error distribution of our method. Reprinted with permission from Jiandong Tian and Yandong Tang, "Wavelength-sensitive-function controlled reflectance reconstruction," Optics Letters, Vol. 38, Issue 15, pp. 2818–2820, 2013. © The Optical Society

reflectance tends to be a uniform distribution as illustrated on the upper right plot of Fig. 2.15. However, sensor response in each channel usually has different wavelengths as illustrated on the left plot of Fig. 2.15. Here, we aim to make the reconstructed reflectance more accurate near the peaks of sensor sensitivities to achieve better imaging or color reproduction performance. The error distribution is expected to appear as the shape shown in the lower right plot of Fig. 2.15, which is a schematic drawing of our idea. We expect to achieve more accurate reconstruction at wavelengths where sensors have high sensitivities. In other words, the higher the sensitivity, the lower the reconstruction errors.

Suppose that the sensor response functions of an imaging device are denoted by $f_i(\lambda)$, $i = 1, \cdot\cdot d$. Let $Q(\lambda) = \sum_{i=1}^{d} f_i(\lambda)$ be a one-dimensional wavelength-sensitive function (WSF). After sampling $Q(\lambda)$ at the resolution of n columns, we can generate a vector $\mathbf{q}$. Further, we build a WSF matrix $\mathbf{W}$ as

$$\mathbf{W} = [\mathbf{q}\ \mathbf{q} \cdots \mathbf{q}]_{m \times n}^{T} \tag{2.31}$$

Our basis functions are obtained by

$$\min_{\mathbf{b}_j} \left\| \mathbf{W} \circ (\mathbf{R} - \hat{\mathbf{R}}) \right\|_2^2 \quad j = 1,2 \ldots d \tag{2.32}$$

Here $\circ$ denotes the Hadamard product ($\mathbf{C} = \mathbf{A} \circ \mathbf{B} : c_{ij} = a_{ij} b_{ij}$). Let $\widetilde{\sigma_j}$, $\widetilde{\mathbf{u}_j}$ and $\widetilde{\mathbf{v}_j}$ denote the singular values, left singular vectors, and right singular vectors of $\mathbf{W} \circ \mathbf{X}$, respectively. Similar to Eq. (2.30), we have

$$\mathbf{W} \circ \mathbf{R} = \sum_{j=1}^{n} \widetilde{\sigma_j} \widetilde{\mathbf{u}_j} \widetilde{\mathbf{v}_j^T} + \mathbf{W} \circ \mathbf{h} \otimes \bar{\mathbf{r}} \tag{2.33}$$

Then the new reconstructed reflectance is obtained by

$$\hat{\mathbf{R}}_1 = \mathbf{D} \circ (\sum_{j=1}^{d} \widetilde{\sigma}_j \widetilde{\mathbf{u}}_j \widetilde{\mathbf{v}_j^T} + \mathbf{W} \circ \mathbf{h} \otimes \bar{\mathbf{r}}) \tag{2.34}$$

where $\mathbf{D}$ is a matrix satisfying $d_{ij} = 1/w_{ij}$. In Eq. 2.34, for any $\beta > 0$, we have,

$$\frac{\mathbf{D}}{\beta} \circ (\sum_{j=1}^{d} \widetilde{\beta\sigma_j} \widetilde{\mathbf{u}}_j \widetilde{\mathbf{v}_j^T} + \beta \mathbf{W} \circ \mathbf{h} \otimes \bar{\mathbf{r}}) = \mathbf{D} \circ (\sum_{j=1}^{d} \widetilde{\sigma}_j \widetilde{\mathbf{u}}_j \widetilde{\mathbf{v}_j^T} + \mathbf{W} \circ \mathbf{h} \otimes \bar{\mathbf{r}}) \tag{2.35}$$

Therefore, $\hat{\mathbf{R}}_1$ is invariant to scale variation of $\mathbf{W}$, and there is no need to consider the value range of $Q(\lambda)$.

In our experiment, we take XYZ color matching functions (400–700 nm) to test our method, since XYZ color space is the standard space in color, imaging, and related sciences. Figure 2.16 shows the XYZ color matching functions (left) and the WSF (right) generated by adding the three matching functions.

To the best of our knowledge, the reflectance spectral data introduced in [3] are the most complete data. They contain 1995 different surface reflectances, including the 24 Macbeth ColorChecker patches, 1269 Munsell chips, 120 Dupont paint chips, and 577 natural objects. Therefore, this reflectance dataset is employed to test our method.

By performing calculation on this reflectance dataset, we obtain that the SVDs of $\mathbf{W} \circ \mathbf{R}$ and $\mathbf{R}$ satisfy

$$\| \sum_{j=d+1}^{n} \widetilde{\sigma}_j \widetilde{\mathbf{u}}_j \widetilde{\mathbf{v}}_j^T \|_2^2 < \| \mathbf{W} \circ \sum_{j=d+1}^{n} \sigma_j \mathbf{u}_j \mathbf{v}_j^T \|_2^2 \tag{2.36}$$

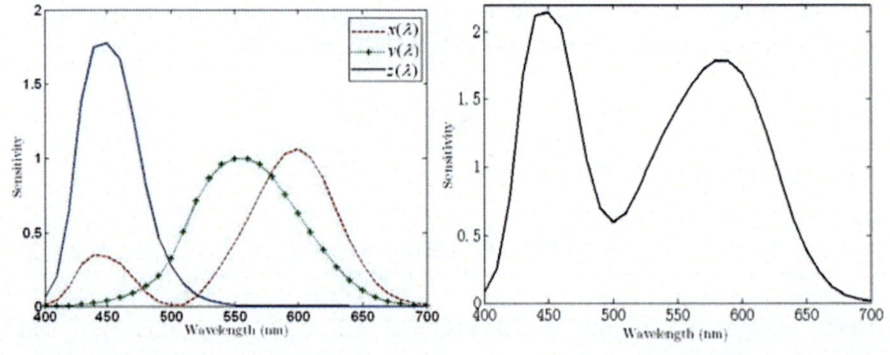

Fig. 2.16 XYZ color matching functions (left) and the WSF (right) generated by adding three matching functions. Reprinted with permission from Jiandong Tian and Yandong Tang, "Wavelength-sensitive-function controlled reflectance reconstruction," Optics Letters, Vol. 38, Issue 15, pp. 2818–2820, 2013. © The Optical Society

In the following, we will prove that the reconstructed reflectance $\hat{\mathbf{R}}_1$ obtained by our method and $\hat{\mathbf{R}}_0$ obtained by the PCA method satisfy the following inequality:

$$\left\|\mathbf{W} \circ (\mathbf{R} - \hat{\mathbf{R}}_1)\right\|_2^2 < \left\|\mathbf{W} \circ (\mathbf{R} - \hat{\mathbf{R}}_0)\right\|_2^2 \tag{2.37}$$

Proof of the inequality 2.37:

$$\begin{aligned}
\left\|\mathbf{W} \circ (\mathbf{R} - \hat{\mathbf{R}}_1)\right\|_2^2 &= \left\|\mathbf{W} \circ \mathbf{R} - \mathbf{W} \circ \hat{\mathbf{R}}_1\right\|_2^2 \\
&= \left\|\sum_{j=1}^{n} \widetilde{\sigma}_j \widetilde{\mathbf{u}}_j \widetilde{\mathbf{v}}_j^T + \mathbf{W} \circ \mathbf{H} \otimes \overline{\mathbf{r}} - \mathbf{W} \circ [\mathbf{D} \circ (\sum_{j=1}^{d} \widetilde{\sigma}_j \widetilde{\mathbf{u}}_j \widetilde{\mathbf{v}}_j^T + \mathbf{W} \circ \mathbf{H} \otimes \overline{\mathbf{r}})]\right\|_2^2 \\
&= \left\|\sum_{j=d+1}^{n} \widetilde{\sigma}_j \widetilde{\mathbf{u}}_j \widetilde{\mathbf{v}}_j^T\right\|_2^2 < \|\mathbf{W} \circ \sum_{j=d+1}^{n} \sigma_j \mathbf{u}_j \mathbf{v}_j^T \|_2^2 \\
&= \left\|\mathbf{W} \circ (\mathbf{R} - \hat{\mathbf{R}}_0)\right\|_2^2
\end{aligned} \tag{2.38}$$

Inequality 2.37 shows that our reconstructed reflectance is more accurate than PCA-based reconstructed reflectance at wavelengths where sensor sensitivities are high.

The comparison of our results, the PCA-based results, and the original reflectance is plotted in Fig. 2.17. We also superimpose the WSF on the plot to illustrate the sensitivity peaks.

Figure 2.17 clearly shows that our reconstructed results are more accurate than the PCA-based reconstructed results near the wavelengths where WSF is high. It also indicates that our results can achieve better imaging or color reproduction performance due to the small reconstruction errors at the sensitivity peaks. To evaluate the results quantitatively, we calculate the mean-square errors (MSEs) of $\left\|\mathbf{W} \circ (\mathbf{R} - \hat{\mathbf{R}}_0)\right\|_2^2$ and $\left\|\mathbf{W} \circ (\mathbf{R} - \hat{\mathbf{R}}_1)\right\|_2^2$ for the 1995 reflectances. To compare the performance of imaging or color reproduction for each reconstructed spectral reflectance, we also calculate the CIE XYZ tristimulus values under illuminant D65 and convert them to CIE Lab values. We then calculate the color difference between the actual and reconstructed Lab values. In Table 2.3, we statistically summarize these two results. Table 8.2 shows that our method achieves the overall smaller MSEs in terms of mean, minimum, and maximum errors. Our method also achieves more accurate color reproduction performance because of its smaller errors at the sensitivity peaks. For the statistically significant mean values, our method reduces MSE and Lab errors by 47% and 55%, respectively, when compared to the PCA-based method.

In this section, we propose a method to selectively control the WSF influence on the reflectance reconstruction process. Our reconstructed results are more accurate than the classical PCA-based results near the wavelengths corresponding to the peaks of the WSF. Consequently, our reconstructed results lead to more accurate color reproduction performance. These results further imply that our method will bring better performance in related applications, such as scene classification, imaging simulation, robot vision, and computer graphics.

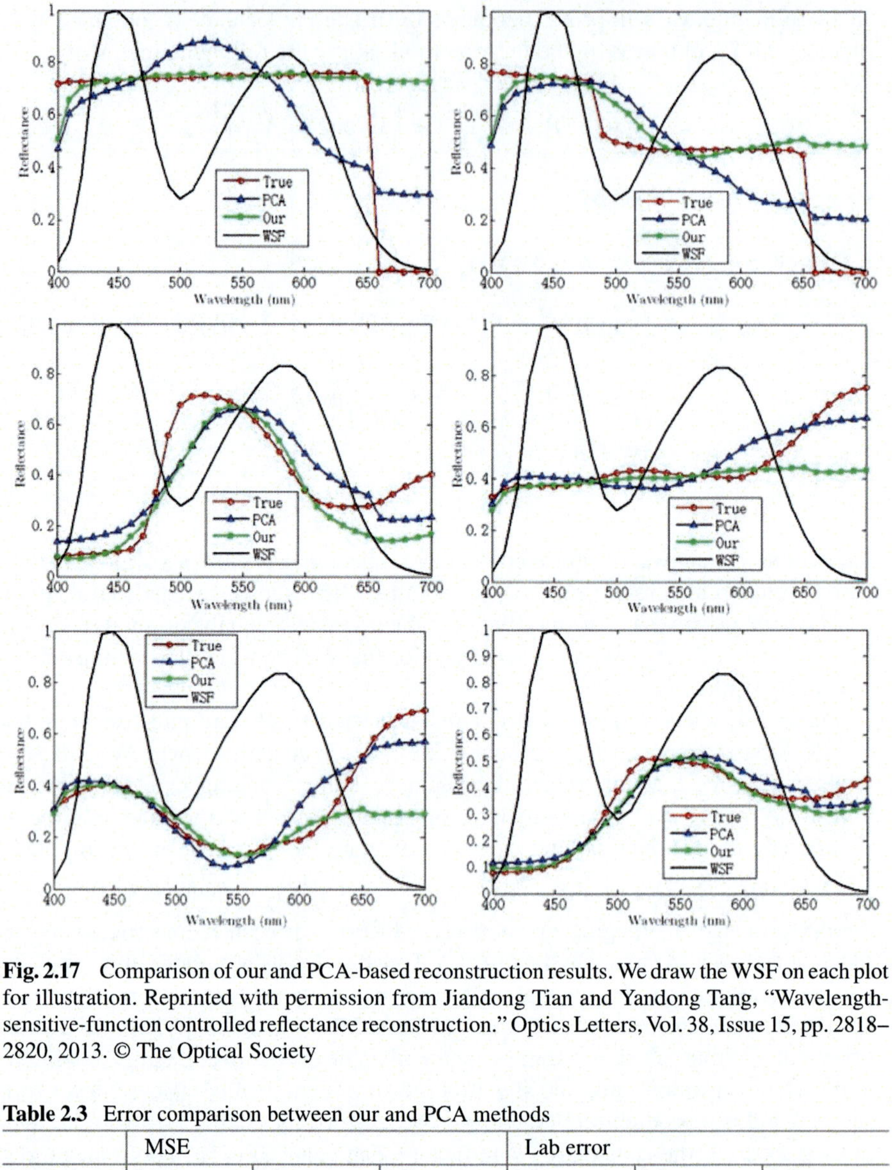

Fig. 2.17 Comparison of our and PCA-based reconstruction results. We draw the WSF on each plot for illustration. Reprinted with permission from Jiandong Tian and Yandong Tang, "Wavelength-sensitive-function controlled reflectance reconstruction." Optics Letters, Vol. 38, Issue 15, pp. 2818–2820, 2013. © The Optical Society

Table 2.3 Error comparison between our and PCA methods

	MSE			Lab error		
Methods	Mean	Max	Min	Mean	Max	Min
PCA	1.5×10^{-3}	6.9×10^{-2}	1.1×10^{-5}	29.45	1564.40	0.60
OURS	8.0×10^{-4}	6.0×10^{-2}	7.0×10^{-6}	13.24	113.76	0.45

Appendix

Our proposed total absorption coefficient

λ	τ′	λ	τ′	λ	τ′	λ	τ′	λ	τ′	λ	τ′
400	0.0270	450	0.0200	500	0.0190	550	0.0362	600	0.0572	650	0.0370
401	0.0258	451	0.0201	501	0.0212	551	0.0357	601	0.0537	651	0.0334
402	0.0244	452	0.0205	502	0.0255	552	0.0358	602	0.0524	652	0.0506
403	0.0249	453	0.0194	503	0.0285	553	0.0361	603	0.0507	653	0.0333
404	0.0259	454	0.0194	504	0.0398	554	0.0356	604	0.0498	654	0.0320
405	0.0251	455	0.0190	505	0.0296	555	0.0355	605	0.0486	655	0.0387
406	0.0236	456	0.0185	506	0.0307	556	0.0359	606	0.0482	656	0.0348
407	0.0248	457	0.0207	507	0.0311	557	0.0366	607	0.0475	657	0.0342
408	0.0267	458	0.0206	508	0.0301	558	0.0373	608	0.0469	658	0.0296
409	0.0264	459	0.0206	509	0.0287	559	0.0382	609	0.0462	659	0.0268
410	0.0260	460	0.0212	510	0.0271	560	0.0390	610	0.0452	660	0.0255
411	0.0246	461	0.0202	511	0.0248	561	0.0402	611	0.0448	661	0.0222
412	0.0260	462	0.0216	512	0.0234	562	0.0412	612	0.0440	662	0.0202
413	0.0265	463	0.0219	513	0.0225	563	0.0432	613	0.0432	663	0.0204
414	0.0247	464	0.0195	514	0.0213	564	0.0441	614	0.0424	664	0.0195
415	0.0230	465	0.0173	515	0.0211	565	0.0443	615	0.0417	665	0.0189
416	0.0230	466	0.0185	516	0.0211	566	0.0441	616	0.0411	666	0.0182
417	0.0240	467	0.0195	517	0.0216	567	0.0455	617	0.0403	667	0.0177
418	0.0245	468	0.0181	518	0.0226	568	0.0494	618	0.0397	668	0.0167
419	0.0253	469	0.0180	519	0.0231	569	0.0571	619	0.0390	669	0.0164
420	0.0255	470	0.0191	520	0.0231	570	0.0550	620	0.0387	670	0.0159
421	0.0254	471	0.0186	521	0.0231	571	0.0502	621	0.0383	671	0.0155
422	0.0253	472	0.0192	522	0.0236	572	0.0543	622	0.0375	672	0.0152
423	0.0258	473	0.0216	523	0.0244	573	0.0530	623	0.0372	673	0.0148
424	0.0251	474	0.0215	524	0.0271	574	0.0519	624	0.0368	674	0.0145
425	0.0226	475	0.0194	525	0.0289	575	0.0519	625	0.0363	675	0.0142
426	0.0223	476	0.0183	526	0.0285	576	0.0503	626	0.0370	676	0.0139
427	0.0242	477	0.0182	527	0.0297	577	0.0492	627	0.0396	677	0.0135
428	0.0230	478	0.0191	528	0.0294	578	0.0475	628	0.0566	678	0.0133
429	0.0218	479	0.0207	529	0.0302	579	0.0464	629	0.0434	679	0.0130
430	0.0238	480	0.0207	530	0.0308	580	0.0457	630	0.0436	680	0.0129
431	0.0229	481	0.0201	531	0.0313	581	0.0448	631	0.0373	681	0.0129
432	0.0219	482	0.0187	532	0.0312	582	0.0446	632	0.0395	682	0.0130
433	0.0236	483	0.0179	533	0.0309	583	0.0443	633	0.0364	683	0.0129
434	0.0253	484	0.0182	534	0.0303	584	0.0448	634	0.0344	684	0.0132
435	0.0247	485	0.0194	535	0.0299	585	0.0451	635	0.0345	685	0.0127
436	0.0220	486	0.0188	536	0.0297	586	0.0466	636	0.0333	686	0.0127
437	0.0200	487	0.0192	537	0.0299	587	0.0499	637	0.0328	687	0.1496
438	0.0231	488	0.0215	538	0.0311	588	0.0608	638	0.0308	688	0.1041
439	0.0240	489	0.0213	539	0.0315	589	0.0792	639	0.0305	689	0.0931
440	0.0206	490	0.0188	540	0.0330	590	0.0787	640	0.0298	690	0.0768
441	0.0194	491	0.0181	541	0.0372	591	0.0700	641	0.0295	691	0.0566
442	0.0213	492	0.0180	542	0.0387	592	0.0774	642	0.0287	692	0.0382
443	0.0260	493	0.0191	543	0.0361	593	0.0690	643	0.0292	693	0.0420
444	0.0260	494	0.0200	544	0.0378	594	0.0713	644	0.0288	694	0.0509
445	0.0250	495	0.0200	545	0.0380	595	0.0770	645	0.0300	695	0.0420
446	0.0236	496	0.0195	546	0.0402	596	0.0642	646	0.0365	696	0.0447
447	0.0245	497	0.0180	547	0.0384	597	0.0667	647	0.0447	697	0.0213
448	0.0231	498	0.0173	548	0.0373	598	0.0636	648	0.0507	698	0.0213
449	0.0201	499	0.0174	549	0.0365	599	0.0615	649	0.0476	699	0.0363
										700	0.0372

References

1. K. Ansari, S. Amirshahi, S. Moradian, Recovery of reflectance spectra from CIE tristimulus values using a progressive database selection technique. Color. Technol. **122**(3), 128–134 (2006)
2. F. Ayala, J.F. Echávarri, P. Renet, A.I. Negueruela, Use of three tristimulus values from surface reflectance spectra to calculate the principal components for reconstructing these spectra by using only three eigenvectors. J. Opt. Soc. Am. A **23**(8), 2020–2026 (2006)
3. K. Barnard, L. Martin, B. Funt, A. Coath, A data set for color research. Color. Res. Appl. **27**(3), 147–151 (2002)
4. S. Bergner, M.S. Drew, T. Möller, A tool to create illuminant and reflectance spectra for light-driven graphics and visualization. ACM Trans. Graph. **28**(1), 1–11 (2009)
5. A. Berk, L.S. Bernstein, D.C. Robertson, *Modtran: A Moderate Resolution Model for Lowtran* (Technical report, SpectralSciences Inc., Burlington, 1987)
6. R.E. Bird, C. Riordan, Simple solar spectral model for direct and diffuse irradiance on horizontal and tilted planes at the earth's surface for cloudless atmospheres. J. Climate Appl. Meteorol. **25**(1), 87–97 (1986)
7. I.E. Commission et al., Multimedia systems and equipment-colour measurement and management-part 2-1: colour management-default RGB colour space-srgb. IEC 61966-2-1 (1999)
8. G.D. Finlayson, M.S. Drew, C. Lu, Entropy minimization for shadow removal. Int. J. Comput. Vision **85**(1), 35–57 (2009)
9. C. Gueymard et al., *SMARTS2: A Simple Model of the Atmospheric Radiative Transfer of Sunshine: Algorithms and Performance Assessment* (Florida Solar Energy Center Cocoa, FL, 1995)
10. M. Iqbal. *An Introduction to Solar Radiation* (Elsevier, 2012)
11. J. Jiang, D. Liu, J. Gu, S. Süsstrunk, What is the space of spectral sensitivity functions for digital color cameras? in *2013 IEEE Workshop on Applications of Computer Vision (WACV)* (IEEE, 2013), , pages 168–179
12. D.B. Judd, D.L. MacAdam, G. Wyszecki, H. Budde, H. Condit, S. Henderson, J. Simonds, Spectral distribution of typical daylight as a function of correlated color temperature. Josa **54**(8), 1031–1040 (1964)
13. O.-S. Kwon, C.-H. Lee, K.-H. Park, Y.-H. Ha, Surface reflectance estimation using the principal components of similar colors. J. Imaging Sci. Technol. **51**(2), 166–174 (2007)
14. Z.L., G.P., T.Y, A new way to study water-vapor absorption coefficient. Mar. Sci. Bull. **7**(2), 67–71 (2005)
15. B. Leckner, The spectral distribution of solar radiation at the earth's surface–elements of a model. Sol. Energy **20**(2), 143–150 (1978)
16. C.-H. Lee, B.-J. Moon, H.-Y. Lee, E.-Y. Chung, Y.-H. Ha, Estimation of spectral distribution of scene illumination from a single image. J. Imaging Sci. Technol. **44**(4), 308–320 (2000)
17. L.T. Maloney, B.A. Wandell, Color constancy: a method for recovering surface spectral reflectance. J. Opt. Soc. Am. A **3**(1), 29–33 (1986)
18. J.L. Nieves, J. Hernandez-Andres, E. Valero, J. Romero, Spectral-reflectance linear models for optical color-pattern recognition. Appl. Opt. **43**(9), 1880–1891 (2004)
19. J.P.S. Parkkinen, J. Hallikainen, T. Jaaskelainen, Characteristic spectra of Munsell colors. J. Opt. Soc. Am. A **6**(2), 318–322 (1989)
20. M.S. Peercy, Linear color representations for full speed spectral rendering, in *Proceedings of the 20th Annual Conference on Computer Graphics and Interactive Techniques* (1993), pages 191–198
21. R. Perez, R. Seals, J. Michalsky, All-weather model for sky luminance distribution–preliminary configuration and validation. Sol. Energy **50**(3), 235–245 (1993)
22. J.H. Pierluissi, C.-M. Tsai, New lowtran models for the uniformly mixed gases. Appl. Opt. **26**(4), 616–618 (1987)

23. A.J. Preetham, P. Shirley, B. Smits, A practical analytic model for daylight, in *Proceedings of the 26th Annual Conference on Computer Graphics and Interactive Techniques* (1999), pages 91–100
24. A. Robles-Kelly, Learning a gaussian basis for spectra representation aimed at reflectance classification, in *CVPR 2011 Workshops* (IEEE, 2011), pages 88–95
25. R. Schroeder, J. Davies, Significance of nitrogen dioxide absorption in estimating aerosol optical depth and size distributions. Atmos. Ocean **25**(2), 107–114 (1987)
26. N. Shimano, Recovery of spectral reflectances of objects being imaged without prior knowledge. IEEE Trans. Image Process. **15**(7), 1848–1856 (2006)
27. J. Tian, Y. Tang, Linearity of each channel pixel values from a surface in and out of shadows and its applications (2011), pages 985–992
28. J. Tian, J. Sun, Y. Tang, Tricolor attenuation model for shadow detection. IEEE Trans. Image Process. **18**(10), 2355–2363 (2009)

Chapter 3
Imaging Modeling and Camera Sensitivity Recovery

Abstract Imaging modeling refers to the modeling and simulation of the raw data generation and the processing process from raw data to sRGB images. In this chapter, we model the imaging process, the main steps of which are: white balance, color space conversion (camera color space to XYZ space conversion, XYZ space to sRGB space conversion), gamma correction, Gamut mapping, and Tone mapping. Besides, we also present a method to recover camera spectral sensitivity. The experimental results show that our method is accurate, simple, and practical.

3.1 Overview of Digital Imaging

The process of image formation can be roughly divided into two parts: the camera imaging and the post-processing. During the whole process, the main factors affecting the imaging are the spectral power distribution of the light source, the spectral response function of the camera, the spectral reflectance of the object, and the post-processing of the camera, as shown in Fig. 3.1.

Light emitted from the light source passes through the air and reaches an object and then enters the camera through the reflection of the target object's surface. The camera collects the light reflected from the target surface and performs photoelectric conversion through the image sensor to generate the raw data. The process can be described as follows:

$$F_H^{\mathrm{RAW}} = \int_{400}^{700} E(\lambda) S(\lambda) Q_H(\lambda) \mathrm{d}\lambda; \qquad H = R, G, B \tag{3.1}$$

Among them, $E(\lambda)$ represents the spectral power distribution of the light source, $S(\lambda)$ represents the spectral reflectance of the object, and $Q_H(\lambda)$ represents the spectral sensitivity functions of the camera. After the above process, the camera

J. Tian, *All Weather Robot Vision*, Research on Intelligent Manufacturing,
https://doi.org/10.1007/978-981-16-6429-8_3

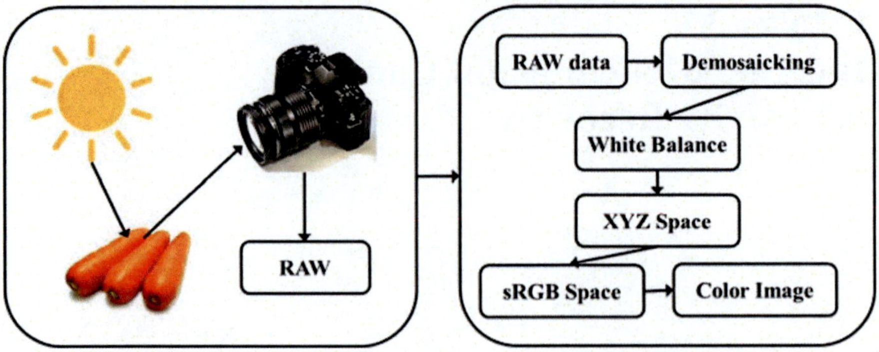

Fig. 3.1 Imaging process can be divided into two parts: raw image formation and post-processing

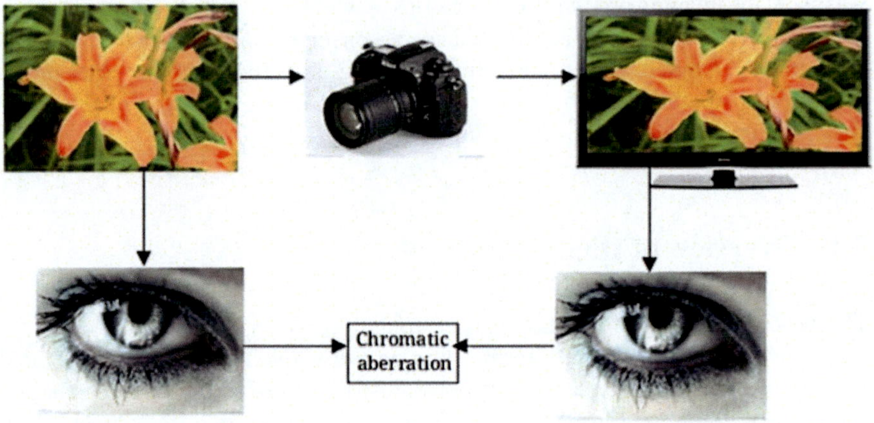

Fig. 3.2 Color space conversion between imaging device and display device to control color difference

generates the most original data, called raw data. We call the above process as the Out-camera model of imaging. At this time, the generated data are only camera sensing data and do not have any color information.

After the raw data are generated, in order to realize the reproduction of the scene color, the raw data must undergo a color interpolation to generate the three-channel data. After the above processing, although the three-channel information is generated to make the image have colors, the color of the image is generated in the color space of the camera, and there is a big gap between the color space for cameras and the scene color for human eyes. The camera color space must be converted to other color spaces to achieve the approximation with the actual scene color, as shown in Fig. 3.2. To realize the accurate reproduction of the scene, several color space conversions are usually required. First, we need to convert the camera color space to the standard CIE color space. Then, to display the image on different devices, it is necessary to go through the conversion from the standard color space to the color space of the display device, and finally perform post-processing and rendering operations on the image. This process is generally referred to as the in-camera model.

3.2 Simulation of Digital Imaging Process

The main task of a digital camera is to record the data of the scene. From the perspective of an observer, the goal of the digital camera is to accurately reproduce the scene, that is, the image obtained by the camera is consistent with the scene seen by the observer. The factors affecting imaging can be divided into two categories: the influence of the external environment of the camera and the influence of the internal processing process of the camera after imaging. The first kind of influencing factors is mainly light source, spectral reflectance of the scene, and spectral response curves of the image sensor. The second factors are mainly the camera internal processing algorithms. This section will analyze and model the two out-camera and in-camera imaging stages. On the basis of this, the simulation calculation of the whole process of imaging is realized.

3.2.1 Color Image Formation

The color image formation process for a single spatial location on a sensor array is given as follows. A beam from a light source illuminants an object and is reflected by the object's surface. A digital imaging device then collects the reflected beam. In this process, part light will enter the device through some filters. An optical detector (CCD/CMOS) is arranged to measure incident light. Then, the response value is outputted by the sensor and the raw data are created. This procedure is shown in Fig. 3.3.

To create a color image, three color bands are required. Therefore, a digital device needs three different sensors to acquire the red, green, and blue parts of the spectrum. However, such a scheme has some drawbacks: expensive cost and additional complexity. Generally, a commercial digital device only employs one sensor and overlays with a color filter array to solve such problems. The most popular array pattern is Bayer pattern [1] as shown in Fig. 3.4.

When the raw data are created by a sensor as shown in Fig. 3.5, a full-color image is reconstructed from the raw data with Bayer pattern. The sensor then outputs the spectral response data of each channel.

When such measured data are saved into an image, the raw image is created. In fact, this imaging process is not finished for many digital devices. In order to realize accurate color reproduction and create visually pleasing pictures, a series of transformations is usually applied. The process is shown in Fig. 3.5. First, the raw data of three channels are converted into the XYZ color space by camera characterization. Then, when the raw image is outputted by a display device, another transformation is applied in order to realize XYZ color space to another color space. Most of the

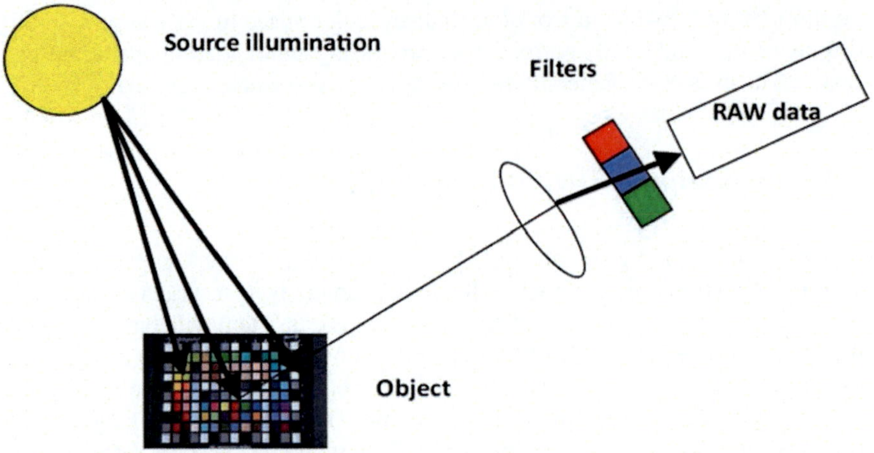

Fig. 3.3 Raw image formation process. Reprinted with permission from Denglu Wu, Jiandong Tian, Bingfeng Li, Yaonan Wang, and Yandong Tang, "Recovering sensor spectral sensitivity from raw data," Journal of Electronic Imaging 22(2), 023032, 2013. ©SPIE

Fig. 3.4 Bayer pattern: the different filter pattern. Reprinted with permission from Denglu Wu, Jiandong Tian, Bingfeng Li, Yaonan Wang, and Yandong Tang, "Recovering sensor spectral sensitivity from raw data," Journal of Electronic Imaging 22(2), 023032, 2013. ©SPIE

monitors and cameras employ sRGB color space, and a gamma correction is applied as usual for color space transformation from XYZ to sRGB. The transformation consists of two steps. The first step is a linear transformation, which calculates the sRGB tristimulus values from the CIEXYZ tristimulus values. The linear transformation is given by

$$\begin{bmatrix} R_{\text{linear}} \\ G_{\text{linear}} \\ B_{\text{linear}} \end{bmatrix} = \begin{bmatrix} 3.2406 & -1.5372 & -0.4986 \\ -0.9689 & 1.8758 & 0.0415 \\ 0.0557 & -0.2040 & 1.0570 \end{bmatrix} \begin{bmatrix} X \\ Y \\ Z \end{bmatrix}, \tag{3.2}$$

where R_{linear}, G_{linear}, and B_{linear} are the intermediate parameters. Note that these values are not the final results. The second step is transforming the linear values into sRGB data. Let X be one of R_{linear}, G_{linear}, and B_{linear}. The gamma correction function is as follows:

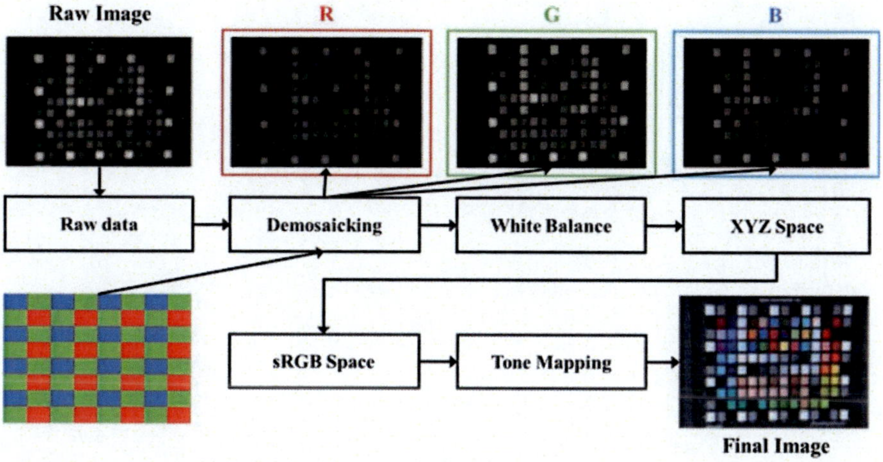

Fig. 3.5 Color image formation and process. Reprinted with permission from Denglu Wu, Jiandong Tian, Bingfeng Li, Yaonan Wang, and Yandong Tang, "Recovering sensor spectral sensitivity from raw data," Journal of Electronic Imaging 22(2), 023032, 2013 ©SPIE

$$\text{gamma}_{\text{sRGB}}(X) = \begin{cases} 12.92X & \text{if} X \leq 0.0031308 \\ 1.055X^{\frac{1}{2.4}} - 0.055 & \text{if} X > 0.0031308 \end{cases} \tag{3.3}$$

From the above process, it can be seen that many transformations are applied from raw data to sRGB data. The above process ensures accurate color reproduction on a display device.

3.2.2 *Imaging Simulation*

The process of camera imaging can be divided into two stages: camera's raw data imaging stage and camera's raw data post-processing stage. To simulate the raw data imaging stage, it is necessary to know the light source spectrum, the spectral reflectance of the target, and the camera response function.

3.2.2.1 Raw Data Simulation

According to the imaging formula, the SPD of light source is a significant factor in the imaging process. When light emitted from a light source irradiates the target, partial light reflected by the target surface enters the camera. To simulate the camera imaging process, we use the Macbeth ColorChecker Chart as the target. First, we use a spectrometer to measure the SPD of the light source and sample the measured SPD at 10nm intervals. The SPD of the light source after sampling is shown in Fig. 3.6. The spectral reflectance of the 24 colored patches of the ColorChecker Chart is shown in Fig. 3.7.

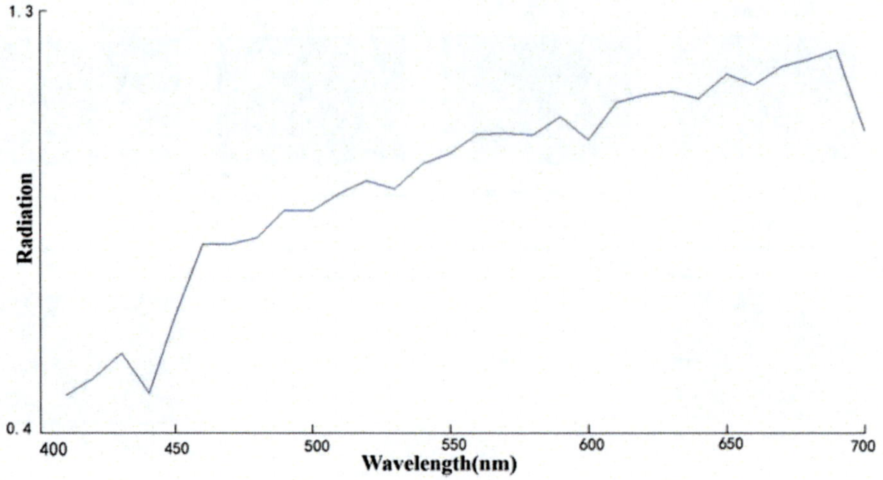

Fig. 3.6 Light source's SPD for calculating imaging simulation with sampling interval at 10 nm

In order to simulate camera imaging, we must know the camera response functions. In this experiment, we use Nikon D100 digital camera, and its spectral response functions are shown in Fig. 3.8.

The spectral reflectance and the camera response functions are sampled at 10nm intervals, and then the raw data is calculated via the imaging formula (3.1–3.3). After that, the simulated raw data and the actual raw data are normalized. The data (the first 12 colored patches of ColorChecker) are shown in Tables 3.1, 3.2, and 3.3.

3.2.2.2 Raw Data Post-processing Simulation

After the above process is completed, we generate raw data of the target scene. But this only completes out-camera part of the whole imaging process. Considering the manufacturing cost, the common consumer-grade cameras adopt single CCD structure, and the generated raw data are single channel gray data only. We take Canon 5D Mark II camera as an example to simulate the post-processing of the camera, and the gray-scale raw data obtained are shown in the left side of Fig.1.14. In order to obtain the response value of the three-channel camera, a Bayer demosaicking interpolation method is applied to the generated raw data to obtain the three-channel response data, as shown in the right side of Fig. 3.9.

After generating the three-channel color image, a third-order diagonal matrix is required for white balance correction. However, the color space of the original camera response values is device dependency. This image is very different from the final image (e.g.,JPEG) produced by the camera post-processing or the scene observed by human eyes directly. Therefore, the color images generated above are device dependent, which means that different cameras will generate different response values

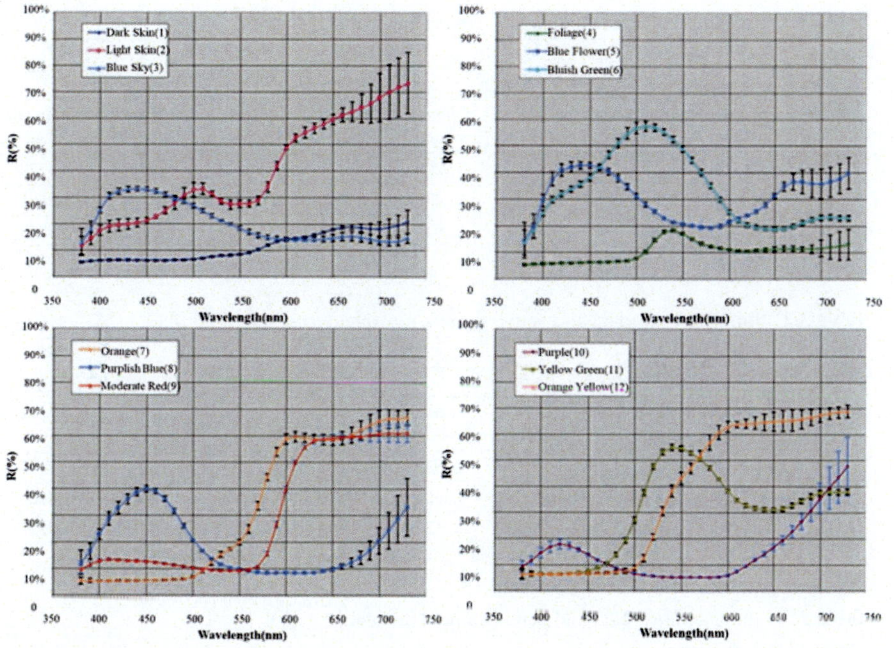

Fig. 3.7 Spectral reflectance of some color patches in ColorChecker, where the horizontal axis represents the wavelength, and the vertical axis represents the reflectivity

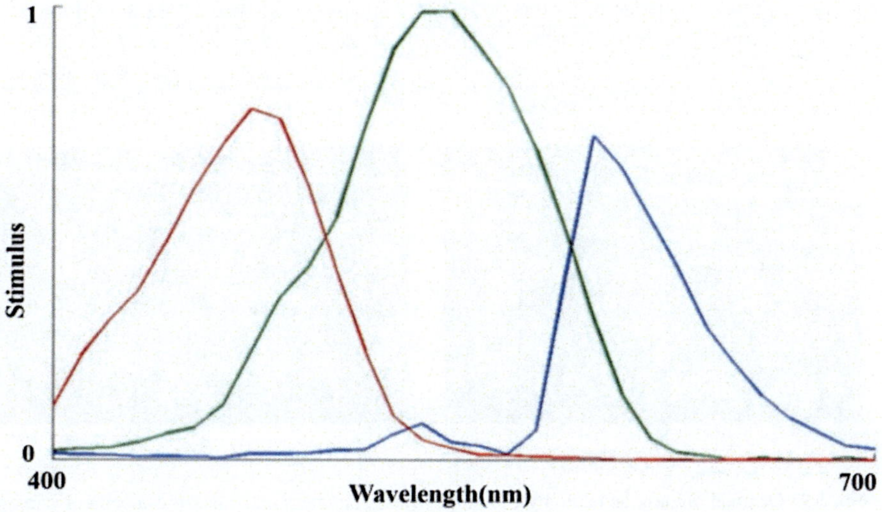

Fig. 3.8 Spectral response functions of Nikon D100 camera, where the horizontal axis represents the wavelength, and the vertical axis represents response rate

Table 3.1 Simulated raw data and actual camera raw data

Channel	Color patch							
	Dark Skin		Light skin		Blue sky		Foliage	
	Actual value	Simulated value	Actual value	Simulated value	Actual value	Simulated value	Actual value	Simulated value
R	0.6114	0.6227	2.0659	2.0299	0.6510	0.6893	0.4876	0.5229
G	0.7460	0.7486	2.6571	2.5470	1.8342	1.9460	1.1672	1.2472
B	0.2869	0.2840	1.1632	1.1036	1.3543	1.4424	0.0318	0.3248

Table 3.2 Simulated raw data and actual camera raw data

Channel	Color patch							
	Blue flower		Bluish green		Orange		Purple blue	
	Actual value	Simulated value	Actual value	Simulated value	Actual value	Simulated value	Actual value	Simulated value
R	1.2048	1.1589	1.1610	1.2330	2.2888	2.3135	0.4031	0.4442
G	2.1531	2.4026	4.2531	4.3321	1.7903	1.7846	1.2341	1.3354
B	1.6960	1.8797	2.0622	2.9537	0.2915	0.2939	1.4177	1.5292

Table 3.3 Simulated raw data and actual camera raw data

Channel	Color patch							
	Moderate red		Purple		Yellow green		Orange yellow	
	Actual value	Simulated value	Actual value	Simulated value	Actual value	Simulated value	Actual value	Simulated value
R	1.7292	1.8120	0.3738	0.4363	1.6308	1.6984	2.5751	2.5434
G	1.0454	1.0814	0.5246	0.6137	3.8089	3.8012	2.8962	2.8849
B	0.5361	0.5774	0.4839	0.5722	0.6853	0.6910	0.3759	0.3843

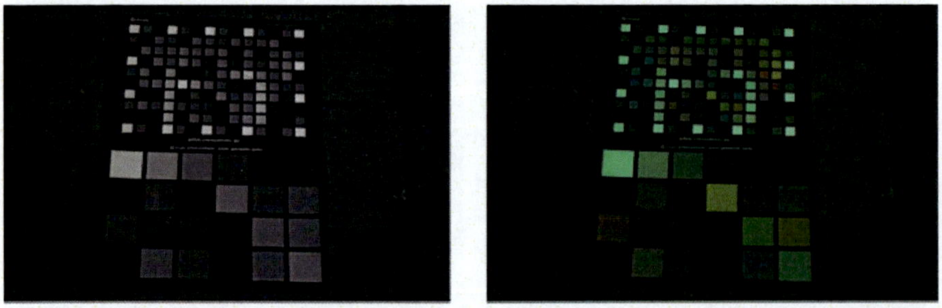

(a) (b)

Fig. 3.9 Original raw image taken by Canon 5D Mark II camera and the demosaiced color image of the same scene. In order to capture the device-independent response values, the above data should be converted from camera color space to XYZ device-independent color space. This process is named as camera characterization. In general, camera characterization is achieved by a conversion matrix from camera color space to XYZ color space. The camera response functions should be consistent with the CIE XYZ color matching functions or sRGB color matching functions if the camera has a good color reproduction ability. However, camera manufacturers rarely publish the characterization matrix of the camera due to commercial reasons.

After the color raw data are converted from the camera color space to the device-independent XYZ color space utilizing the camera characterization matrix, it is also necessary further to convert the data from the XYZ color space to the sRGB color space. In the process of the color space conversion, the conversion matrices from raw camera color space to XYZ color space and from XYZ color space to sRGB color space are usually combined and named as the color conversion matrix. Currently, the converted sRGB data is linear, which is different from the final data generated by the camera.

3.2.3 Camera Radiometric Calibration

There are several steps should be done to convert the raw data to the linear sRGB data, among which the most critical calculation is the transformation matrix from the raw color space to the XYZ color space. Since different cameras have different transformation matrices, this matrix is device-dependent and is called the camera characterization matrix. Due to commercial reasons, this characterization matrix is difficult to obtain. In addition, the purpose of many cameras is to produce pleasing photographs, so some specific effects will be applied except the common imaging processing [4, 8]. To obtain more accurate simulation results, we need to calculate the characterization matrix of camera via some specific methods, such as [6].

A series of calculation process needs to be done from raw data to the final processed camera data (JPEG). This process can be expressed by the formula as follows:

$$I_x = f(e_x) \tag{3.4}$$

where f refers to the mapping function between the raw data e_x and the final output image I_x. x denotes the pixel location in the image. Equation (3.4) can be extended to three-channel and can be written as follows:

$$\begin{bmatrix} I_{rx} \\ I_{gx} \\ I_{bx} \end{bmatrix} = \begin{bmatrix} f_r(e_{rx}) \\ f_g(e_{gx}) \\ f_b(e_{bx}) \end{bmatrix} \tag{3.5}$$

$$\mathbf{e_x} = \begin{bmatrix} e_{rx} \\ e_{gx} \\ e_{bx} \end{bmatrix} = \mathbf{TE_x} \tag{3.6}$$

where $\mathbf{T}$ is a 3×3 transformation matrix, which includes the color transformation matrix that transforms the camera color space ($\boldsymbol{E}_x$) to the sRGB color space ($\boldsymbol{e}_x$) and the white balance matrix. The mapping function f is nonlinear due to camera manufacturing reasons such as dynamic range compression (gamut compression) and gamma correction. Obtaining function f is the key to get other parameters.

In general, the camera calibration here is used to obtain the function f. There are many calibration methods, e.g., polynomial model [9], PCA-based model [3], statistics-based model [7], and probability-based model [10]. Firstly, the ColorChecker is used to obtain the image pairs with different exposure conditions of the camera, and then the pixel values of the colored patches are extracted from the obtained images pairs and expressed in coordinates, which is called the brightness transfer function (BTF). The above process can be expressed as follows:

$$\frac{f^{-1}(\mathbf{I}'_{\mathbf{x}})}{f^{-1}(\mathbf{I}_{\mathbf{x}})} = k' \tag{3.7}$$

where $(\mathbf{I}, \mathbf{I}')$ stands for the image pairs with different exposure conditions. From formula (3.7), it can be concluded as follows:

$$\mathbf{I}_{\mathbf{x}} = \tau_k(\mathbf{I}_{\mathbf{x}}) = f(k' f^{-1}(\mathbf{I}_{\mathbf{x}})) \tag{3.8}$$

where τ_k is the BTF, f is the camera mapping function, and k' is the exposure ratio. BTF describes how the intensity values of the image change with exposures when given the mapping function. It can be derived from Eq. (3.8) that if function f is known, the BTF should be the same for all exposed images. It is worth noting that the BTF is constant even if $\mathbf{T}$ is considered. For example:

$$\frac{f^{-1}(\mathbf{I}'_{\mathbf{cx}})}{f^{-1}(\mathbf{I}_{\mathbf{cx}})} = k' \frac{t'_c \mathbf{E}_{\mathbf{x}}}{t_c \mathbf{E}_{\mathbf{x}}} = k' \; if \; t_c = t'_c \tag{3.9}$$

Therefore, [6] proposed a new imaging formula to calculate the camera imaging process.

$$\begin{bmatrix} I_{rx} \\ I_{gx} \\ I_{bx} \end{bmatrix} = \begin{bmatrix} f_r(e_{rx}) \\ f_g(e_{gx}) \\ f_b(e_{bx}) \end{bmatrix} \tag{3.10}$$

where

$$\begin{bmatrix} e_{rx} \\ e_{gx} \\ e_{bx} \end{bmatrix} = h(\mathbf{T}_{\mathbf{s}} \mathbf{T}_{\mathbf{w}} \mathbf{E}_{\mathbf{x}}) \tag{3.11}$$

$\mathbf{E}_{\mathbf{x}} = \left[E_{rx}, \; E_{gx}, \; E_{bx}\right]$ is the radiometric value, which is recorded with raw format in a digital camera. We could know that matrix $\mathbf{T}_{\mathbf{w}}$ is used to white balance the raw data firstly. Then the white-balanced raw data is transformed into sRGB space by matrix $\mathbf{T}_{\mathbf{s}}$. We can notice that the biggest difference between the imaging model of formula (3.10) and the general one is the introduced function h. The function h is mainly used to implement the gamut mapping for data in sRGB space. When the color range of a camera exceeds the sRGB gamut, the gamut mapping function h will transform the color inside the gamut.

In order to calculate the parameters h, $\mathbf{T_w}$, $\mathbf{T_s}$, and f, we need to capture the final images and raw data of the ColorChecker in different exposure conditions. In this experiment, we take the X-Rite ColorChecker (XCC) to calculate the parameters mentioned above.

3.2.4 Simulation of In-Camera Post-processing

3.2.4.1 Calculation of Camera Mapping Function

The images and raw data under various exposure conditions are taken by the camera, as shown in Fig. 3.10. Then the pixel values of color patches are extracted from the final sRGB and raw images. The camera mapping function f is calculated by means of formula (3.12) based on PCA method [3]. The result is shown in Fig. 3.11.

$$\frac{f^{-1}(\mathbf{I}'_{\mathbf{cx}})}{f^{-1}(\mathbf{I}_{\mathbf{cx}})} = k' \tag{3.12}$$

Fig. 3.10 Color patches in X-RITE 24 and 140 ColorChecker Chart.

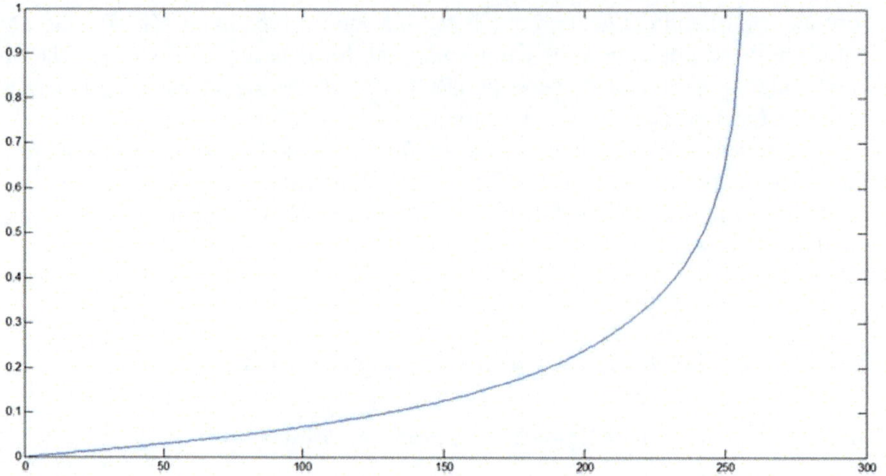

Fig. 3.11 Camera mapping function result (X-axis: intensity of the image)

3.2.4.2 Calculation of Color Space Transformation Matrix

When the camera response function f is known, the linear sRGB value can be obtained by the inverse process. After obtaining the linear sRGB value, we can use the following formula to minimize the error function and calculate the matrix $\mathbf{T_w}, \mathbf{T_s}$:

$$\sum_{i=1}^{M}\sum_{j=1}^{N}\left\|\mathbf{T_s^{-1}X_{ij}} - \mathbf{T_{w_i}E_{ij}}\right\|^2 \tag{3.13}$$

where M is the number of white balance settings, N is the amount of pixels in the image used for estimation, and the $\mathbf{X_{ij}}$ is the sRGB value linearized by the inverse process ($X_{ij} = \left[f_r^{-1}(I_{r,ij}), f_g^{-1}(I_{g,ij}), f_b^{-1}(I_{r,ij})\right]^T$). The transformation matrix of Canon 5D Mark II is calculated by formula (3.13):

$$\mathbf{T} = \mathbf{T_s T_w} = \begin{bmatrix} 4.1313 & -1.1764 & 0.3016 \\ -0.6465 & 1.3894 & -0.4627 \\ -0.1923 & -0.2275 & 2.0422 \end{bmatrix}$$

3.2.4.3 Calculation of Gamut Mapping Function

After the completion of above steps, the gamut mapping function can be calculated according to Eq. (3.14). A nonparametric calculation method based on radial basis function (RBF) is applied to obtain the gamut mapping function h. There are many kinds of radial basis functions. We calculate it by means of the following models (Buhmann 2003; Carr 2001).

Table 3.4 RMS error of raw and JPEG (sRGB) conversion of XCC colorchecker

Raw2JPG	0.0007014	0.00084525	0.0007720349
JPG2Raw	0.0007049	0.00081956	0.0007754838

$$h^{-1}(\mathbf{X}) = p(\mathbf{X}) + \sum_{i=1}^{N} \lambda_i \phi(\|\mathbf{X} - \mathbf{X_i}\|) \tag{3.14}$$

where $\mathbf{X} = \left[f_r^{-1}(\mathbf{I_r}), f_g^{-1}(\mathbf{I_g}), f_b^{-1}(\mathbf{I_b})\right]^T$, λ_i is the weight of the basis function ϕ. $\phi(r) = r$ is selected as the base function. For $p(\mathbf{X})$, Let $p(\mathbf{X}) = \mathbf{c^T}\tilde{\mathbf{X}}$, where $c = [c_1, c_2, c_3, c_4]$ and $\tilde{\mathbf{X}} = [1, \mathbf{X}^T]^T$.

Given sRGB-raw image pair and matrix $\mathbf{T_w}$, $\mathbf{T_s}$, the point $\mathbf{X_i}$ in the raw image can be calculated as $\mathbf{X_i'} = h^{-1}(\mathbf{X_i}) = \mathbf{T_s T_w E_i}$. Given set of point pairs $(\mathbf{X_i}, \mathbf{X_i'})$ in the image pair, the parameters $\lambda = [\lambda_1, \lambda_2, \ldots, \lambda_N]^T$ and $\mathbf{c}$ of RBF in equation (3.15) can be calculated as follows:

$$\begin{pmatrix} \mathbf{D} - 8N\pi\rho\mathbf{I} & \tilde{\mathbf{P}} \\ \tilde{\mathbf{P}}^T & \mathbf{0}_{4\times4} \end{pmatrix} \begin{pmatrix} \lambda \\ \mathbf{c} \end{pmatrix} = \begin{pmatrix} \mathbf{P'} \\ \mathbf{0}_{4\times3} \end{pmatrix} \tag{3.15}$$

where $\mathbf{D}$ is the $N \times N$ matrix with $D_{ij} = \|x_i - x_j\|$ as the matrix element, $\tilde{\mathbf{P}}$ is the $N \times 4$ matrix with $\mathbf{X_i'}$ as the i-th row, and $\mathbf{P'}$ is the $N \times 3$ matrix with $\mathbf{X_i}'^T$ as the ith row. The parameter ρ is smooth coefficient of RBF.

3.2.4.4 Imaging Simulation Results

After the above calculation, all parameters from raw data to sRGB data can be estimated. We can compute the RMS error from raw space to sRGB space and vice versa. The results are shown in Table 3.4.

It can be seen from Table 3.4 that the simulated image is a little different from the camera captured image. This might be caused by the calibration matrix from camera color space to XYZ space is not very accurate.

3.3 Camera Spectral Sensitivity Recovery

It can be seen from formula (3.1) that the problem of estimating the camera response function $Q_H(\lambda)$ can be attributed to the case where the spectral power distribution of the light source $E(\lambda)$, the spectral reflectance of the target $S(\lambda)$, and the three-channel raw data are known. For convenience, we denote red channel, green channel, and blue channel with 1, 2, and 3, respectively. Let $Q_i(\lambda)(i = 1, 2, 3)$ be the spectral sensitivity of the i'th channel. $E(\lambda)$ is the spectral power distribution of the scene illuminant, $S_j(\lambda)$ is the surface reflection imaged at pixel j, and I_j^i is the spectral response data of the $i'th$ channel at pixel j. It is given by

$$\mathbf{I}_j^H = \int_{\lambda_{\min}}^{\lambda_{\max}} E(\lambda)S_j(\lambda)Q_i(\lambda)d\lambda,\ j = 1, 2, ..., M, \tag{3.16}$$

where M is the total number of pixels. To perform the calculation, we discretize Eq. (3.16) as follows:

$$\mathbf{I}_j^H = \sum_{k=1}^{N} E(k\Delta\lambda)S_j(k\Delta\lambda)Q_i(k\Delta\lambda),\ j = 1, 2, ..., M, \tag{3.17}$$

where $\Delta\lambda$ is the sample interval and N is the total number of sample points. Let $R_j^k = E(k\Delta\lambda)S_j(k\Delta\lambda)$, $\mathbf{R}_j = [R_j^1, R_j^2, ..., R_j^k, ..., R_j^N]$, $Q_H^k = Q_H(k\Delta\lambda)$, and $\mathbf{Q}_H = [Q_H^1, Q_H^2, ..., Q_H^k, ..., Q_H^N]$, then Eq. (3.17) can be converted into a vector form:

$$\mathbf{I}_j^H = \mathbf{R}_j\mathbf{Q}_H^T,\ j = 1, 2, ..., M. \tag{3.18}$$

Given a set of spectrum and the corresponding spectral response values, Eq. (3.18) can be written as follows with matrix form:

$$\mathbf{I} = \mathbf{RQ}, \tag{3.19}$$

where $\mathbf{I}^H = [\mathbf{I}_1^H, \mathbf{I}_2^H, ..., \mathbf{I}_j^H, ..., \mathbf{I}_M^H]^T$, the $i'th$ column of the matrix $\mathbf{I}$ is $\mathbf{I}^i$. The $j'th$ row of matrix $\mathbf{R}$ is $\mathbf{R}_j$. The $i'th$ column of the matrix $\mathbf{Q}$ is $\mathbf{Q}_i^T$.

In general, to recover the camera spectral sensitivity, the spectral power distribution of light reflected from the color chart checker to the camera and three-channel spectral response values need to be obtained. Specifically, matrix **I** and **R** are known in Eq. (3.19). These data can be collected by a spectrometer and a camera with raw support, respectively.

The images of X-Rite Color Checker SG (XCC SG) were captured by Canon 5D MARK II camera and Nikon D100 camera, respectively. Both raw and sRGB files of XCC SG were recorded. Then the raw files can be decoded with software DCRAW[2]. Three-channel spectral response values are calculated through demosaicing. The raw image and sRGB image are given in Fig. 3.12. We can see that the raw image is clearly different from the sRGB image.

Then, we extract three-channel spectral response values in different patches from the raw image. The average spectral response value within a patch is computed by using the center pixels for each color patch. The boundary patches are not used. This calculation procedure can reduce the impact of image noise. For each channel, 96 spectral response values are obtained. Thus, the size of the matrix I is 96×3.

The reflectance data were acquired by Avantes spectrometer from 400 to 700 nm. According to Eq. (3.17), we sample the spectrum by 10-nm intervals for each

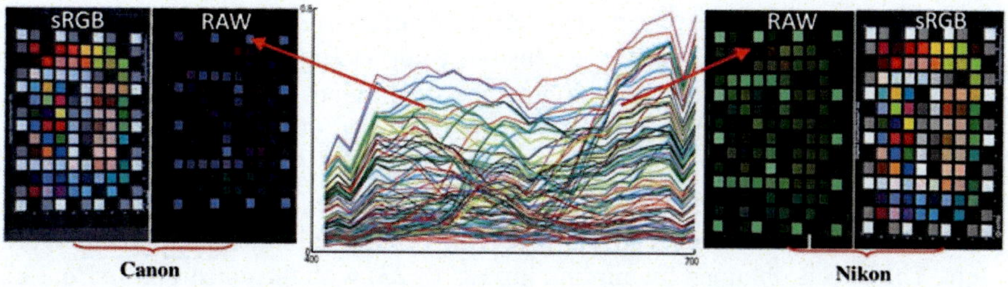

Fig. 3.12 Left: sRGB image and raw image by Cannon. Middle: the sampled reflectance spectra of the XCC SG. Right: sRGB image and raw image by Nikon

color patch, i.e., $\Delta\lambda = 10$ nm. The sample results are shown in Fig. 3.12 (middle). The matrix **R** for the spectrum is created. Its size is 96×30.

When matrix **I** and **R** are known, we estimate matrix **Q** from the following equation:

$$\mathbf{Q} = \mathbf{R}^p\mathbf{I}, \tag{3.20}$$

where $\mathbf{R}^p$ is the pseudoinverse of **R** and equals to $(\mathbf{R}^T\mathbf{R})^{-1}\mathbf{R}^T$. Then, we can easily recover the SSS from Eq. 3.20. This method can get the excellent results without the noise. However, it is always unstable in the calculation process because of noise.

In order to yield the reliable results, we propose a new objective function to solve this problem as follows:

$$F_1(Q_i) = \left[\sum_{j=1}^{P}(I_j^i - \widehat{I}_j^i)^2\right]^{1/2}, \qquad j = 1, 2, \ldots, P \tag{3.21}$$

$$F_2(Q_i) = \left[\sum_{k=2}^{N}(Q_i^k - Q_i^{k-1})^2\right]^{1/2}, \qquad k = 1, 2, \ldots, N, \tag{3.22}$$

$$Q_i^* = \underset{Q_i}{\operatorname{argmin}}\, F_t[Q_i], \qquad t = 1, 2; \qquad i = 1, 2, 3, \tag{3.23}$$

$$\text{s.t. } Q_i^k \subset [0, 1], \tag{3.24}$$

where I_j^i is the actual average pixel value of an image patch for the $i'th$ channel; $\widehat{I}_j^i$ is the estimated pixel value corresponding to such a patch for the $i'th$ channel. P is the total number of the XCC SG image patches, i.e., $P = 96$. N is the number of samples, i.e., $N = 30$.

The objective function $F_1[Q_i]$ measures the error between the estimated pixel values and the actual pixel values of all image patches for each channel. We minimize the total error by Eq. (3.21) to get the Q_i.

Due to the fact that the spectral sensitivity of most sensors varies smoothly, we add the term $F_2[Q_i]$ to smooth the sensitivity curves for each channel. Specifically, we enforce such terms by smoothing neighboring values Q_i^k and Q_i^{k-1}.

Generally, the SSS is larger than zero, i.e., $Q_i^k \geq 0$. In addition, the recovery result is invariant to uniform scaling of all sensitivities. Thus, the max value of the sensitivity Q_i^k is 1. We can take these as a constraint, i.e., $Q_i^k \subset [0, 1]$.

In this way, the SSS recovery problem is converted into an optimization problem. The SSS is obtained by minimizing Eq. (3.23) with the constraint in Eq. (3.24) for each channel, and the multi-object genetic optimization is used to get its solution in the next section.

Compared with other methods in the literature, the cost function in our method does not need a basic function to express the SSS. The optimization cost function decreases the error between the measured camera response and the estimated response of the camera under the same input. Moreover, there is no parameter to be determined by the user for the cost function.

We present our experimental results to evaluate our method for SSS recovery. Three-channel spectral response values are created, and a total number of 96 reflectance spectrums from the XCC SG is sampled. The SSS of each channel is obtained by optimizing the cost function in Eq. (3.23) with the following parameter values which are fixed for our all experiments.

To recover the SSS, the multi-object genetic optimization in MATLAB is used to solve the minimization problem. The parameters are set as follows. A double precision number represents each gene in the chromosome, and a bit mutation method is used. In all the experiments, the population type is set as double vector. The population size is 2000. The creation function is set to be constraint-dependent. The scaling function is set to be rank scaling. The selection rate is created by stochastic uniform. The reproduction options determine how the genetic algorithm creates children at each new generation. We use the default settings. In other words, the elite count is set to 2 and the crossover is 0.8. The mutation function is set to be constraint dependent. The crossover is set to be scattered.

We tested our recovered method on two different camera sensors: Canon 5D Mark II and Nikon D100. In order to evaluate the effectiveness of our method, we also obtained the ground truth using monochromatic lighting as used in some state-of-the-art methods. The ground truth is shown as the dashed curves in Fig. 3.13.

To evaluate the effectiveness of our method under different illumination conditions, we test our methods under four common illuminations: lamp, sun, blue sky, and cloudy sky. The recovered results by our method are shown as the dotted curves in Fig. 3.13. The first row shows the results of Canon 5D Mark II. The second row shows the results of Nikon D100. Observing the above results, we can see that both of the recovered results are reasonably smooth. The estimated curves are very close to the ground truth. However, we note that the estimated curves of blue and green channels are smooth and match closely to the actual one. The error of the calculated SSS of the red channel is somewhat larger than that of blue and green channels. This may be due to the introduced noise in the image and reflectance spectrum.

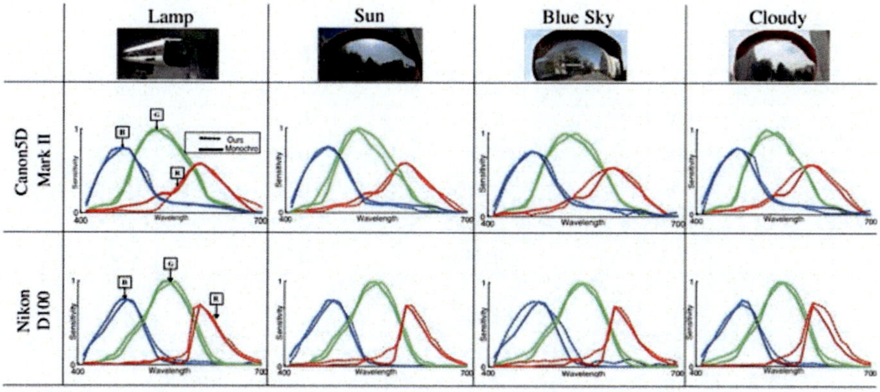

Fig. 3.13 Estimated spectral sensitivities of cameras under different illuminations. The results of our method are similar to the ground truth. Reprinted with permission from Denglu Wu, Jiandong Tian, Bingfeng Li, Yaonan Wang, and Yandong Tang, "Recovering sensor spectral sensitivity from raw data", Journal of Electronic Imaging 22(2), 023032, 2013. ©SPIE

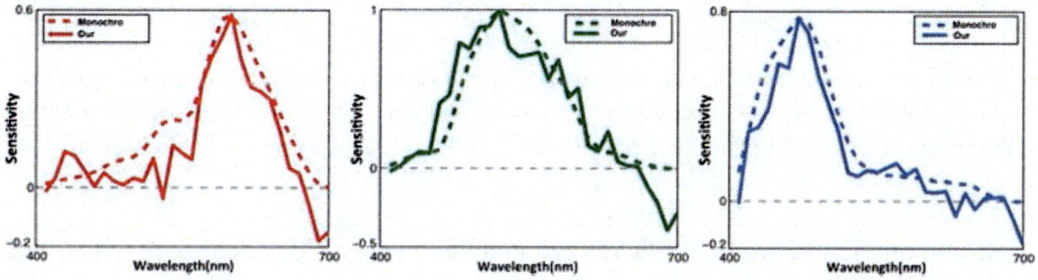

Fig. 3.14 Recovered results without the smoothness and positivity constraints. Reprinted with permission from Denglu Wu, Jiandong Tian, Bingfeng Li, Yaonan Wang, and Yandong Tang, "Recovering sensor spectral sensitivity from raw data," Journal of Electronic Imaging 22(2), 023032, 2013. ©SPIE

Besides, another observation is that the recovered results have less high frequency. It is due to the fact that the smooth constrained term is added into the cost function and the high frequency is eliminated. To examine this in detail, we test our method with the following settings on Canon 5D Mark II under the blue sky. The SSS is calculated by our method without the smoothness and positivity constraint terms. Specifically, the cost function is set to Eq. (3.21) and the constraint term is set to the spectral sensitivity $Q_i^k < 1$. The results are shown in Fig. 3.14. We can see that the curves have sharp peaks and the spectral sensitivity is negative among some ranges. In contrast, the results shown in Fig. 3.13 agree with all constraints. They are smooth and positive. The difference indicates the utility of the smooth constraints for SSS recovery.

To quantify the accuracy of the recovered results, we calculate the average values of root mean square (RMS) errors over three channels between the recovered

Table 3.5 Comparison results

			RMSE	
Method	Sun	Blue	Cloudy	Lamp
Hai et al.	0.1241	0.1063	0.0996	0.0981
	0.1354	0.0877	0.0938	0.0829
	0.1219	0.0752	0.0942	0.0813
Ours	0.0526	0.0545	0.0639	0.0517
	0.0519	0.0471	0.0531	0.0456

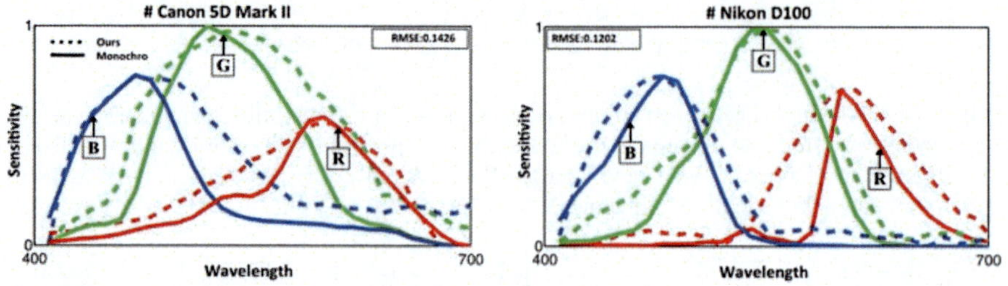

Fig. 3.15 Recovered results and RMS errors by sRGB data. Reprinted with permission from Denglu Wu, Jiandong Tian, Bingfeng Li, Yaonan Wang, and Yandong Tang, "Recovering sensor spectral sensitivity from raw data," Journal of Electronic Imaging 22(2), 023032, 2013. ©SPIE

data and the ground truth. The RMS error can describe the correlation between actual and estimated spectral sensitivity curves.

The RMS errors of the spectral sensitivity curves shown in Fig. 3.13 are calculated. The results are also presented in Table 3.5. From this table, it can be seen that the recovered data are very close to the ground truth. These results show that our method performs well on different sensors under different illumination conditions.

Furthermore, to compare the accuracy between the raw data and sRGB data for SSS recovery, we also recover the SSS by our methods using sRGB data. The results are shown in Fig. 3.15. The shape of recovered curves is also consistent with the ground truth.

Meanwhile, the average values of RMS errors are also calculated. We note that the RMS errors recovered by sRGB data are relatively larger than the RMS errors recovered by raw data. The primary reason is the noise. In process of the recovered results, the noise mainly comes from spectrum measurement errors and the transformation errors between the raw data and sRGB data. Especially, the transformations introduce significant noise and cause the error for results. It is shown by the experimental results in Fig. 3.15. Therefore, it validates that the raw data can be used to recover the SSS accurately.

For comparing the accuracy of different methods, we compared our method with the method [5] on different illumination conditions. Their method is based on the fact that SSS of different cameras is various and should not deviate a lot from

Table 3.6 Root mean square error of PE, POCS, and our methods

			RMSE	
Method	Sun	Blue	Cloudy	Lamp
PE	0.1335	0.1169	0.1243	0.0942
	0.1487	0.1275	0.1358	0.1026
POCS	0.0821	0.0786	0.0612	0.0538
	0.0836	0.0651	0.0783	0.0604
Ours	0.0526	0.0545	0.0639	0.0517
	0.0519	0.0471	0.0531	0.0456

each other. Thus, SSS can be expressed by a limited number of parameters and basic functions. In [5], nine Fourier basic functions are used in the calculation. The comparison results are shown in Table 3.5. Each line of Table 3.5 represents the average RMS errors between the recovered results and the ground truth on different cameras under different illumination conditions. Our method is significantly better than the method in [5]. Through these comparisons, it can be seen that the SSS can be accurately recovered by our method.

We also compared our method with principal eigenvector (PE) method and projecting onto convex sets (POCS) method12 on two cameras (Canon 5D Mark II and Nikon D100) under different illuminant conditions. The average RMS errors are also calculated between the recovered data and the ground truth for different methods. Table 3.6 shows the comparison results. The performance with our method is better than the PE and POCS methods.

In the above two experiments, the SSS is recovered by our method with raw data and the other methods with sRGB data. For comparison of the accuracy of different methods with the raw data, we also recovered the SSS by PE and POCS methods. As an example, SSS is calculated on Nikon D100 camera under different illumination conditions. The average RMS errors are calculated for three channels. The results are shown in Fig. 3.16.

Note that average RMS errors in PE and POCS methods by raw data are relatively smaller than PE and POCS methods by sRGB data. It is also verified that SSS is recovered accurately by raw data.

We can also see that the performance with our method by sRGB is close to PE method. The performance with POCS method by sRGB data is more accurate than that of our method by sRGB data. That is due to the fact that the POCS method uses many constraints to reduce the effect of noise, while our methods have limited constraints for dealing with the noise and focus on recovering SSS from the raw data. Thus, in case of "recovering the SSS from sRGB data," the performance with our method is not much better.

However, when the raw data are used for SSS recovery, the performance with our method improves radically while the PE and POCS do not. In Fig. 3.16, we can see that our method outperforms the POCS method. Although the difference of

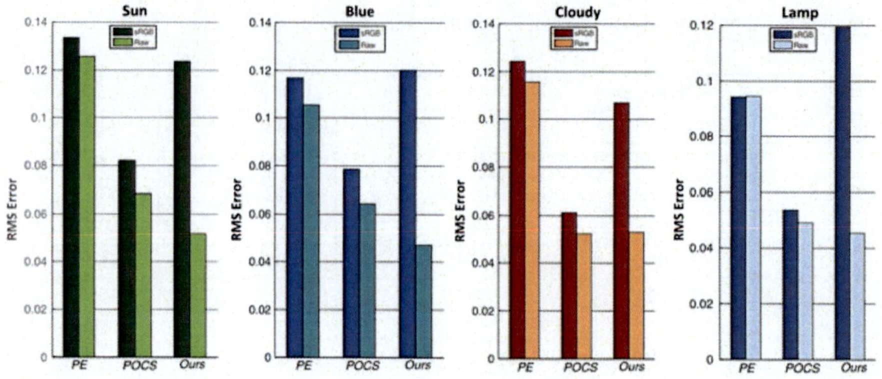

Fig. 3.16 Average RMS errors of different methods with sRGB and raw data. Reprinted with permission from Denglu Wu, Jiandong Tian, Bingfeng Li, Yaonan Wang, and Yandong Tang, "Recovering sensor spectral sensitivity from raw data," Journal of Electronic Imaging 22(2), 023032, 2013. ©SPIE

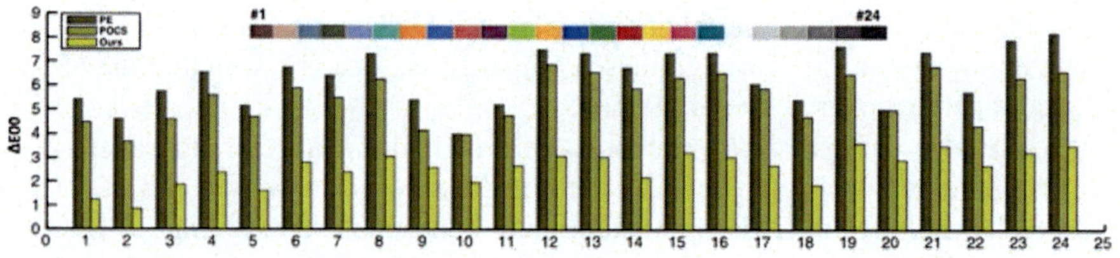

Fig. 3.17 ΔE_{00} for different methods

RMS errors is small, our method is simple and without the parameters determined by experiments or prior knowledge.

In order to evaluate colorimetric performance of different methods, the average color difference CIEDE2000 (ΔE_{00}) is calculated between these estimated CIEXYZ and the original CIEXYZ values of the color set for the given illumination.

As an example, we calculated ΔE_{00} for different methods using the estimated SSS (recovered in Table 3.6) and the actual SSS under the blue sky. We choose 24 patches in XCC SG chart as the test target. Figure 3.17 shows the results of ΔE_{00} calculated by different methods. We can see that the performances with our method are better than other methods.

Compared with the previous methods, our method is simple and practical. No prior parameter for the cost function is set by experiments and prior knowledge in our algorithm. By only taking a single image of a color chart and measuring the illumination using a spectrometer, the SSS can be recovered accurately.

The effectiveness of our method is demonstrated with experiments using real images taken under various illumination conditions. The results show that the recovered sensor curves are very close to the ground truth for different devices. Additionally, some comparison experiments are presented. We find that the SSS recovery from the raw data is more accurate than that from the sRGB color space.

References

1. B.E. Bayer, Color Imaging Array (1976)
2. D. Coffin, Dcraw: Decoding raw digital photos in linux. http://cybercom.net/dcoffin/dcraw 81 (2008)
3. M.D. Grossberg, S.K. Nayar, Determining the camera response from images: what is knowable? IEEE Trans. Pattern Anal. Mach. Intell. **25**(11), 1455–1467 (2003)
4. M.D. Grossberg, S.K. Nayar, Modeling the space of camera response functions. IEEE Trans. Pattern Anal. Mach. Intell. **26**(10), 1272–1282 (2004)
5. S. Han, Y. Matsushita, I. Sato, T. Okabe, Y. Sato. Camera spectral sensitivity estimation from a single image under unknown illumination by using fluorescence (2012), pp. 805–812
6. S.J. Kim, H.T. Lin, Z. Lu, S. Süsstrunk, S. Lin, M.S. Brown, A new in-camera imaging model for color computer vision and its application. IEEE Trans. Pattern Anal. Mach. Intell. **34**(12), 2289–2302 (2012)
7. S. Kuthirummal, A. Agarwala, D.B. Goldman, S.K. Nayar, Priors for large photo collections and what they reveal about cameras, in *European Conference on Computer Vision* (Springer, 2008), pp. 74–87
8. S. Lin and L. Zhang. Determining the radiometric response function from a single grayscale image. In *2005 IEEE Computer Society Conference on Computer Vision and Pattern Recognition (CVPR'05)*, volume 2, pages 66–73. IEEE, 2005
9. T. Mitsunaga, S.K. Nayar. Radiometric self calibration, in *Proceedings. 1999 IEEE Computer Society Conference on Computer Vision and Pattern Recognition (Cat. No PR00149)*, vol. 1 (IEEE, 1999), pp. 374–380
10. C. Pal, R. Szeliski, M. Uyttendaele, N. Jojic, Probability models for high dynamic range imaging, in *Proceedings of the 2004 IEEE Computer Society Conference on Computer Vision and Pattern Recognition, 2004. CVPR 2004*, vol. 2 (IEEE, 2004), pp. II–II

Chapter 4
Shadow Modeling and Detection

Abstract Shadows are the common phenomena in most outdoor scenes, which bring many problems in image processing and computer vision. In this chapter, we present two novel methods focusing on modeling and extracting shadows from a single outdoor image, i.e., tricolor attenuation model and tricolor linear model. The tricolor attenuation model (TAM) describes the attenuation relationship between shadow and its non-shadow background, and it is derived based on image formation theory. The parameters of the TAM are fixed by using the spectral power distribution (SPD) of daylight and skylight, which are estimated according to Planck's blackbody irradiance law. Based on the TAM, a multi-step shadow detection algorithm is proposed to extract shadows. The shadow linearity model is a linear ratio model which is derived from the perspective of physical imaging. The parameters of the linearity model are also fixed by the SPD of daylight and skylight, which are estimated according to our SPD calculation method proposed in Chap. 2, and they do not depend on the reflection curve of an object and the model can be applied to different scenes. Besides, we find additional new physical properties of shadows based on the tricolor linear model and take these shadow properties as features to design an effective shadow detection method. Finally, we conduct an extensive evaluation of eight widely adopted shadow features for shadow detection. Through sets of experiments, several useful and interesting conclusions are obtained.

4.1 Introduction

Shadows, the common phenomena in most outdoor scenes, bring many problems to robot vision and image processing. Researchers have already recognized the necessity of shadow processing and conducted a lot of research. We found two new models through the combination of atmospheric physics and imaging optics and proposed the tricolor attenuation model and the linearity model of shadows in an image. The tricolor attenuation model describes the tri-channel attenuation relationship between shadow and non-shadow background in an image. The shadow linearity model is a linear ratio model which is derived from the perspective of physical imaging. We also proved that the parameters of the linearity model do not depend on the

J. Tian, *All Weather Robot Vision*, Research on Intelligent Manufacturing,
https://doi.org/10.1007/978-981-16-6429-8_4

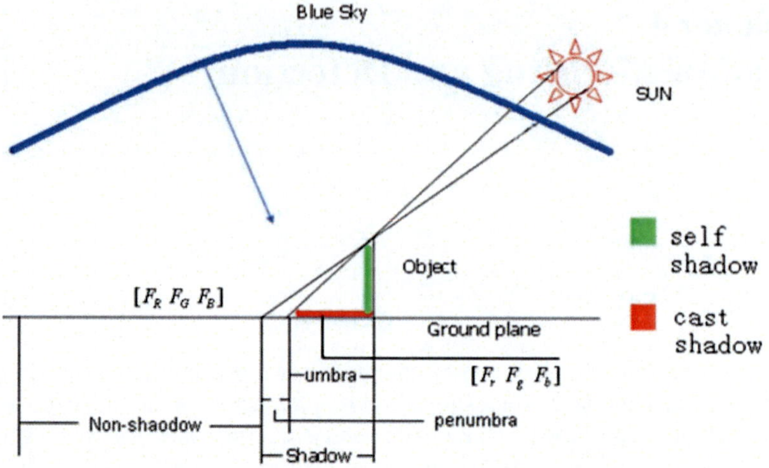

Fig. 4.1 Physical explanation of shadow formation © [2009] IEEE. Reprinted, with permission, from [Jiandong Tian, Jing Sun, and Yandong Tang, "Tricolor Attenuation Model for Shadow Detection," IEEE Transactions on Image Processing, Vol. 18, No. 10, pp. 2355–2363, 2009]

Table 4.1 Illumination and light intensity of shadow and non-shadow

	Illumination	Light intensity
Non-shadow	Daylight	High
Penumbra	Part of direct sunlight + Diffused skylight	Middle
Umbra	Diffused skylight	Low

reflection curve of an object, i.e., no matter how complex the scene is, the pixel ratio of different objects covered by shadows to their non-shadow surrounding regions is a fixed value, and it has proved that the value is uniquely determined by the sun angle (zenith angle). Based on this model, the methods of intrinsic-image deriving and intrinsic illumination decomposition are presented. Light source is a necessary factor for imaging, as shown in Fig. 4.1 and Table 4.1. Shadow can be further divided into umbra and penumbra regions. The illumination on umbra region is the diffused skylight; that on penumbra is diffused skylight and a small part of direct sunlight. The illumination on non-shadow region is daylight (diffused skylight and direct sunlight). Since skylight is a component of daylight, light intensity in shadow must be lower than that of the corresponding non-shadow background. The illumination and light intensity in shadow and non-shadow regions are given in Table 4.1.

4.2 Tricolor Attenuation Model of Shadows

Denoting $[F_{\mathrm{R}}\ F_{\mathrm{G}}\ F_{\mathrm{B}}]$ as a tricolor vector of a pixel value in a color image F, $[F_{\mathrm{NS_R}}\ F_{\mathrm{NS_G}}\ F_{\mathrm{NS_B}}]$ as a pixel value vector in a non-shadow background region, $[F_{\mathrm{S_R}}\ F_{\mathrm{S_G}}\ F_{\mathrm{S_B}}]$ as a pixel value vector in the corresponding shadow region, $[\Delta R\ \Delta G\ \Delta B]$

as the value attenuation vector in the shadow and non-shadow region, the relationship between $[F_{\mathrm{NS_R}}\ F_{\mathrm{NS_G}}\ F_{\mathrm{NS_B}}]$ and $[F_{\mathrm{S_R}}\ F_{\mathrm{S_G}}\ F_{\mathrm{S_B}}]$ is

$$\begin{cases} F_{\mathrm{S_R}} = F_{\mathrm{NS_R}} - \Delta R \\ F_{\mathrm{S_G}} = F_{\mathrm{NS_G}} - \Delta G \\ F_{\mathrm{S_B}} = F_{\mathrm{NS_B}} - \Delta B \end{cases} \tag{4.1}$$

Formula (4.1) implies that if ΔR, ΔG, ΔB are different, the disparities of the three channels of a shadow region are expected to be different from those of the corresponding non-shadow background region. Taking $\Delta R > \Delta G > \Delta B$ as an example, if we subtract B channel from R channel, we have

$$\begin{aligned} & F_{\mathrm{S_R}} - F_{\mathrm{S_B}} \\ & = F_{\mathrm{NS_R}} - \Delta R - (F_{\mathrm{NS_B}} - \Delta B) \\ & = F_{\mathrm{NS_R}} - F_{\mathrm{NS_B}} + (\Delta B - \Delta R) < F_{\mathrm{NS_R}} - F_{\mathrm{NS_B}}. \end{aligned} \tag{4.2}$$

Obviously, in this example, the disparity of shadow region is lower than that of the non-shadow region. Therefore, our model originates from the idea that if we subtract the minimum attenuated channel from the maximum attenuated channel, the results in shadow regions will be lower than the results in non-shadow regions. This is very useful for shadow identification. In fact, in most cases, ΔR, ΔG, ΔB are different. The key problem is how to find the maximum and the minimum attenuated channels.

The imaging calculation can be expressed by Formula (4.3), where $E(\lambda)$ is the SPD of the illumination, $S(\lambda)$ is the reflection spectrum, $Q_H(\lambda)$ is the response curve of the R, G, and B channel of a camera.

$$F_H = \eta \int E(\lambda) S(\lambda) Q_H(\lambda) \mathrm{d}\lambda,\ H = \{R, G, B\}. \tag{4.3}$$

If the camera has narrowband sensitivity; i.e., the camera sensor response properties can be approximated by Dirac delta function: $Q_H(\lambda) = q_H \delta(\lambda - \lambda_H)$, then Formula (4.3) is reduced to a simpler form

$$F_H = \sigma E(\lambda_H) S(\lambda_H) q_H. \tag{4.4}$$

Denoting the SPD of the illumination on non-shadow region as E_1 and that on shadow region as E_2, the difference between shadow and non-shadow is

$$\sigma \begin{bmatrix} E_1(\lambda_{\mathrm{R}}) S(\lambda_{\mathrm{R}}) q_{\mathrm{R}} \\ E_1(\lambda_{\mathrm{G}}) S(\lambda_{\mathrm{G}}) q_{\mathrm{G}} \\ E_1(\lambda_{\mathrm{B}}) S(\lambda_{\mathrm{B}}) q_{\mathrm{B}} \end{bmatrix} - \sigma \begin{bmatrix} E_2(\lambda_{\mathrm{R}}) S(\lambda_{\mathrm{R}}) q_{\mathrm{R}} \\ E_2(\lambda_{\mathrm{G}}) S(\lambda_{\mathrm{G}}) q_{\mathrm{G}} \\ E_2(\lambda_{\mathrm{B}}) S(\lambda_{\mathrm{B}}) q_{\mathrm{B}} \end{bmatrix} = \begin{bmatrix} \Delta R \\ \Delta G \\ \Delta B \end{bmatrix}. \tag{4.5}$$

Dividing both sides of (4.5) by $\Delta B(\Delta B \neq 0)$, we get

$$\begin{cases} \dfrac{\Delta R}{\Delta B} = \dfrac{\sigma S(\lambda_R) q_R [E_1(\lambda_R) - E_2(\lambda_R)]}{\sigma S(\lambda_B) q_B [E_1(\lambda_B) - E_2(\lambda_B)]} \\ \dfrac{\Delta G}{\Delta B} = \dfrac{\sigma S(\lambda_G) q_G [E_1(\lambda_G) - E_2(\lambda_G)]}{\sigma S(\lambda_B) q_B [E_1(\lambda_B) - E_2(\lambda_B)]} \end{cases}. \tag{4.6}$$

From (4.4), we have

$$\begin{cases} \dfrac{\sigma S(\lambda_R) q_R}{\sigma S(\lambda_B) q_B} = \dfrac{F_{NS_R}}{F_{NS_B}} \cdot \dfrac{E_1(\lambda_B)}{E_1(\lambda_R)} \\ \dfrac{\sigma S(\lambda_G) q_G}{\sigma S(\lambda_B) q_B} = \dfrac{F_{NS_G}}{F_{NS_B}} \cdot \dfrac{E_{1(\lambda_B)}}{E_1(\lambda_G)} \end{cases}. \tag{4.7}$$

Substituting Formula (4.7) into (4.6), we have

$$\begin{cases} \dfrac{\Delta R}{\Delta B} = \dfrac{F_{NS_R}}{F_{NS_B}} \cdot \dfrac{E_1(\lambda_B)}{E_1(\lambda_R)} \cdot \dfrac{E_1(\lambda_R) - E_2(\lambda_R)}{E_1(\lambda_B) - E_2(\lambda_B)} = m \cdot \dfrac{F_{NS_R}}{F_{NS_B}} \\ \dfrac{\Delta G}{\Delta B} = \dfrac{F_{NS_G}}{F_{NS_B}} \cdot \dfrac{E_1(\lambda_B)}{E_1(\lambda_G)} \cdot \dfrac{E_1(\lambda_G) - E_2(\lambda_G)}{E_1(\lambda_B) - E_2(\lambda_B)} = n \cdot \dfrac{F_{NS_G}}{F_{NS_B}} \end{cases} \tag{4.8}$$

where

$$\begin{cases} m = \dfrac{E_1(\lambda_B)}{E_1(\lambda_R)} \cdot \dfrac{E_1(\lambda_R) - E_2(\lambda_R)}{E_1(\lambda_B) - E_2(\lambda_B)} \\ n = \dfrac{E_1(\lambda_B)}{E_1(\lambda_G)} \cdot \dfrac{E_1(\lambda_G) - E_2(\lambda_G)}{E_1(\lambda_B) - E_2(\lambda_B)} \end{cases}. \tag{4.9}$$

Then, the vector $[\Delta R \ \Delta G \ \Delta B]$ can be represented by

$$\begin{bmatrix} \Delta R \\ \Delta G \\ \Delta B \end{bmatrix} = \begin{bmatrix} \Delta R / \Delta B \cdot \Delta B \\ \Delta G / \Delta B \cdot \Delta B \\ 1 \cdot \Delta B \end{bmatrix} = \begin{bmatrix} m \cdot F_{NS_R} / F_{NS_B} \\ n \cdot F_{NS_G} / F_{NS_B} \\ 1 \end{bmatrix} \cdot \Delta B. \tag{4.10}$$

The maximum channel and the minimum channel in $[\Delta R \ \Delta G \ \Delta B]$ equal to those in $[m \cdot F_{NS_R}/F_{NS_B} \ \ n \cdot F_{NS_G}/F_{NS_B} \ \ 1]$ because of $\Delta B \geq 0$. Next, we discuss how to get the value of m and n.

Denoting E_{sun} as the SPD of direct sunlight and E_{sky} as that of diffuse skylight. Non-shadow region is lighted by daylight (sunlight plus skylight). Therefore, $E_1 = E_{sun} + E_{sky} = E_{day}$; shadow region is lighted by skylight plus partially sunlight and can be denoted as

$$E_2 = \alpha E_{sun} + E_{sky}. \tag{4.11}$$

Here, α is a proportional factor. In umbra region, α equals to 0; in non-shadow, α equals to 1; in penumbra region, $\alpha \in (0\ 1)$.

Rewriting (4.9) , we get

$$\begin{cases} m = \dfrac{E_{\text{day}}(\lambda_{\text{B}})}{E_{\text{day}}(\lambda_{\text{R}})} \cdot (1-\alpha)\dfrac{E_{\text{sun}}(\lambda_{\text{R}})}{E_{\text{sun}}(\lambda_{\text{B}})} \\ n = \dfrac{E_{\text{day}}(\lambda_{\text{B}})}{E_{\text{day}}(\lambda_{\text{G}})} \cdot (1-\alpha)\dfrac{E_{\text{sun}}(\lambda_{\text{G}})}{E_{\text{sun}}(\lambda_{\text{B}})} \end{cases}. \tag{4.12}$$

We will use blackbody irradiance (Wyszecki 1982) to approximate the SPDs of daylight and direct sunlight. The blackbody formula determines the SPD by color temperature T

$$E(\lambda, T) = c_1 \lambda^{-5} (e^{c_2/T\lambda} - 1)^{-1}. \tag{4.13}$$

where $c_1 = 2\pi hc^2$ and $c_2 = \frac{hc}{k}$, in which:

c is the velocity of light: $3.0 \times 10^8\ \text{m} \cdot \text{s}^{-1}$;
h is the Plank constant: $6.63 \times 10^{-34}\ \text{J} \cdot \text{s}$;
k is the Boltzmann constant: $1.38 \times 10^{-23}\ \text{J} \cdot \text{K}^{-1}$.

Therefore, we obtain

$$c_1 = 3.74 \times 10^{-16} (\text{W} \cdot \text{m}^2)$$

$$c_2 = 1.43 \times 10^{-2} (\text{m} \cdot \text{K}).$$

Normally, taking T_{day}=6500 K for daylight, $T_{\text{sun}} = 5500$ K for direct sunlight, and setting the wavelengths at red, green, and blue equal to 650, 550, and 440 nm, respectively, thus we have

$$\frac{E(\lambda_{\text{B}}, T_{\text{day}})}{E(\lambda_{\text{R}}, T_{\text{day}})} = \left(\frac{\lambda_{\text{B}}}{\lambda_{\text{R}}}\right)^{-5} \cdot \frac{e^{\frac{c_2}{T_{\text{day}}\lambda_{\text{R}}}} - 1}{e^{\frac{c_2}{T_{\text{day}}\lambda_{\text{B}}}} - 1} = 1.36 \tag{4.14}$$

similarly

$$\frac{E(\lambda_{\text{B}}, Tv)}{E(\lambda_{\text{G}}, T_{\text{day}})} = 1.11; \frac{E(\lambda_{\text{R}}, T_{\text{sun}})}{E(\lambda_{\text{B}}, T_{\text{sun}})} = 0.97; \frac{E(\lambda_{\text{G}}, T_{\text{sun}})}{E(\lambda_{\text{B}}, T_{\text{sun}})} = 1.07.$$

Substituting them into (4.12) and setting $\alpha = 0$ for simplicity, it is easy to get: $m = 1.31$ and $n = 1.19$. The parametric values of m and n are just approximations because they are calculated from the correlated color temperatures of standard daylight and direct sunlight that may not be same as the actual color temperatures under which the experimental images are taken.

4.3 Tricolor Linear Model of Shadows

In the previous section, based on Planck's law of black body radiation, we calculated the spectral power distribution (SPD) of direct sunlight and sunlight (scattered skylight plus direct sunlight) and estimated the parameters of the tricolor attenuation model. However, there is a certain error between black body approximation and the actual SPD of outdoor light sources. At the same time, we also assume that the camera's spectral response function can be approximated by the dirac delta function, but the actual camera usually does not meet this assumption. In this section, we will derive the tricolor linear model of shadows based on actual SPD and sRGB camera responses.

Figure 4.2 shows the image formation procedure with outdoor illumination. In this chapter, we will consider the response characteristics in the sRGB space because we are modeling and analyzing actual captured images. In sRGB space, the ideal spectral response function of the camera is the color matching function (CMF) in sRGB space. In fact, the response curve of most cameras (the combined result of Bayer filter and camera color correction matrix) is consistent with the sRGB color matching function [5, 10].

In clear weather that shadows most probably take place, the spectral power distributions (SPD) of daylight and skylight show strong regularities, and they are mainly controlled by sun angles, as shown in Fig. 4.3. We will combine the camera imaging calculation formula to study the relationship between the images under daylight and skylight.

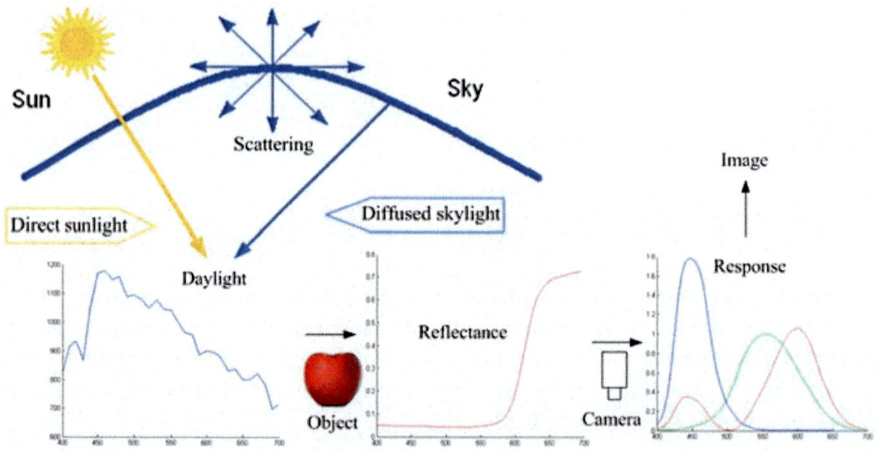

Fig. 4.2 Illustration of the image formation procedure. © [2011] IEEE. Reprinted, with permission, from [Jiandong Tian and Yandong Tang, "Linearity of Each Channel Pixel Values from a Surface in and out of Shadows and Its Applications," IEEE Conference on Computer Vision and Pattern Recognition, pp. 985–992, 2011]

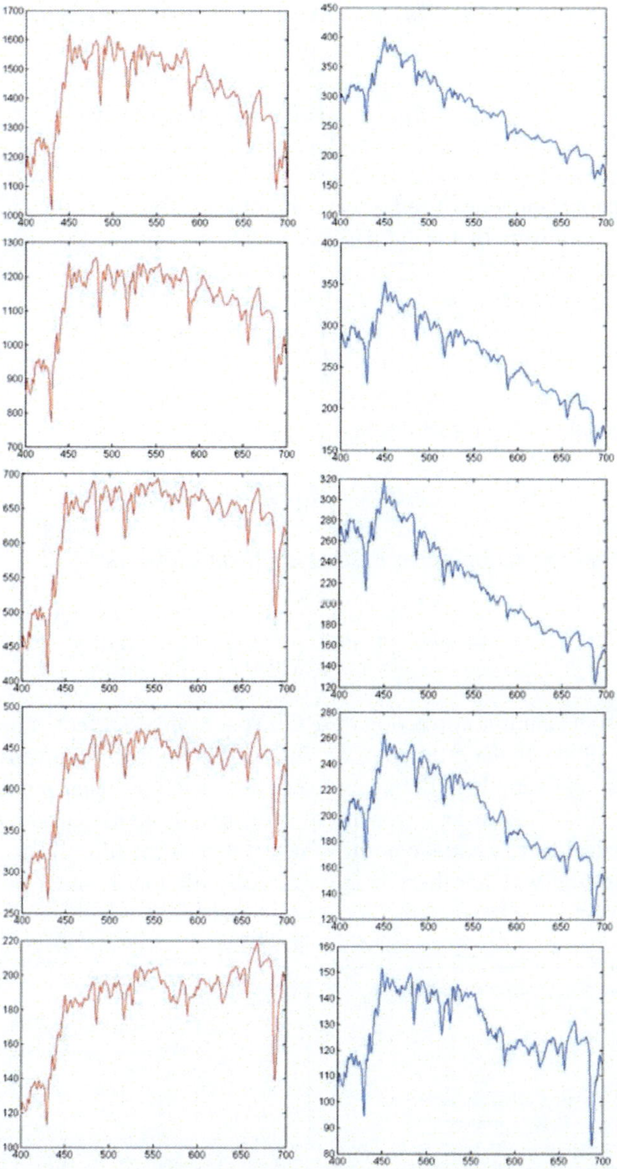

Fig. 4.3 SPD of daylight (left) and that of skylight (right). The x-coordinate denotes wavelength (nm) and the y-coordinate denotes irradiance power ($W \cdot m^{-2} \cdot nm^{-1}$). From top to bottom are of sun angles equal to 20, 40, 60, 70, 80°, respectively © [2011] IEEE. Reprinted, with permission, from [Jiandong Tian and Yandong Tang, "Linearity of Each Channel Pixel Values from a Surface in and out of Shadows and Its Applications," IEEE Conference on Computer Vision and Pattern Recognition, pp. 985–992, 2011]

If $Q_H(\lambda)$ means sRGB CMF, the linear RGB tristimulus values in sRGB color space are

$$F_H^L = \eta \cdot w_H \cdot \int_{400}^{700} E(\lambda)S(\lambda)Q_H(\lambda)\mathrm{d}\lambda, \tag{4.15}$$

where H denotes any one of red, green, and blue channel. η is the factor to approximate camera exposure that is defined as

$$\eta = 100 / \int_{400}^{700} E(\lambda)Q_{\mathrm{G}}(\lambda)\mathrm{d}\lambda, \tag{4.16}$$

w_H is white balance factor satisfying $w_{\mathrm{G}} = 1$ and

$$w_{\mathrm{R}} F_{\mathrm{R}}^L = w_{\mathrm{G}} F_{\mathrm{G}}^L = w_{\mathrm{B}} F_{\mathrm{B}}^L. \tag{4.17}$$

The gamma correction on linear RGB is applied as follows:

$$F_H = 255 \cdot [1.055 \cdot \left(\frac{F_H^L}{100}\right)^{(1/2.4)} - 0.055]. \tag{4.18}$$

For details of the gamma correction in sRGB space, please refer to the literature (IEC Publication 1966). In the following, we will use the 60-degree sun angle SPD shown in Fig. 4.4 as an example to derive and verify the tricolor linear model of shadows.

Because illumination and camera spectral responses have much less randomness than reflectance, we make point-wise product of the illumination and CMFs in Eq. (4.15) to find whether there is any characteristic that is independent of surface reflectance.

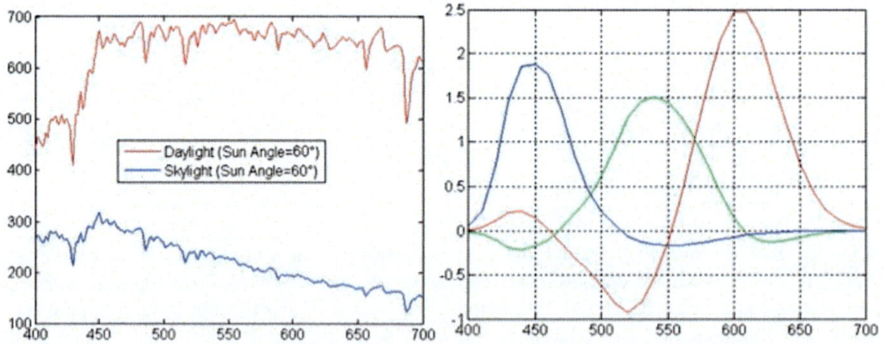

Fig. 4.4 SPD of daylight and skylight at sun angle of 60° and sRGB CMFs © [2011] IEEE. Reprinted, with permission, from [Jiandong Tian and Yandong Tang, "Linearity of Each Channel Pixel Values from a Surface in and out of Shadows and Its Applications," IEEE Conference on Computer Vision and Pattern Recognition, pp. 985–992, 2011]

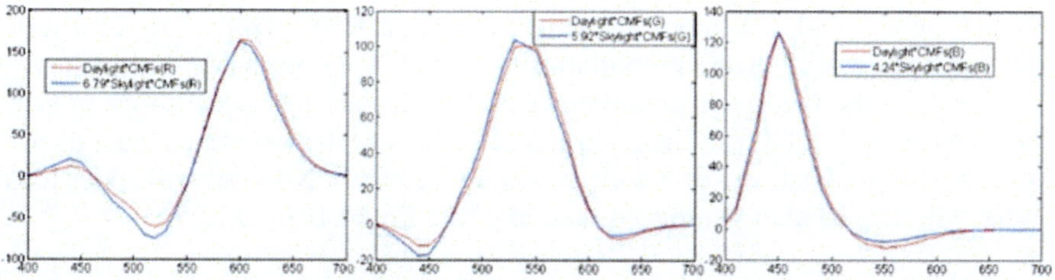

Fig. 4.5 Point-wise products of daylight with CMFs can be well-approximated by those of skylight with CMFs multiplying a constant in each channel. The constants are derived by Eq. (4.19) (we assume half skylight can illuminate a shadow in this paper) © [2011] IEEE. Reprinted, with permission, from [Jiandong Tian and Yandong Tang, "Linearity of Each Channel Pixel Values from a Surface in and out of Shadows and Its Applications," IEEE Conference on Computer Vision and Pattern Recognition, pp. 985–992, 2011]

As can be seen from Fig. 4.5, though the SPD of daylight and skylight are quite different, the point-wise product of daylight with CMFs (H) can be well-approximated by that of skylight with CMFs (H) multiplied by a constant. That is

$$E_{\text{day}}(\lambda) \cdot Q_H(\lambda) = K_H \cdot E_{\text{sky}}(\lambda) \cdot Q_H(\lambda), \tag{4.19}$$

where $Q_H(\lambda)$ is sRGB CMF, and K_H is a constant independent with wavelength. The constants can be derived from

$$K_H = \arg\min \sum_{\lambda=400}^{700} \left| E_{\text{day}}(\lambda) \cdot Q_H(\lambda) - K_H \cdot E_{\text{sky}}(\lambda) \cdot Q_H(\lambda) \right|. \tag{4.20}$$

If we rewrite Eq. (4.15), for a surface in shadow area, we have

$$F_H^L = \eta \cdot w_H \cdot \int_{400}^{700} E_{\text{sky}}(\lambda) S(\lambda) Q_H(\lambda) \mathrm{d}\lambda. \tag{4.21}$$

For the same surface in non-shadow area, we have

$$F_H^L = \eta \cdot w_H \cdot \int_{400}^{700} E_{\text{day}}(\lambda) S(\lambda) Q_H(\lambda) \mathrm{d}\lambda. \tag{4.22}$$

If we divide Eq. (4.22) by Eq. (4.21) and combine Eq. (4.19), we have

$$\frac{F_H^L}{f_H^L} = \frac{\eta \cdot w_H \cdot \int_{400}^{700} E_{\text{day}}(\lambda) S(\lambda) Q_H(\lambda) \mathrm{d}\lambda}{\eta \cdot w_H \cdot \int_{400}^{700} E_{\text{sky}}(\lambda) S(\lambda) Q_H(\lambda) \mathrm{d}\lambda} = K_H. \tag{4.23}$$

So far we can find that, in each channel, the linear RGB pixel values of a surface illuminated by daylight are proportional to those of the same surface illuminated by skylight. Furthermore, the proportional coefficients are just equal to the constants determined by Eq. (4.20). The good characteristic of the proportional coefficients is that they are independent of wavelength and reflectance. Now, we will take a look at what will happen after gamma correction. From Eq. (4.18), we have

$$\begin{cases} F_H = 39.5 \cdot {F_H^L}^{(1/2.4)} - 14, \\ f_H = 39.5 \cdot {f_H^L}^{(1/2.4)} - 14, \end{cases} \tag{4.24}$$

Combined with Eq. (4.23), we have

$$\begin{aligned} F_H &= 39.5 \cdot {F_H^L}^{(1/2.4)} - 14 \\ &= 39.5 \cdot K_H^{(1/2.4)} \cdot f_H^{L(1/2.4)} - 14 \\ &= K_H^{(1/2.4)} \cdot (39.5 \cdot f_H^{L(1/2.4)} - 14) + (K_H^{(1/2.4)} - 1) \cdot 14 \\ &= K_H^{(1/2.4)} \cdot f_H + (K_H^{(1/2.4)} - 1) \cdot 14. \end{aligned} \tag{4.25}$$

Rewrite Eq. (4.25) to get the linear model inside and outside the shadow area of a certain channel H, we have

$$\begin{cases} F_H = \kappa_1 f_H + \mu_1, \\ \kappa_1 = K_H^{(1/2.4)}, \\ \mu_1 = (K_H^{(1/2.4)} - 1) \cdot 14. \end{cases} \tag{4.26}$$

The above formula shows that after gamma correction the proportional relationship becomes a linear polynomial relationship, and the parameters of the linear polynomial are determined by the proportional coefficients.

We employ the data that contain 1995 different surface reflectances to simulate four images based on Eqs. (4.15)–(4.18). The four images include one linear image in daylight, one linear image in skylight, one gamma-corrected image in daylight, and one gamma-corrected image in skylight. The 1995 surface reflectance curves and the gamma-corrected image in daylight are shown in Fig. 4.6. Using the four simulated images, in each channel, for each surface, we take the pixel value in daylight and that in skylight as one counterpart. So for one paired images, in each channel, we have 1995 counterparts. The fitting results of these counterparts are shown in Fig. 4.7. The first row shows results of the linear images, and the second row shows results of the gamma-corrected images. From the results, we can find that these surfaces in daylight and in skylight show strong statistical linear property. Furthermore, we also use two images captured by a real camera to test whether they satisfy the linearity. Figure 4.8 shows two real images of Xrite ColorChecker (XCC), which contains 24 most common colors in our daily life. The two images were captured at 60° sun angle using Canon 5D Mark II digital camera equipped with a 50 mm len,

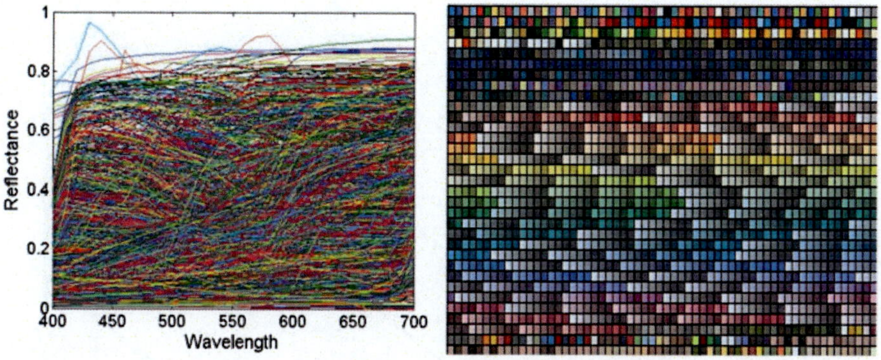

Fig. 4.6 1995 surface reflectance curves and the simulated image in daylight at sun angle of 60° © [2011] IEEE. Reprinted, with permission, from [Jiandong Tian and Yandong Tang, "Linearity of Each Channel Pixel Values from a Surface in and out of Shadows and Its Applications," IEEE Conference on Computer Vision and Pattern Recognition, pp. 985–992, 2011]

and they were captured under identical conditions: the same exposure, same white balance, and same geometric view. The simulated and true captured pixel values of 24 XCC surface counterparts are plotted in Fig. 4.7. The third row shows results of simulated gamma-corrected RGB images of 24 XCC surfaces, and the fourth row shows the results of real-captured RGB images of 24 XCC surfaces. We can observe that linear features exist in the real image. To summarize, in each channel of linear RGB images, pixel values of a same surface in shadow region and in non-shadow region have a proportional relationship. After gamma correction, the proportional relationship becomes a linear polynomial relationship. Both the proportional and polynomial coefficients are intrinsically determined by Eq. (4.20), i.e., determined by daylight and skylight and have nothing to do with reflectance. This is also the biggest advantage of our linear model. Because the model parameters are independent of the reflection spectrum, the model can be applied to different scenes in the physical world.

4.4 Shadow Detection Based on Tricolor Attenuation Model

Based on the proposed tricolor attenuation model, we introduce our multi-step algorithm for shadow detection. Before detection, the calculating regions of the algorithm should be determined. It is meaningless to calculate $[m \cdot F_{NS_R}/F_{NS_B} \quad n \cdot F_{NS_G}/F_{NS_B} \quad 1]$ pixel-wisely because shadow and non-shadow will certainly not be the same pixel. Calculating it on the whole image is also not proper, because we assume that the shadow and its corresponding non-shadow background have the same reflective surface when we derive the tricolor attenuation model. As shown in Fig. 4.9a, region ① is non-shadow background and region ② is shadow region. Red

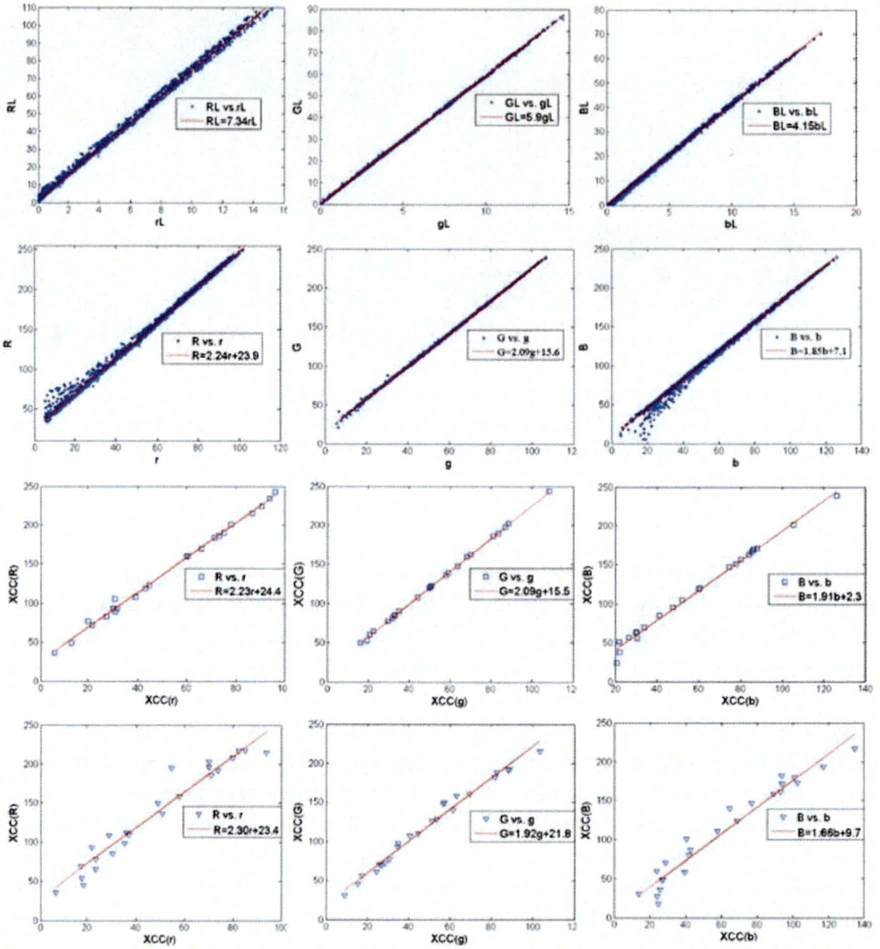

Fig. 4.7 Pixel values in daylight versus those in skylight show strong linearity in R, G, B channels, respectively © [2011] IEEE. Reprinted, with permission, from [Jiandong Tian and Yandong Tang, "Linearity of Each Channel Pixel Values from a Surface in and out of Shadows and Its Applications," IEEE Conference on Computer Vision and Pattern Recognition, pp. 985–992, 2011]

color is the domain color (in region ①, $R > G > B$). If we subtract B channel from R channel, as shown in Fig. 4.9b, not only the shadow region but also the blue region will be apparent from domain color region; that is, region ② and region ③ are both darker than region ① after channel subtraction. Then the blue region may be falsely detected as shadows. Similarly, as shown in Fig. 4.10, if a same shadow falls on different reflecting surface, calculating it on the whole image may also cause false shadow detection. Therefore, we need to segment the original image into subregions with similar color before shadow detection, which aims to segment the shadow and non-shadow on the same reflecting surface into one subregion. Because the

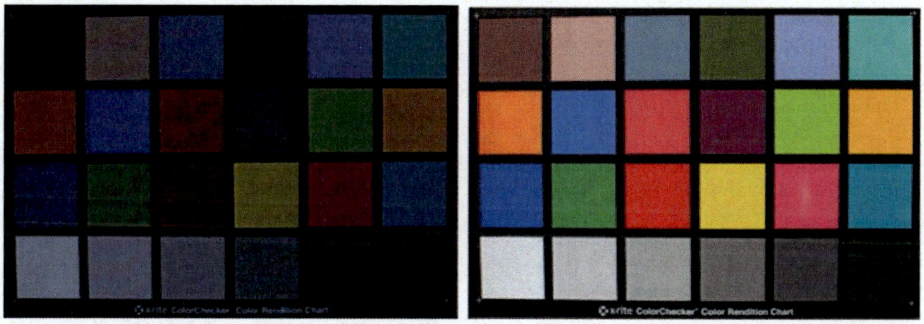

Fig. 4.8 Real images captured at 60° sun angle. Left image is captured in skylight and right in daylight © [2011] IEEE. Reprinted, with permission, from [Jiandong Tian and Yandong Tang, "Linearity of Each Channel Pixel Values from a Surface in and out of Shadows and Its Applications," IEEE Conference on Computer Vision and Pattern Recognition, pp. 985–992, 2011]

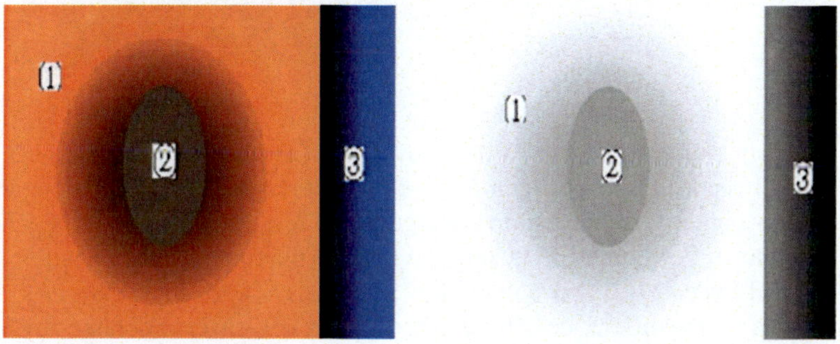

(a) (b)

Fig. 4.9 Subtraction on whole image. **a** red color is the domain color; (b): the subtractive image using R-B. © [2009] IEEE. Reprinted, with permission, from [Jiandong Tian, Jing Sun, and Yandong Tang, "Tricolor Attenuation Model for Shadow Detection," IEEE Transactions on Image Processing, Vol. 18, No. 10, pp. 2355–2363, 2009]

preprocessing of segmentation is easily affected by shadows, it is better to segment the image in a space which is less sensitive to shadows. Here, we adopt Formula (4.27) to transform an RGB image to a gray one that is not very sensitive to shadows.

$$Y = \log\left(\frac{\max[F_R F_G F_B]}{\min[F_R F_G F_B] + 1}\right). \tag{4.27}$$

The reasons to use the transformation for computing the shadow-invariant image are listed below:

(1) The color ratios (chromaticity) do not vary so obviously as pixel intensity when illumination changes.
(2) Constant color ratios such as F_R/F_B are not appropriate. Taking $F_{NS_R} > F_{NS_G} > F_{NS_B}$ for an example, we get $\Delta R > \Delta G > \Delta B$ from the tricolor attenuation

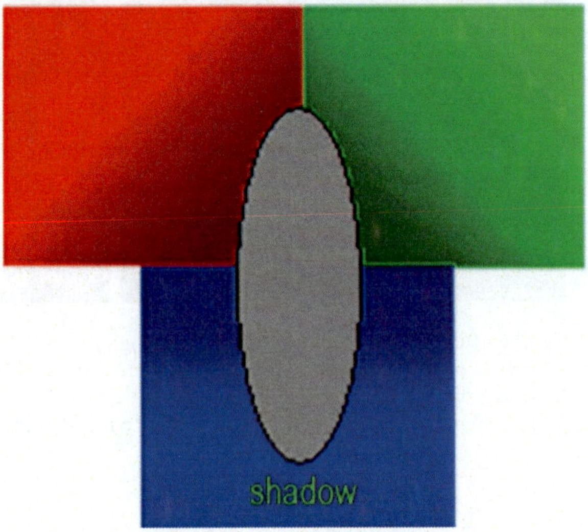

Fig. 4.10 Same shadow falls on different color surface. © [2009] IEEE. Reprinted, with permission, from [Jiandong Tian, Jing Sun, and Yandong Tang, "Tricolor Attenuation Model for Shadow Detection," IEEE Transactions on Image Processing, Vol. 18, No. 10, pp. 2355–2363, 2009]

model of shadows (Eq. 4.10). However, the inequality $F_{S_R} < F_{S_G} < F_{S_B}$ may happen according to Formula (4.1). The difference between shadow and non-shadow obtained by color ratios is large due to $F_{NS_R}/F_{NS_B} > 1$ and $F_{S_R}/F_{S_B} < 1$. In contrast, Formula (4.27) using the ratio of maximum channel and minimum channel instead of fixed color ratio, so, $F_{NS_R}/F_{NS_B} > 1$ and $F_{S_B}/F_{S_R} > 1$. In most cases, the difference between shadow and non-shadow obtained by Formula (4.27) is smaller than that obtained by color ratios.

(3) The logarithm operation is used for compressing the dynamic range to overcome the over-segmentation of the watershed segment algorithm.

Our multi-step algorithm for shadow detection is as follows:

(1) To transform the original color image F into the gray image Y by formula (4.27).
(2) To segment Y into subregions with similar color by the well-known watershed algorithm, that is, $Y = \cup Y^i$, where $Y^i \cap Y^j = \Phi\ (if\ i \neq j)$, and i is the segmented region number.
(3) For each subregion, to calculate the mean value of $[F_R^i\ F_G^i\ F_B^i]$ by

$$[\overline{F_R^i}\ \overline{F_G^i}\ \overline{F_B^i}] = \frac{1}{M}\left(\sum_{k \in F^i}^{k=1\ldots M.} [F_R^{i,k}\ F_G^{i,k}\ F_B^{i,k}]\right). \tag{4.28}$$

where $F_R^{i,k}$ denotes the kth pixel in the ith region of F in R channel, and M is the number of pixels in region i.

(4) To calculate the mean value of $[F_{NS_R}^i\ F_{NS_G}^i\ F_{NS_B}^i]$ by Formula (4.28). As shown in formula (4.10), $[\Delta R\ \Delta G\ \Delta B]$ is determined by $F_{NS_R/F_{NS_B}}$ and F_{NS_G}/F_{NS_B}. However, we do not know where is non-shadow and where is shadow before detection. We take pixels whose values are larger than the mean value of region i as the

non-shadow background in this step. Then $[m \cdot \overline{F^i_{\mathrm{NS_R}}}/\overline{F^i_{\mathrm{NS_B}}} \quad n \cdot \overline{F^i_{\mathrm{NS_G}}}/\overline{F^i_{\mathrm{NS_B}}} \quad 1]$ is calculated.

(5) To subtract the minimum channel from the maximum one. Namely, if $m \cdot \overline{F^i_{\mathrm{NS_R}}}/\overline{F^i_{\mathrm{NS_B}}} > n \cdot \overline{F^i_{\mathrm{NS_G}}}/\overline{F^i_{\mathrm{NS_B}}} > 1$ in F^i, we get $X^i = F^i_{\mathrm{R}} - F^i_{\mathrm{B}}$, and vise versa.

(6) To binarizate X^i by the following threshold based on the observation that shadows are often darker than the mean value of X^i

$$T = \frac{1}{M}\left(\sum_{k \in X^i}^{k=1\cdots\cdots M.} X^i_k\right) \tag{4.29}$$

where M is the number of pixels in region i. An initial shadow result in F^i can be obtained by

$$S^i = \{(x, y)|(x, y) \subset F^i, X^i(x, y) < T\}. \tag{4.30}$$

(7) To verify the shadow. Shadows gotten in each subregion may not be real ones due to false detection. After judging each detected S^i by Formula (4.31), shadow region S in F can be obtained. S is initialized by the first S^i gotten by Formula (4.30)

$$S = \begin{cases} S \cup S^i & \text{if } \overline{F^{\mathrm{NS}^i}_{[\mathrm{R\,G\,B}]}} - \overline{F^{\mathrm{S}^i}_{[\mathrm{R\,G\,B}]}} \subset [k_1 \cdot L \quad k_2 \cdot L] \\ S & \text{if } \overline{F^{\mathrm{NS}^i}_{[\mathrm{R\,G\,B}]}} - \overline{F^{\mathrm{S}^i}_{[\mathrm{R\,G\,B}]}} \not\subset [k_1 \cdot L \quad k_2 \cdot L] \end{cases} \tag{4.31}$$

where $L = [m \cdot \overline{F^{\mathrm{NS}^i}_{\mathrm{R}}}/\overline{F^{\mathrm{NS}^i}_{\mathrm{B}}} \quad n \cdot \overline{F^{\mathrm{NS}^i}_{\mathrm{G}}}/\overline{F^{\mathrm{NS}^i}_{\mathrm{B}}} \quad 1] \cdot \Delta B$, the coefficients k_1 and k_2 are empirically set to 0.8 and 1.2, respectively. The two coefficients are adopted because m and n are not accurate values.

(8) To obtain more accurate boundaries of shadows. Shadows detected by previous steps are based on the subtractive image X_i. However, subtractive operation blurred image X because of high correlation among R, G, and B components. This may cause inaccurate boundaries of shadows. Therefore, another constraint is imposed to improve the details; that is, the pixel values in shadow regions are often darker. Applying this constraint helps to improve the details of the test results. The final result of detected shadows is denoted as Formula (4.32).

$$\text{Shadow} = \{(x, y)|S(x, y) \cap F^i_{[\mathrm{RGB}]}(x, y) < [\overline{F^i_{\mathrm{R}}} \quad \overline{F^i_{\mathrm{G}}} \quad \overline{F^i_{\mathrm{B}}}]\}. \tag{4.32}$$

where $F^i_{[\mathrm{R\,G\,B}]}(x, y)$ denotes the tricolor vector at location (x, y) in i th region of original image F.

Figure 4.11 shows the results of our shadow detection method on a simple image, and the comparison results with other method. Image (a) is an image with church shadow. Our method and literature (Finlayson 2007) have achieved similar results, but the method in (Finlayson 2007) needs two images of the same scene. As shown in Fig. 4.12, our algorithm is used for dealing with images obtained from real complex

(a) (b) (c)

Fig. 4.11 Shadow detection results of a simple image. **a** Original image; **b** result of our method; **c** result of the method (Finlayson 2007) © [2009] IEEE. Reprinted, with permission, from [Jiandong Tian, Jing Sun, and Yandong Tang, "Tricolor Attenuation Model for Shadow Detection," IEEE Transactions on Image Processing, Vol. 18, No. 10, pp. 2355–2363, 2009]

scenes. Image (a) is a top view image of a tree cluster. Both the big cast shadow in the middle and the small shadow in the bottom can be detected out and are free of affection of the complicated texture in the background. Image (c) is a scene that contains shadows of a man and of a tree. Image (d) shows the extracted shadows. Though there is a little noise on the grass, the shadows on the ground are detected out. Because the shadow on the upper portion of the trunk is encompassed by leaves and is lighted by only part of skylight, which violates the light environments of our model (shadow lighted by clear skylight) and further causes the changes of parameter values of m and n, the shadow is missed in our detecting results. Image (e) has shadows on the road and on the soil. As shown in image (f), the weak penumbra shadows on the road, the self shadow on the right side of the road, and the shadows on the soil are all detected, except that there are some pixels misclassified as shadows.

4.5 Shadow Detection Based on Tricolor Linear Model

As a natural phenomenon, a shadow has physical properties that should be adopted in shadow detection. As shown in Fig. 4.13, humans may identify a shadow by exploring the contrast near edges. Without contrast, it is difficult to determine whether the color checker is in the shadow or not. Therefore, the contrast between the shadow and its background should be used to perform the shadow detection task.

In Sect. 4.2, we proposed the tricolor attenuation model that describes the attenuation relationship between a shadow and its non-shadow background in three color channels. In Sect. 4.3, we further deduced a linear model that describes the pixel values of surfaces in a shadow region and a non-shadow region having a linear relationship. In last section, we describe a shadow detection algorithm based on tricolor

Fig. 4.12 Results of complex scenes. Image (**a**), (**c**), (**e**) original images; image (**b**), (**d**), (**f**) detected shadows © [2009] IEEE. Reprinted, with permission, from [Jiandong Tian, Jing Sun, and Yandong Tang, "Tricolor Attenuation Model for Shadow Detection," IEEE Transactions on Image Processing, Vol. 18, No. 10, pp. 2355–2363, 2009]

(a) Non-obvious shadow without the context of contrast.

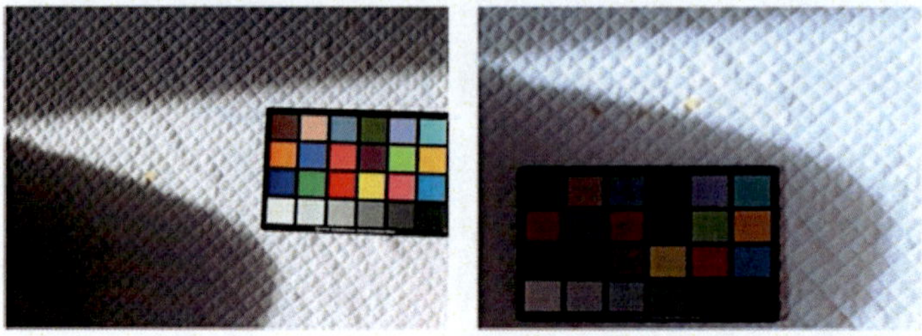

(b) Obvious shadow with the context of contrast.

Fig. 4.13 Illustration of non-obvious and obvious shadows. It is much easier for us to identify shadows by exploring the contrast near edges. The two checkerboards shown in (**a**) were captured under identical imaging environment with the two pictures shown in (**b**) Reprinted from [Jiandong Tian, Xiaojun Qi, Liangqiong Qu, and Yandong Tang, "New Spectrum Ratio Properties and Features for Shadow Detection," Pattern Recognition, Vol. 51, No. 3, 2016, pp. 85–96], Copyright 2016, with permission from Elsevier

attenuation model. In this section, we aim to find additional new physical properties of shadows based on tricolor linear model and then take these shadow properties as features to design an effective shadow detection method.

From Eq. 4.24, we have,

$$\left(\frac{F_H^L}{f_H^L}\right)^{(1/2.4)} = \frac{F_H + 14}{f_H + 14} \tag{4.33}$$

Based on Eq. 4.23, we can simplify Eq. 4.33 to the following,

$$K_H = \left(\frac{F_H + 14}{f_H + 14}\right)^{2.4} \tag{4.34}$$

Equation 4.34 shows that spectrum ratios can be directly calculated from the gamma-corrected sRGB pixel values by: $K_H = \left(\frac{F_H+14}{f_H+14}\right)^{2.4}$.

Certainly, we can also convert the gamma-corrected sRGB values to linearized values by inverse gamma correction and calculate K_H by $K_H = \frac{F_H^L}{f_H^L}$. The

Table 4.2 Spectrum ratios and their differences at representative sun angles

	20°	30°	40°	50°	60°	70°	80°
K_R	12.11	10.26	9.49	8.59	6.79	4.79	3.18
K_G	10.40	9.19	8.52	7.40	5.92	4.25	2.86
K_B	8.10	7.28	7.19	5.94	4.24	3.58	2.53
$K_R - K_G$	1.71	1.07	0.97	1.19	0.87	0.54	0.32
$K_G - K_B$	2.30	1.91	1.33	1.46	1.68	0.67	0.33

advantage of taking the spectrum ratios as physical properties of shadows is that they are independent of wavelength and reflectance.

4.5.1 New Shadow Properties

Similar to calculating K_H in 60° , we can further calculate K_H in other representative sun angles ranging from 20° to 80°. We measured SPDs of daylight and skylight at representative sun angles by an Avantes USB 2.0 spectrometer for one year. We measured them one or two times each week for one year at different times (sun angle) and under different weather conditions. To eliminate the influence of noise and little errors of the measurements, we used mean values of the measurements for the K_H calculation. For most situations that shadow often takes place (e.g., clear weather, little overcast, and big clouds that do not cover the sun), SPDs of daylight and skylight are relatively stable. But sometimes, their amplitude varies noticeably; fortunately, the variation in amplitude does not affect the K_H according to Eq. 4.23. The value of K_H for the sun angle less than 20° or greater than 80° is not computed since in our and nearby cities, we can only observe sun angles between 20° and 80°. We used these measured SPDs only to find general rules of K_H, our proposed shadow features are not sensitive to small SPDs variation caused by factors such as time, location, and seasons, since our shadow features and detection method do not rely on the exact values of K_H . Table 4.2 lists the spectrum ratios of daylight and skylight for sun angles from 20° to 80°. To facilitate future discussion, we also list the difference between K_R and K_G and the difference between K_G and K_B for each sun angle.

Table 4.2 clearly shows the four following new physical shadow properties.

Property 4.1 The values of K_H decrease as sun angle increases.

Property 4.2 The values of $K_H (H = R, G, B)$ satisfy $K_R > K_G > K_B$.

Property 4.3 The differences between K_H ($K_R - K_G$ and $K_G - K_B$) are not very small, and these differences normally decrease as the sun angle increases. The differences between K_H satisfy $K_R - K_G < K_G - K_B$.

Property 4.4 The value of K_H at any angle is less than twice of the value of K_H at the next angle, which is 10° larger than the current investigated angle. That is, $K_H^x < 2K_H^{x+10}$, where K_H^x is the value of the K_H at the angle of x.

These referenced shadow properties will be utilized to generate the three shadow verification criteria in our proposed shadow detection algorithm.

4.5.2 *Shadow Detection Algorithm*

Based on the four new physical shadow properties, we propose a novel shadow detection algorithm for an outdoor color image. Our shadow detection algorithm consists of three steps: multi-threshold edge extraction, K_H calculation, and shadow edge classification. The multi-threshold edge extraction step functions as the preprocessing step, which effectively breaks the connection between shadow edges and object edges. The K_H calculation step employs the dilated regions along the detected edges to compute the values of K_H. The shadow edge classification step employs the three criteria derived from the new shadow properties to classify each edge as shadow edge or non-shadow edge (e.g., object edge). In the following, we explain each of the three steps in detail.

4.5.2.1 Multi-threshold Edge Extraction

We employ the Canny edge detector to extract edges from an image due to its superior performance in extracting thin edges and its robustness to noises. We should avoid long edges since the shadow edge and the object edge may be connected together, which will affect the shadow verification step. To this end, we adopt the multiple thresholds (T = 0.9, 0.8,..., and 0.2) when we employ the Canny edge detector and use the subtraction of Canny results with two adjacent thresholds as the edge images.

4.5.2.2 K_H Calculation

We compute K_H by Eq. 4.34

$$K_H = \left(\frac{F_H + 14}{f_H + 14}\right)^{2.4} \tag{4.35}$$

where F_H and f_H denote the gamma-corrected pixel values of the non-shadow and shadow counterparts, respectively. To overcome noise and obtain fast computing, for each edge, we calculate the mean pixel values on its both sides in each channel to determine the spectrum ratios.

Proposition 4.1 The "mean" operation does not change the spectrum ratios computed from the shadow and non-shadow paired pixel elements within the area covered by the mean filter.

Proof We rewrite Eq. 4.34 as follows:

$$F_H + 14 = {K_H}^{1/2.4}(f_H + 14) \tag{4.36}$$

Suppose we have n shadow and non-shadow counterparts on two sides of an edge, we have,

$$\sum_{i=1}^{n} F_H^i + 14n = {K_H}^{1/2.4}\left(\sum_{l=1}^{n} f_H^i + 14n\right) \tag{4.37}$$

Dividing n on both sides, we have

$$\frac{1}{n}\sum_{i=1}^{n} F_H^i + 14 = {K_H}^{1/2.4}\left(\frac{1}{n}\sum_{i=1}^{n} f_H^i + 14\right) \tag{4.38}$$

It can then be rewritten as follows:

$$\overline{F_H} + 14 = {K_H}^{1/2.4}(\overline{f_H} + 14) \tag{4.39}$$

Finally we have

$$K_H = \left(\frac{\overline{F_H} + 14}{\overline{f_H} + 14}\right)^{2.4} \tag{4.40}$$

□

Comparing Eqs. 4.34 and 4.40, it clearly demonstrates that the averaged pixel values in non-shadow vs. in shadow region have the same spectrum ratios as its individual non-shadow vs. shadow pixels within the region. In other words, the "mean" operation does not change the spectrum ratios. This conclusion also verifies the claim about Eq. 4.23 that "the spectrum ratios are independent of the wavelength and reflectance."

For each detected edge, we enlarge the edge regions by applying a binary dilation operation with the size of six pixels on the both sides. Figure 4.14 shows the dilated edge regions for the edge, where the blue area represents the shadow region and the red area represents the non-shadow region.

We calculate the average pixel values in each dilated region along the edge in each color channel. We then take the one with the larger average pixel value as the non-shadow side and the smaller one as the shadow side to compute K_H using Eq. 4.40.

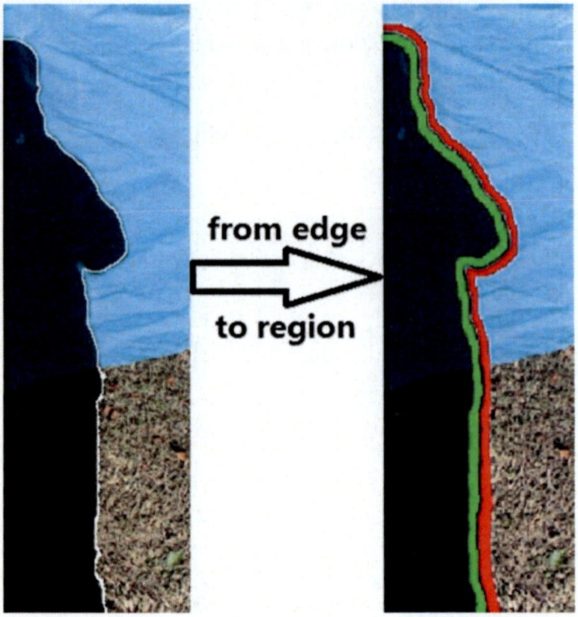

Fig. 4.14 Illustration of the shadow and non-shadow regions along the Canny edge after applying a binary dilation operation Reprinted from [Jiandong Tian, Xiaojun Qi, Liangqiong Qu, and Yandong Tang, "New Spectrum Ratio Properties and Features for Shadow Detection," Pattern Recognition, Vol. 51, No. 3, 2016, pp. 85–96], Copyright 2016, with permission from Elsevier

4.5.2.3 Shadow Edge Classification

We finally use the calculated K_H and the aforementioned shadow properties to differentiate shadow edges from object edges. Specifically, we classify the edge as the shadow edge if all of the following three verification criteria are satisfied:

Criterion 4.1 $\tau_H \cdot K_H^{80} < K_H < \eta_H \cdot K_H^{20}$
where K_H^{80} and K_H^{20} denote K_H at the sun angles of 80° and 20°, respectively. τ_H and η_H are two predefined tolerance coefficients to decide the lower and upper bounds for the value of K_H. In our system, we adopt these two coefficients since we have to consider all possible values of K_H for sun angles ranging from 0° to 90°. Table 4.2 shows that the value of K_H decreases as the sun angle increases, and Property 4.4 shows the relationship between the values of K_H at two adjacent angles. Based on Property 4.4, we compute two predefined tolerance coefficients, τ_H and η_H, as follows:

(1) $\tau_H = 2^{-1}$ since only one angle (i.e., 90°) is greater than 80°when considering the step size of 10°.
(2) $\eta_H = 2^2$ since two angles (i.e., 10°and 0°) are less than 20°when considering the step size of 10°.

This criterion ensures that a valid range for all values of K_H under different sun angles is established.

Criterion 4.2 $K_{\mathrm{R}} > K_{\mathrm{G}} > K_{\mathrm{B}}$.

This criterion directly corresponds to the Property 4.2.

Criterion 4.3 $\begin{cases} K_{\mathrm{R}} - K_{\mathrm{G}} > \varepsilon \wedge K_{\mathrm{G}} - K_{\mathrm{B}} > \varepsilon \;\; if \;\; K_{\mathrm{R}} > K_{\mathrm{R}}^{80} \\ K_{\mathrm{R}} - K_{\mathrm{G}} > \frac{\varepsilon}{2} \wedge K_{\mathrm{G}} - K_{\mathrm{B}} > \frac{\varepsilon}{2} \;\; if \;\; K_{\mathrm{R}} \le K_{\mathrm{R}}^{80} \end{cases}$

Parameter ε is the lower bound for the difference between K_{R} and K_{G} and the difference between K_{G} and K_{B} for sun angle less than 90° (i.e., $K_{\mathrm{R}} > K_{\mathrm{R}}^{80}$). Since we cannot obtain the differences among three K_H values for sun angle of 90° in our city and nearby cities, the lower bound for this angle is set to be a half of ε based on the shadow Property 4.4.

Table 4.2 shows that $K_{\mathrm{R}}^{80} - K_{\mathrm{G}}^{80}$ is the smallest difference among three K_H values for sun angles ranging from 20° to 80°. The shadow Property 4.3 also summarizes that the difference increases as the sun angle decreases. Therefore, we set ε as:

$$\varepsilon = \frac{K_{\mathrm{R}}^{80} - K_{\mathrm{G}}^{80}}{2} \tag{4.41}$$

Our experiments show that these simple predefined coefficients can achieve good shadow detection performance. Coefficients obtained by other methods (e.g., fitting K_H with different sun angles, and obtaining two coefficients in each channel) do not offer better performance.

It should be noted that Criterion 4.2 is a special case of Criterion 4.3. When $\varepsilon = 0$, Criterion 4.3 becomes Criterion 4.2. Here, we list them as two separate criteria due to the following reasons: (1) Criterion 4.2 does not involve any parameter and has simple expression; (2) Criterion 4.2 is powerful and robust since it can correctly classify about 85% of non-shadow edges. Thus in some applications that only need coarse shadow detection, Criterion 4.2 can be separately used for fast detection.

4.5.2.4 Algorithm Pseudocode and Illustration

The detail of our shadow detection algorithm is presented as follows:
Algorithm **input:** an outdoor color image
Algorithm **output:** shadow edge detected image

```
loop (T = 0.9, 0.8, ··· , 0.2)
1: Do 'Edge Extraction' Step
     if T == 0.9 then
   edge ← Canny(T)
     else
   edge ← Canny(T) − Canny(T + 0.1)
     endif
2: Do 'K_H Calculation' Step
3: Do 'Edge Classification' Step
End loop
```

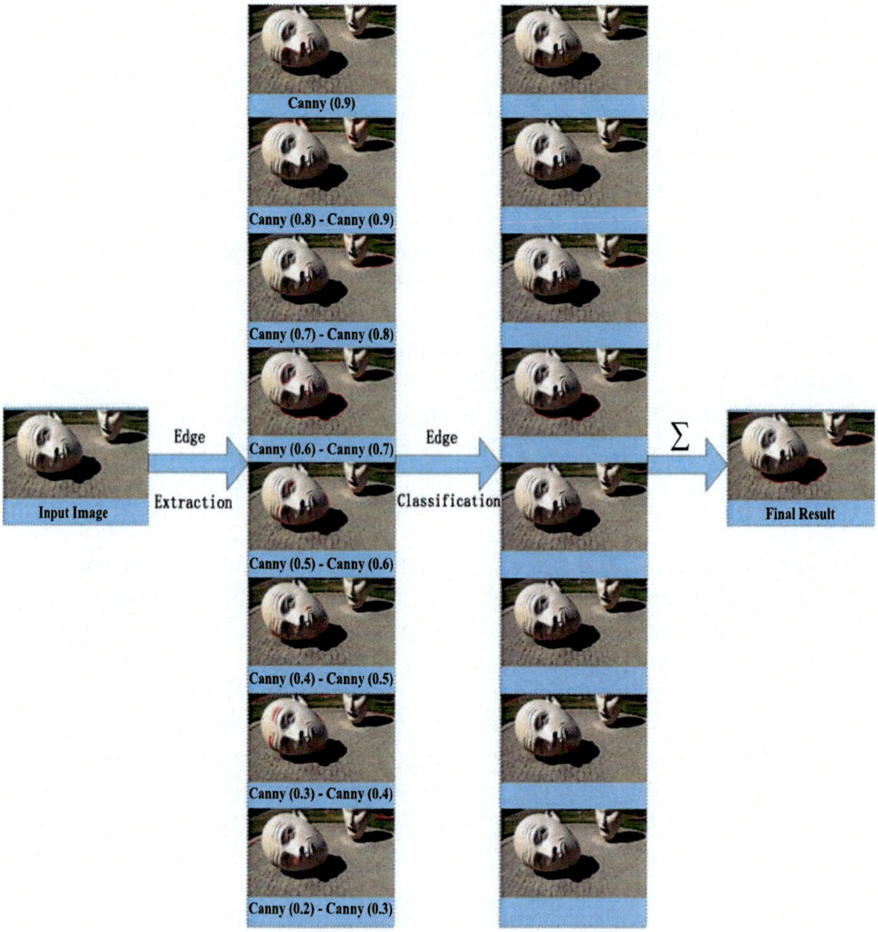

Fig. 4.15 Shadow edges preserved after shadow classification on Canny edge image. Reprinted from [Jiandong Tian, Xiaojun Qi, Liangqiong Qu, and Yandong Tang, "New Spectrum Ratio Properties and Features for Shadow Detection," Pattern Recognition, Vol. 51, No. 3, 2016, pp. 85–96], Copyright 2016, with permission from Elsevier

where $Canny(T)$ denotes the results of Canny edge detector on an image with threshold T. The left part in Fig. 4.15 shows one original image and its edge images obtained by the multi-threshold edge detector. Two kinds of edges, namely shadow edges and non-shadow edges, are clearly presented in the edge image. The right part in Fig. 4.15 shows shadow edges satisfying all three criteria and the final shadow detection result. It clearly shows that all non-shadow edges are successfully removed after classification on each edge image. The final result is obtained by adding all shadow edges in different Canny edge images.

4.5.3 Experiments

In this section, we compare our proposed shadow detection algorithm with two state-of-the-art shadow detection algorithms [9, 16] on various outdoor single-color images in terms of shadow detection results, quantitative results, running time, and results on images without shadows. In our experiments, we use images containing shadows with different scenes and reflectance and images that do not contain shadows to test the shadow detection results of the three methods. Figure 4.16 shows shadow detection results obtained by our method and two state-of-the-art methods [9, 16] on seven representative images, which cover a large selection of shadow scenarios. The two images shown in the top two rows contain shadows on surfaces with different reflectance. The image shown in the third row contains noisy shadow. The image shown in the fourth row is a textureless aerial image. The image shown in the fifth row contains shadows on different surfaces such as plastic and grass. The two images shown in the last two rows contain more complex contents. The shadow detection results of our proposed method, [9, 16] are presented in the second, third, and fourth columns, respectively. The results and comparisons are summarized as follows:

(1) For the first two images with different reflectance, our method detects more accurate shadows than Lalonde's method and Guo's method. Both methods miss some shadow edges along the contour of the bicycle and falsely detect some grass and skin as shadow edges.
(2) For the next three images, our method detects a majority of shadow edges with a few misclassified shadow edges. For example, it detects edges along the curbside in the fourth image as shadows. In comparison, both other methods miss many shadow edges. In detail, Lalonde's method misses some shadow edges of the wheel in the third image and the tree in the fifth image. Guo's method misses most shadows in the third image, mistakenly classifies the edges on the curbside edges in the fourth image as shadow edges, and misses the human body shadow in the fifth image.
(3) For the last two images, our method successfully detects all the shadow edges with a few misclassified shadow edges. Lalonde's method has false cast shadow detection on the house in the sixth image and on the ground in the last image. Guo's method gives more false positives. Figure 4.16 shows that our proposed shadow detection method achieves better shadow detection results than the other two methods. It should be noted that our method does not require any training and therefore always achieves the same shadow detection results.

To evaluate whether our method generalizes well to other illumination conditions, we conducted extensive experiments on four datasets including Zhu's dataset [28], Lalonde's dataset [16], Guo's dataset [9], and our dataset. Our dataset contains a variety of images, downloaded from the web or real-captured by a Canon EOS 5D Mark II camera, with scenes that other three datasets do not include. Particularly we carefully captured images under different sun angles and different weather conditions (includes clear sky, little overcast, and big clouds that did not cover the sun). The four

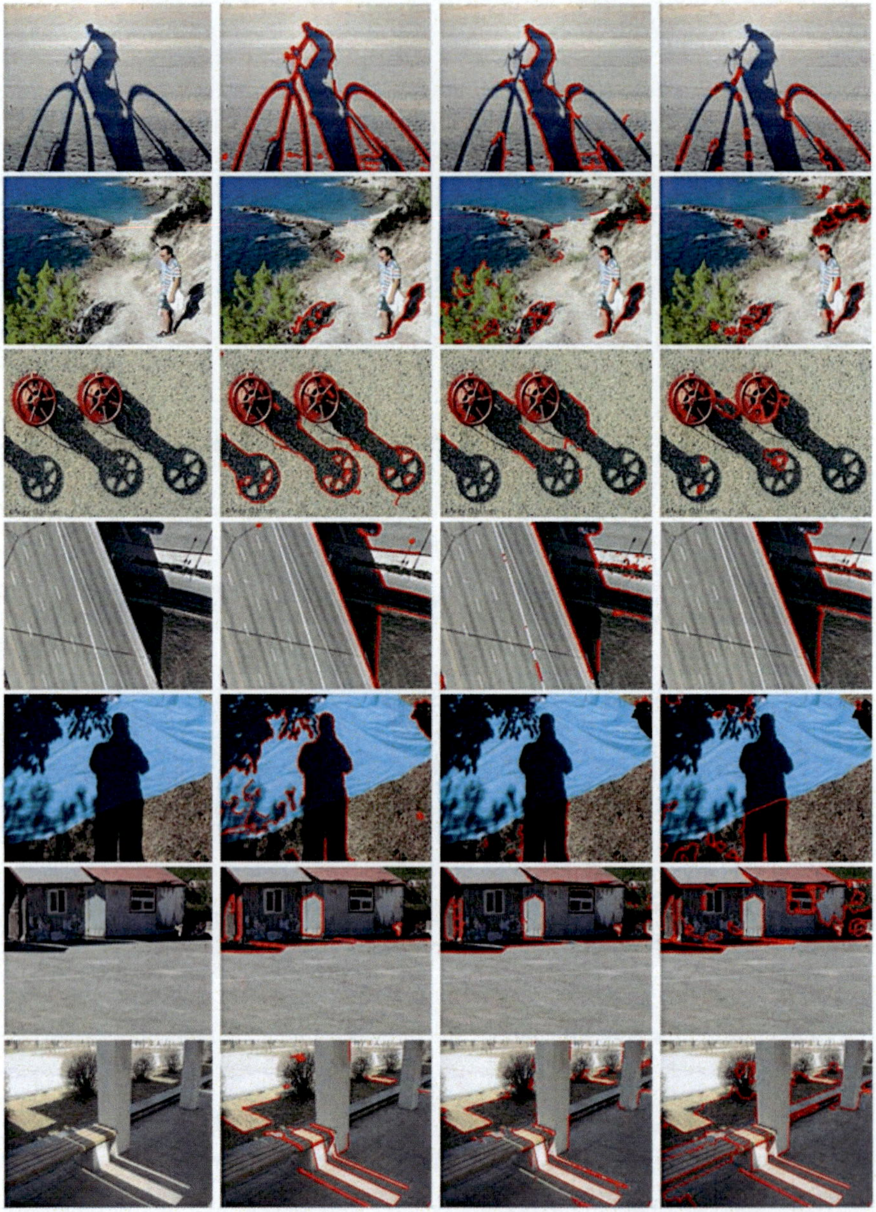

Fig. 4.16 Comparison with the state-of-the-art methods. First column: original images; second column: our results; third column: Lalonde's results; fourth column: Guo's results Reprinted from [Jiandong Tian, Xiaojun Qi, Liangqiong Qu, and Yandong Tang, "New Spectrum Ratio Properties and Features for Shadow Detection," Pattern Recognition, Vol. 51, No. 3, 2016, pp. 85–96], Copyright 2016, with permission from Elsevier

datasets contain 563 real-world images altogether. To the best of our knowledge, we have employed the most extensive images to compare the shadow detection results. We demonstrate some original images with different scenes and our shadow detection results in Fig. 4.17. Specifically, we test our shadow detection method on images with different reflectance materials (e.g., road, grass, sands, wood, and snow) and different occluders (e.g., shadows of people, building, leaves, and animals). Some images also contain one shadow which covers the areas with different reflectance. Figure 4.17 demonstrates that our shadow detection method works well on these different kinds of images. To quantitatively evaluate the effectiveness of our and compared methods, the commonly used precision, recall, and F-measure are computed.

$$\text{Precision} = \frac{\text{TP}}{\text{TP} + \text{FN}} \tag{4.42}$$

$$\text{Recall} = \frac{\text{TP}}{\text{TP} + \text{FP}} \tag{4.43}$$

$$F\text{-measure} = 2 \cdot \frac{\text{Precision} \cdot \text{Recall}}{\text{Precision} + \text{Recall}} \tag{4.44}$$

where TP denotes the true shadows that are correctly identified, FN denotes the true shadows that are erroneously classified as non-shadow, and FP denotes the true non-shadow pixels that are erroneously classified as shadows. The higher precision, recall, and F-measure denote better shadow detection results. All these factors are determined by a direct pixel-wise comparison between the ground truth shadow edge map and the predicted shadow edge map. Considering the fact that many of the shadow boundaries are soft, we allow 2 pixel Euclidean distance error between the ground truth and the detected shadow edge maps to tolerate a slight localization error. Since Guo's method is focused on shadow area detection, we extract its edge maps for a fair result evaluation.

The experimental results by our results and those by [9, 16] on each and the entire dataset are summarized in Table 4.3. It clearly shows that our method achieves the highest precision, recall, and F-measure on all images, indicating that our method is the most robust, and that our proposed four shadow properties and three shadow verification criteria are effective. Additionally, on each dataset, our method outperforms the compared methods in all situations, with the exclusion of the one case that Lalonde's method has 0.01 better precision on Lalonde's dataset. However, unlike Lalonde's and Guo's methods employing a complex classifier to learn several complicated shadow features, our method only requires a simple verification on Canny edges, indicating the effectiveness of the proposed spectrum ratio properties.

Fig. 4.17 More results of our shadow detector. Reprinted from [Jiandong Tian, Xiaojun Qi, Liangqiong Qu, and Yandong Tang, "New Spectrum Ratio Properties and Features for Shadow Detection," Pattern Recognition, Vol. 51, No. 3, 2016, pp. 85–96], Copyright 2016, with permission from Elsevier

4.6 Evaluation of Shadow Features

Previous works for local illumination variation processing (especially shadow) mainly focus on designing an effective classifier for existing illumination-related features, but pay less attention to the analysis of the features themselves. Readers often do not know which feature is more effective or whether these features are complementary. In this section, we present a comprehensive evaluation of eight commonly applied shadow features to investigate the robustness and limitation of each shadow feature. We hope that this evaluation can offer guidance for future shadow processing algorithm.

Shadow features, such as color ratio, texture, and chromaticity, have proved to be quite effective in applications such as shadow detection and shadow removal. A number of categorized methods of shadow features have been proposed in the past studies. Zhu et al. [28] proposed three types of features: shadow-variant features, shadow-invariant features, near-black features. Shadow-variant features describe different characteristics between shadows and non-shadows. Shadow-invariant features exhibit similar behaviors across shadow boundaries. Near-black features distinguish shadows from near-black pixels. Prati et al. [20] mentioned three types of features: spectral, spatial, and temporal features. They claimed that spectral features can be exploited using gray level or color information; spatial features worked at a region level or at a frame level instead of pixel level and can improve results in some detection methods; temporal features were only exploited by some special methods to integrate and improve results. Sanin et al. [22] observed that the choice of features has greater impact on shadow detection results compared to the choice of algorithms. Sanin et al. [22] divided spectral features into intensity, chromaticity, and physical properities and also divided spatial features into geometry and textures.

Some researchers have noted that obtaining better performance is difficult using a single feature; thus, they developed shadow detection algorithms by combining different features. For example, many recent algorithms detect shadows by training a classifier using several shadow feature combinations; however, the choice of features may be heuristic. They may try different combinations of features to determine which combination can achieve the best performance. Due to the lack of works on evaluating each feature, some problems are still unsolved, such as the robustness of each feature, which feature plays the most important role in a detection method, and what is the best performance that current features can achieve. Related to shadow detection,

Table 4.3 Comparison of quantitative results in terms of precision, recall, and F-measure (P/R/F)

Method	Dataset				All Images
	Tian's	Guo's	Lalonde's	Zhu's	
Lalonde's	0.59/0.72 /0.61	0.49/0.69/0.55	0.37/0.63/0.40	0.35/0.66/0.43	0.45/0.68/0.50
Guo's	0.48/0.48/0.40	0.53/0.57/0.51	0.13/0.33/0.17	0.35/0.54/0.36	0.37/0.48/0.36
Ours	**0.70/0.78/0.70**	**0.69/0.73/0.66**	**0.36/0.66/0.44**	**0.47/0.67/0.50**	**0.56/0.70/0.57**

several works on shadow detection evaluation have been published in recent years. However, no such papers have been reported for shadow feature evaluation.

In this section, we present a survey and comprehensive evaluation of eight widely adopted shadow features for shadow detection. We first give a brief review of the eight commonly adopted shadow features, and then two sets of experiments on a large shadow detection dataset are conducted to quantitatively evaluate different shadow features. One set of experiment is directly conducted on the region labels using simple thresholding techniques to investigate the robustness and limitation of each shadow feature. The other set of experiment is conducted using different classifiers to investigate the performance of each feature with respect to different classifiers.

4.6.1 Features Selected for Evaluation

In this study, we use eight commonly applied shadow features for evaluation. In the following, we will provide a brief review of each feature.

Lower Intensity: This feature describes the fact that shadow regions are darker than the surface upon which they are cast. In outdoor environment, shadows are illuminated by skylight, whereas non-shadow background is illuminated by both skylight and sunlight. Therefore, the intensity of shadow region is lower than that of non-shadow region in every channel. This feature may be the most straightforward shadow feature, and it is generally applied as a rough shadow segmentation step. In [18], the lower intensity rule is used as a preclassifier in their method. In [6], the pixels that are darker than the background become candidate shadow points. The algorithm in [21] is based on the observation that shadows are generally darker than their surroundings in both visible and near-infrared images. This feature is also adopted by the algorithms in [13, 27].

Smoothness: This feature is based on the observation that reducing illumination not only decreases the object intensity but also suppresses local variations on the underlying surfaces. As suggested in [1, 28], we use the method proposed by Forsyth and Fleck in [7] to capture this cue. The method subtracts a Gaussian-smoothed version from the original image, and then it uses the standard deviations to measure the smoothness.

Entropy: To correctly differentiate shadows from near-black objects, Zhu et al. [28] introduced the entropy feature, which describes that shadows have a different entropy value compared to that of near-black objects. The major reason is that most of the darker objects appear textureless. As suggested in [1, 12], in our experiment, we calculate the entropy using:

$$E_i = \sum_{i \in \omega} -p_i \times \log_2 (p_i), \tag{4.45}$$

where ω is a window and p_i is the probability of the histogram counts at pixel i.

Skewness: This cue was also suggested by Zhu et al. [28] based on their observation that the asymmetries in shadows and in non-shadows are different. They found a mean value of 1.77 for shadows and −0.77 for non-shadows in skewness. The methods in [1, 16, 26] also employed the skewness feature to capture intensity distribution differences on both sides of a boundary. In our experiment, we directly use the "skewness" function in MATLAB.

Chromaticity: This feature arises from the fact that cast shadows result in a significant change in intensity without much change in chromaticity. In detail, if a shadow is cast on a background, then the hue and saturation components vary within a certain limit. Cucchiara et al. [4] obtained the following four thresholds via a training step:

$$\begin{array}{l} \alpha \leq \frac{I_F^V(x,y)}{I_B^V(x,y)} \leq \beta \\ \wedge \left| I_F^S(x,y) - I_B^S(x,y) \right| < \tau_S \\ \wedge \left| I_F^H(x,y) - I_B^H(x,y) \right| < \tau_H \end{array} \cdot \tag{4.46}$$

This feature is also applied in [17, 18].

Higher hue: This cue assumes that shadows have higher hue components. Tsai [25] proposed a ratio map as Eq. 4.47 by assuming that the hue of shadow pixels will appear high in HSV color space.

$$R = \frac{\text{Hue} + 1}{\text{Value} + 1}, \ \text{Hue, Value} \in [0, 1]. \tag{4.47}$$

This map was thresholded using Otsu's method to compute the final shadow map. This feature was also applied by [3].

Color ratio: Color ratio is defined as the ratio of shadow/non-shadow pairs. This feature is based on the assumption that color ratios are distributed in a limited interval. However, this assumption does not hold for objects. Therefore, the shadow pixels are discriminated according to whether their color ratios are located in the interval. The most common approach to obtain the area is through a learning step. We find that this feature is the most frequently used. For example, it is used in [2, 11, 14, 15, 23, 24, 26, 28]. In our first experiment, we obtain the interval borderline through a brute force search to pursue the best performance; while in our second experiment, we train different classifiers to locate the interval borderline.

Texture: The principle behind the textural feature is that the texture of shaded surfaces changes little across shadow boundaries. This feature is generally applied to verify the candidate shadows. A shadow candidate is classified as either object or shadow by comparing the texture in the candidate region with the texture in the background region. This feature is also frequently used, such as in [8, 13, 16, 26, 28]. As suggested in [8, 16, 28], we apply the method provided by Martin et al. [19] to represent texture with the texton histogram. The method filters an image with a bank of Gaussian derivative filters consisting of 8 orientations and 3 scales, and then applies clustering to form 128 discrete centers. To compare textures between regions, we use the chi-square distance between their texton histograms. The distances of both sides of a shadows boundary will satisfy a range: The upper limit should meet the texture similarity, and the lower limit should meet the minimal change of illumination. Similar to the color ratio feature, we also used a complete search in experiment 1 and a training technique in experiment 2 to find the interval.

4.6.2 *Feature Evaluation*

To fairly and quantitatively evaluate the shadow features, we constructed an extensive image dataset and conducted two experiments on the dataset. The first experiment is directly conducted on the region labels using simple thresholding techniques to investigate how powerful each feature itself is without drawing support from advanced classifiers. The second experiment is conducted using different classifiers to investigate the performance of each feature with respect to different classifiers.

4.6.2.1 Dataset

At present, three datasets exist for shadow detection, namely Zhu's dataset [28], Lalonde's dataset [16], and Guo's dataset [8]. After carefully studying the existing three datasets, we created a new dataset to supplement the other three datasets with scenes that are not included in terms of three aspects. Firstly, we carefully captured images under different sun angles and different weather conditions (including clear sky, slightly overcast, and large clouds that did not cover the sun). Secondly, we captured some images that also contain one shadow that covers areas with different reflectances, such as asphalt, brick, grass, and sand. Specifically, we captured images using different cameras, including single-lens reflex cameras (Canon EOS 5D Mark II, Canon EOS 60D, and Nikon D100), a consumer camera (Sony DSC-W320), mobile phone cameras (iPhone 5s, Motorola XT883, and Lenovo a788t), and a surveillance camera (BOSCH NBN-832V-P). The overall constructed dataset contains four subdatasets, including Zhu's dataset [28], Lalonde's dataset [16], Guo's dataset [8], and our new dataset which contains 108 real-world shadow images. Compared to each specific subdataset, the overall constructed dataset is large scale. The images in the overall dataset cover a wide variety of lighting conditions, surface materials, scenarios, and camera characteristics.

4.6.2.2 Generation of Shadow Versus Non-shadow Labels

All images in the dataset are labeled with ground-truth shadow edges. For each image, we also extract edges with the "Canny" detector. Then, object edges can be obtained by subtracting the ground-truth shadow edges from the Canny edges. The object edges are divided into two groups. For experiment 1, we randomly selected an equal number of object edges to shadow edges in each image for a fair comparison. In this way, we have 3947 shadow edges and 3947 object edges. For experiment 2, we use all object edges, i.e., 3947 shadow edges and 5050 object edges. This choice is logical since there are typically more object edges than shadow edges in shadow detection. After we obtained the shadow boundaries and non-shadow boundaries, we selected two regions along these boundaries for feature evaluation. For each detected edge, we perform morphological image processing along the boundaries to obtain these two regions; i.e., we enlarge the edge regions by applying a binary dilation operation with size of six pixels on both sides. Figure 4.14 shows the dilated edge regions for the edge, where the red region denotes a non-shadow part, and the green region denotes a shadow part.

4.6.2.3 Directly Evaluate Features on Labels

In this section, we present the results of our evaluation experiments directly on region labels using simple thresholding techniques. We performed this experiment for two reasons. First, we want to investigate how powerful each feature itself is without drawing support from advanced classifiers. Second, some researchers still apply simple thresholding techniques to develop their methods. For quantitative evaluation, we use three metrics. The precision (P), recall (R), and F-measure are computed and compared. These metrics are defined as follows:

$$\text{Precision} = \frac{\text{TP}}{\text{TP} + \text{FN}}, \tag{4.48}$$

$$\text{Recall} = \frac{\text{TP}}{\text{TP} + \text{FP}}, \tag{4.49}$$

$$F\text{-measure} = 2 \cdot \frac{\text{Precision} \cdot \text{Recall}}{\text{Precision} + \text{Recall}}, \tag{4.50}$$

where TP denotes the true shadows that are correctly identified, FN denotes the true shadows that are erroneously classified as non-shadows, and FP denotes the true non-shadow pixels that are erroneously classified as shadows. Higher precision, recall, and F-measure denote better shadow detection results. For features without parameters such as lower intensity, smoothness, and higher hue, TP, FP, and FN can easily be obtained by directly comparing the "large and small" relationship from region labels. For features that depend on some parameters such as chromaticity, color ratio, and texture, we carefully optimized these parameters with the complete search technique to ensure that the best performance is achieved for each feature.

The performances of the shadow features in terms of precision, recall, and F-measure are listed in Table 4.4. From this table, we can find some interesting aspects.

1. On this dataset, the F-measure for all features is not high, which indicates that these features do not have very good robustness. This result may account for why widely used shadow detection algorithms are lacking in real applications, and why most existing methods need to apply multiple features simultaneously to detect shadow regions.

Table 4.4 Performance of shadow features in terms of precision, recall, and *F*-measure

Feature	Precision	Recall	*F*-measure	Rank
Intensity	1.000	0.502	0.668	5
Smoothness	0.792	0.625	0.700	3
Skewness	0.965	0.507	0.665	7
Entropy	0.772	0.652	0.711	2
Chromaticity	0.985	0.504	0.667	6
High Hue	0.877	0.547	0.674	4
Color ratio	0.85	0.676	0.753	1
Texture	0.764	0.540	0.633	8

It was ranked by the *F*-measure

2. For all features, we find that precision is better than recall, which indicates that the true positive is high and that the false positive is also high. This result means that many objects are falsely identified as shadows, which may be because all shadow features are proposed based on observations of what phenomena shadows should have, but it is seldom considered that many non-shadow objects also share the same properties. For example, shadowed regions are often dark, with less texture and little gradient, but some non-shadowed objects may also share similar characteristics.

3. The color ratio feature performs the best. We also find that color ratios are adopted the most frequently by researchers among these evaluated features. The literature [16, 26] uses color ratio in different scales and different color spaces in algorithms. This result may imply that some researchers have realized that color ratios may be more powerful, and this feature plays a more important role in their detection methods.

In Fig. 4.18, we present some statistical characteristics for the three features (i.e., smoothness, skewness, and high hue) by plotting the histograms of dark vs. bright sides. From this figure, we can clearly find that these features indeed have properties that are useful for shadow identification. However, this result does not indicate that the features have good specificity. As shown in Fig. 4.18a, a shadow is indeed a smoothed version of a non-shadow, but Fig. 4.18b shows that many objects are also smoother than their neighbors. Fig. 4.18a and Fig. 4.18b indicates that it is better to use a threshold rather than a simple large or small relationship. We conducted an experiment to verify this observation, and we found that by using a threshold, the precision, recall, and *F*-measure can increase to 0.843, 0.665, and 0.744, respectively. Zhu stated that the mean value of skewness in the shadow regions is greater than zero, whereas the mean value of non-shadow regions is less than zero. As shown in Fig. 4.18c, it is true that the shadow regions and non-shadow regions near the shadow border possess this property; however, the same property also holds for the regions that are near the side of the object border.

Figure 4.18e presents the histogram of the high hue feature, which assumes that the hue of shadow regions is higher than that of non-shadow regions. Tsai et al. [25]

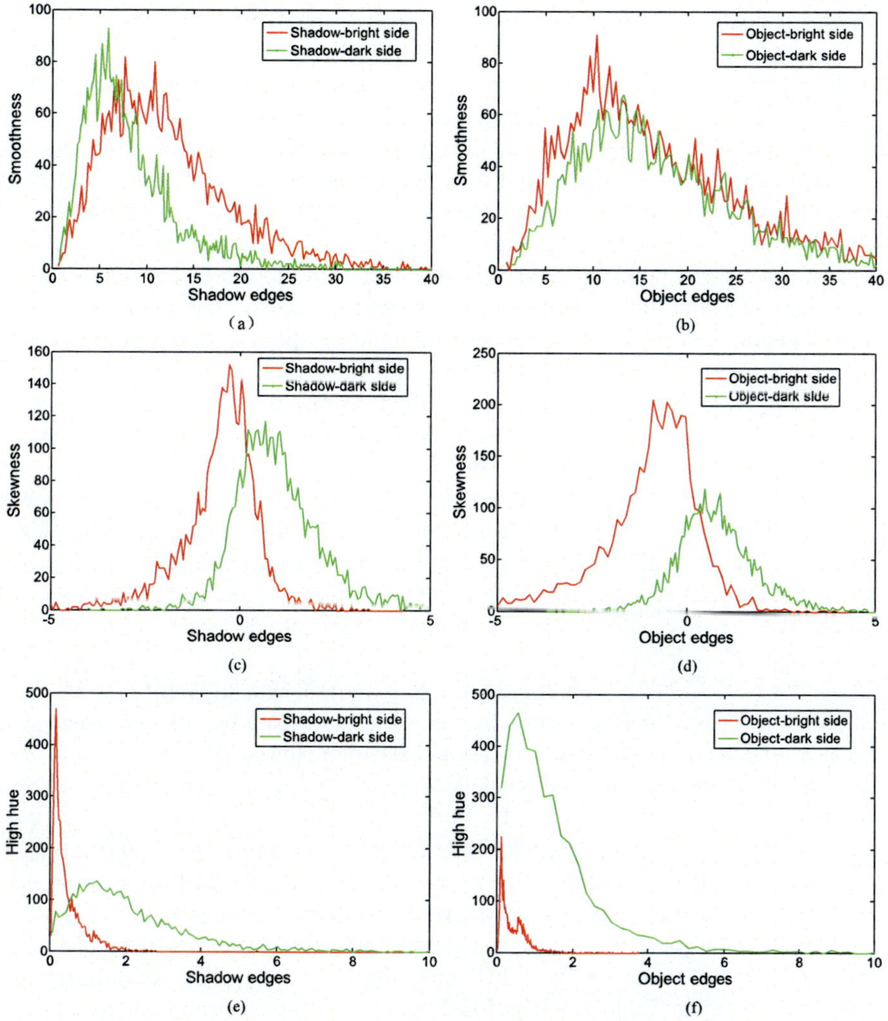

Fig. 4.18 Histograms of dark side versus bright side of shadows and those of objects Reproduced from [Jiandong Tian, Liangqiong Qu, Huijie Fan, Wentao Li, and Yandong Tang, "Evaluation of Shadow Features," IET Computer Vision, Vol.12, Issue 1, 2018, pp. 95–103] by permission of John Wiley & Sons Ltd

proposed a ratio map (Eq. 4.47) and then thresholded this ratio to compute the final shadow map. As shown in Fig. 4.18e, when shadow regions possess higher hue values than non-shadow regions, this method performs well. However, this threshold does not work well for object edges, as shown in Fig. 4.18f. It indicates that pixels satisfying this condition are not guaranteed to belong in shadowed regions and require additional validation.

4.6.2.4 Feature Evaluation via Classifiers

This experiment aims to evaluate the performance of each feature using several widely used classifiers. We applied four different classifiers: naive Bayes, SVM, AdaBoost, and logistic regression. These classifiers have been successfully used in a variety of robot vision tasks. Specifically, we use an estimator for numeric attributes in the naive Bayes classifier. For SVM, we selected LIBSVM with the RBF kernel as the classifier. AdaBoost is an iterative algorithm, and it trains many different weak classifiers and integrates them together efficiently. We selected decision tree as our base classifier and 10 iterations to be performed. Logistic regression is a classic classification method in statistical learning, and its spline is a sigmoid curve. We selected BFGS to update our parameters in each iteration. In this chapter, aiming at estimating features more precisely rather than the effect of different classifiers, we only split the database into two parts: 66% of the database serves as the training set, and the remainder serves as the test set.

The performances in terms of precision (P), recall (R), and F-measure (F) of each feature with different classifiers are listed in Table 4.5. By comparing Table 4.4 with Table 4.5, we can find that the feature performance is similar. However, note that for the two features, lower intensity and smoothness, their performance becomes better. These results may due to the classifiers using a threshold rather than a simple "large and small" relationship.

In Fig. 4.19, we draw each feature's ROC curve with different classifiers. As shown, except for the Bayes classifier on color ratio features, the performances of each feature are quite stable with different classifiers. In Fig. 4.20, we draw different features' ROC curves with the same classifier. As shown, the performances of these features are apparently different. By comparing Figs. 4.19 and 4.20, we conclude that for shadow detection, the choice of features plays a more important role than the choice of classifier. From Fig. 4.20, we can also find that the performances of color ratio and smoothness outperform those of other features in most cases.

The relevant quantitative comparison performances (in terms of precision (P), recall (R), and F-measure (F)) of different feature combinations with different classifiers are listed in Table 4.6. The ROC curves of these feature combination with different classifiers are shown in Fig. 4.21 (for simplicity, here we only draw the ROC curve of four kinds of feature combinations with SVM and the AdaBoost classifier). This experiment aims to investigate how well shadow detection can be obtained using state-of-the-art features and classifiers. We can find that combining features can generally enhance the detection performance. However, the degree of the performance enhancement relies on the choice of different feature combinations. Combining color ratio and entropy features can obtain nearly the best performance. The impacts of adding more features are quite limited.

In this section, we conducted an extensive evaluation of the widely used shadow features. To our knowledge, this is the first work to evaluate shadow features. Several useful and interesting conclusions are obtained: The color ratio feature performs the best among these eight investigated shadow features in our experiments; the choice of features has more impact than the choice of classifiers for a shadow detection

Table 4.5 Performances of shadow features in terms of precision (P), recall (R), and *F*-measure (F) with four classifiers (i.e., Naive Bayes, SVM, AdaBoost, and Logistic)

Feature	Naive Bayes			SVM			AdaBoost			Logistic		
Lower intensity	P	R	F	P	R	F	P	R	F	P	R	F
	0.707	0.706	0.706	0.722	0.722	0.722	0.708	0.704	0.702	0.718	0.718	0.718
Smoothness	P	R	F	P	R	F	P	R	F	P	R	F
	0.748	0.748	0.748	0.755	0.752	0.753	0.758	0.759	0.757	0.754	0.750	0.751
Entropy	P	R	F	P	R	F	P	R	F	P	R	F
	0.688	0.686	0.686	0.701	0.699	0.699	0.705	0.681	0.682	0.684	0.686	0.684
Skewness	P	R	F	P	R	F	P	R	F	P	R	F
	0.552	0.570	0.542	0.551	0.569	0.541	0.554	0.550	0.551	0.551	0.575	0.494
Chromaticity	P	R	F	P	R	F	P	R	F	P	R	F
	0.615	0.615	0.614	0.593	0.593	0.592	0.622	0.614	0.610	0.593	0.592	0.591
High hue	P	R	F	P	R	F	P	R	F	P	R	F
	0.587	0.599	0.573	0.604	0.611	0.580	0.614	0.618	0.615	0.596	0.603	0.556
Color ratio	P	R	F	P	R	F	P	R	F	P	R	F
	0.597	0.605	0.595	**0.791**	**0.791**	**0.791**	0.775	0.775	0.775	0.739	0.740	0.747
Texture	P	R	F	P	R	F	P	R	F	P	R	F
	0.573	0.581	0.574	0.580	0.589	0.518	0.573	0.581	0.574	0.596	0.575	0.422

The best results and the second best results for each classifier are shown in red and blue color, respectively

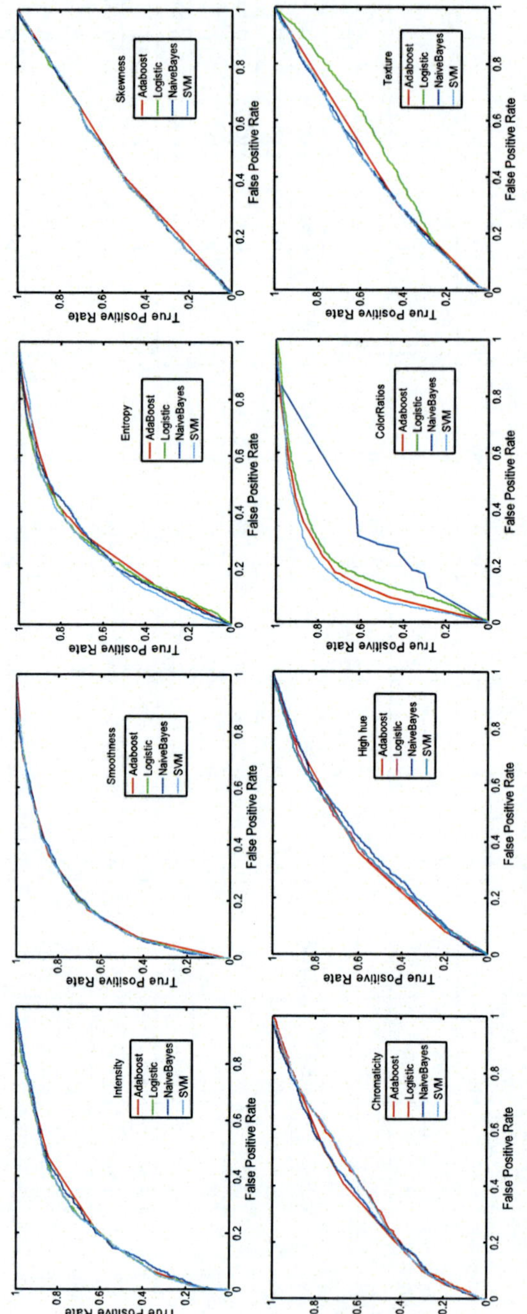

Fig. 4.19 Each feature's ROC curve with different classifiers Reproduced from [Jiandong Tian, Liangqiong Qu, Huijie Fan, Wentao Li, and Yandong Tang, "Evaluation of Shadow Features," IET Computer Vision, Vol.12, Issue 1, 2018, pp. 95–103] by permission of John Wiley & Sons Ltd

Table 4.6 Performances of the combined features in terms of precision (P), recall (R), and F-measure (F) with four classifiers (i.e., Naive Bayes, SVM, AdaBoost, and Logistic)

Combined features	Naive Bayes			SVM			AdaBoost			Logistic		
Color ratio + Smoothness	P	R	F	P	R	F	P	R	F	P	R	F
	0.771	0.772	0.771	0.775	0.772	0.773	0.816	0.817	0.817	0.787	0.783	0.784
Color ratio + Entropy	P	R	F	P	R	F	P	R	F	P	R	F
	0.726	0.722	0.723	0.824	0.824	0.824	0.778	0.777	0.777	0.747	0.745	0.746
Smoothness + Entropy	P	R	F	P	R	F	P	R	F	P	R	F
	0.755	0.751	0.752	0.774	0.770	0.771	0.767	0.767	0.767	0.776	0.771	0.772
Skewness + Texture	P	R	F	P	R	F	P	R	F	P	R	F
	0.572	0.581	0.539	0.593	0.582	0.496	0.586	0.593	0.567	0.600	0.570	0.442
Skewness+Texture+ High hue	P	R	F	P	R	F	P	R	F	P	R	F
	0.639	0.642	0.639	0.622	0.627	0.602	0.634	0.635	0.634	0.623	0.627	0.602
Skewness + Texture + Intensity	P	R	F	P	R	F	P	R	F	P	R	F
	0.711	0.711	0.711	0.727	0.728	0.727	0.733	0.733	0.733	0.717	0.718	0.717
Color ratio + Smoothness + Intensity	P	R	F	P	R	F	P	R	F	P	R	F
	0.799	0.800	0.799	0.792	0.792	0.792	0.833	0.833	0.833	0.795	0.795	0.795
Color ratio + Smoothness + Entropy	P	R	F	P	R	F	P	R	F	P	R	F
	0.795	0.793	0.794	0.784	0.782	0.783	0.828	0.827	0.828	0.796	0.793	0.794
Color ratio + Smoothness + Intensity + Entropy	P	R	F	P	R	F	P	R	F	P	R	F
	0.796	0.794	0.795	0.797	0.797	0.797	0.827	0.826	0.827	0.803	0.802	0.802
All features	P	R	F	P	R	F	P	R	F	P	R	F
	0.800	0.798	0.799	0.811	0.811	0.311	0.840	0.840	0.840	0.819	0.817	0.817
Remove texture	P	R	F	P	R	F	P	R	F	P	R	F
	0.812	0.803	0.804	0.810	0.809	0.309	0.839	0.839	0.839	0.807	0.806	0.806

(continued)

Table 4.6 (continued)

Combined features	Naive Bayes			SVM			AdaBoost			Logistic		
Remove skewness	P	R	F	P	R	F	P	R	F	P	R	F
	0.816	0.808	0.809	0.813	0.813	0.813	0.842	0.842	0.842	0.815	0.814	0.814
Remove chromaticity	P	R	F	P	R	F	P	R	F	P	R	F
	0.804	0.796	0.797	0.804	0.803	0.803	**0.843**	**0.842**	**0.842**	0.813	0.811	0.812
Remove high hue	P	R	F	P	R	F	P	R	F	P	R	F
	0.804	0.794	0.796	0.809	0.809	0.809	0.836	0.837	0.836	0.814	0.813	0.813
Remove texture + Skewness	P	R	F	P	R	F	P	R	F	P	R	F
	0.799	0.798	0.798	0.809	0.809	0.809	0.830	0.830	0.830	0.806	0.805	0.805
Remove Chromaticity + High hue	P	R	F	P	R	F	P	R	F	P	R	F
	0.801	0.799	0.800	0.803	0.803	0.803	0.837	0.836	0.836	0.803	0.803	0.803
Remove skewness + Chromaticity	P	R	F	P	R	F	P	R	F	P	R	F
	0.794	0.792	0.793	0.802	0.802	0.802	0.836	0.835	0.835	0.813	0.811	0.812
Remove texture + High hue	P	R	F	P	R	F	P	R	F	P	R	F
	0.803	0.801	0.802	0.811	0.811	0.811	0.833	0.832	0.832	0.811	0.811	0.811
Remove texture + Skewness + Chromaticity	P	R	F	P	R	F	P	R	F	P	R	F
	0.803	0.801	0.802	0.811	0.811	0.811	0.833	0.832	0.832	0.806	0.805	0.806
Remove texture + Skewness + High hue	P	R	F	P	R	F	P	R	F	P	R	F
	0.801	0.799	0.800	0.809	0.809	0.809	0.809	0.833	0.833	0.807	0.806	0.807
Remove texture + Chromaticity + High hue	P	R	F	P	R	F	P	R	F	P	R	F
	0.798	0.796	0.797	0.797	0.796	0.796	0.832	0.832	0.832	0.803	0.802	0.802
Remove skewness + Chromaticity + High hue	P	R	F	P	R	F	P	R	F	P	R	F
	0.802	0.800	0.801	0.799	0.799	0.799	0.828	0.828	0.828	0.808	0.807	0.807

We have listed the performance of the following types of combinations: combination of 2 features, combination of 3 features, combination of 4 features, combination of all 8 features, combination of 7 features, combination of 6 features, and combination of 5 features. The best results and the second best results for each classifier (among various types of combined features) are shown in red and blue color respectively

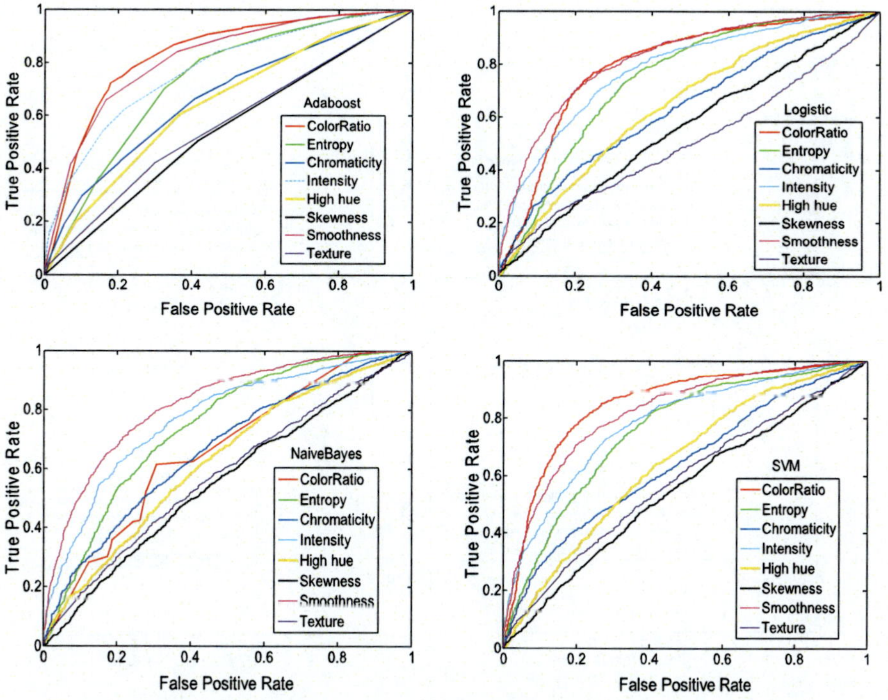

Fig. 4.20 Different features' ROC curve with the same classifier Reproduced from [Jiandong Tian, Liangqiong Qu, Huijie Fan, Wentao Li, and Yandong Tang, "Evaluation of Shadow Features," IET Computer Vision, Vol. 12, Issue 1, 2018, pp. 95–103] by permission of John Wiley & Sons Ltd

algorithm; combining color ratio and entropy features can obtain nearly the best performance. The impacts of adding more features are quite limited.

The main problem of current features is that these features are pure based on image data, but neglecting the physically imaging mechanism. They often impose some assumptions on the shadow image; for example, the intensity shadow features assume that the shadow regions are darker than non-shadow regions. Thus, these pure image data-based shadow features often lack "specificity"; in other words, shadows indeed satisfy the features, but many non-shadow objects also satisfy the features. In our following work, instead of directly designing a purely image data-based shadow feature, we build a universal and practical illumination model from the view of atmosphere transmittance effects and physically imaging mechanism to handle these illumination variations in outdoor environment.

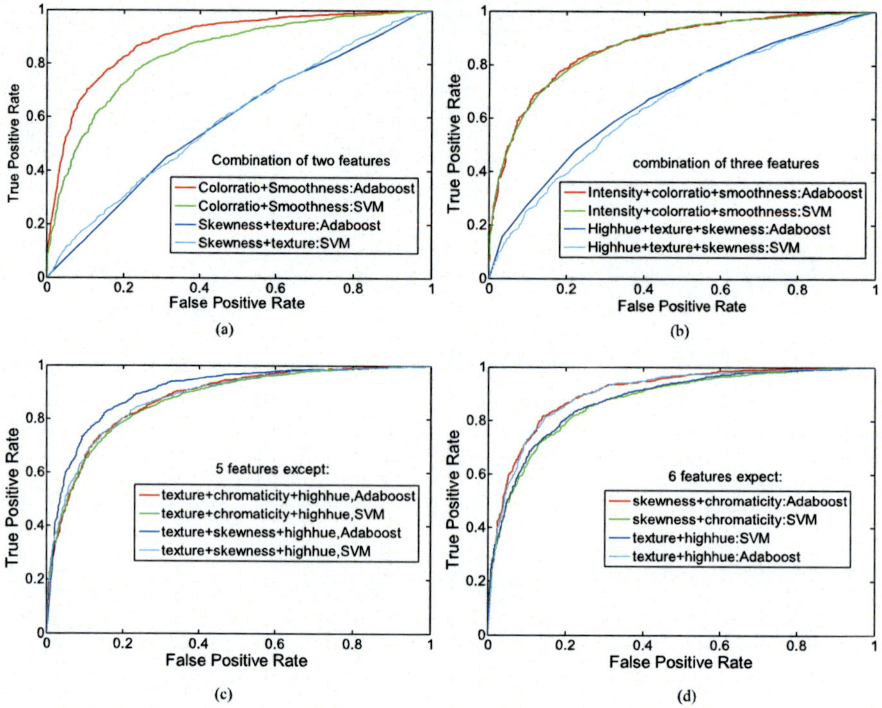

Fig. 4.21 ROC curves of the combined features with SVM and the AdaBoost classifiers Reproduced from [Jiandong Tian, Liangqiong Qu, Huijie Fan, Wentao Li, and Yandong Tang, "Evaluation of Shadow Features," IET Computer Vision, Vol. 12, Issue 1, 2018, pp. 95–103] by permission of John Wiley & Sons Ltd

References

1. Y. Aksoy, A.A. Alatan, Utilization of false color images in shadow detection, in *European Conference on Computer Vision* (Springer, 2012), pp. 472–481
2. K. Barnard, G. Finlayson, Shadow identification using colour ratios, in *Color and Imaging Conference*, vol. 2000 (Society for Imaging Science and Technology, 2000), pp. 97–101
3. K.-L. Chung, Y.-R. Lin, Y.-H. Huang, Efficient shadow detection of color aerial images based on successive thresholding scheme. IEEE Trans. Geosci. Remote Sens. **47**(2), 671–682 (2009)
4. R. Cucchiara, C. Grana, M. Piccardi, A. Prati, Detecting moving objects, ghosts, and shadows in video streams. IEEE Trans. Pattern Anal. Mach. Intell. **25**(10), 1337–1342 (2003)
5. K. Engelhardt, P. Seitz, Optimum color filters for CCD digital cameras. Appl. Opt. **32**(16), 3015–3023 (1993)
6. L.Z. Fang, W.Y. Qiong, Y.Z. Sheng, A method to segment moving vehicle cast shadow based on wavelet transform. Pattern Recogn. Lett. **29**(16), 2182–2188 (2008)
7. D.A. Forsyth, M.M. Fleck, Identifying nude pictures, in *Proceedings Third IEEE Workshop on Applications of Computer Vision. WACV'96* (EEE, 1996), pp. 03–108
8. R. Guo, Q. Dai, D. Hoiem, Single-image shadow detection and removal using paired regions, in *CVPR 2011* (IEEE, 2011), pp. 2033–2040
9. R. Guo, Q. Dai, D. Hoiem, Paired regions for shadow detection and removal. IEEE Trans. Pattern Anal. Mach. Intell. **35**(12), 2956–2967 (2013)

10. H. Haneishi, T. Shiobara, Y. Miyake, Color correction for colorimetric color reproduction in an electronic endoscope. Opt. Commun. **114**(1–2), 57–63 (1995)
11. J.-B. Huang, C.-S. Chen, Moving cast shadow detection using physics-based features, in *2009 IEEE Conference on Computer Vision and Pattern Recognition* (IEEE, 2009), pp. 2310–2317
12. X. Huang, G. Hua, J. Tumblin, L. Williams. What characterizes a shadow boundary under the sun and sky? in *2011 International Conference On Computer Vision*. (IEEE, 2011), pp. 898–905
13. I. Huerta, M. Holte, T. Moeslund, J. Gonzalez, Detection and removal of chromatic moving shadows in surveillance scenarios, in *2009 IEEE 12th International Conference on Computer Vision* (IEEE, 2009), pp. 1499–1506
14. A.J. Joshi, N.P. Papanikolopoulos, Learning to detect moving shadows in dynamic environments. IEEE Trans. Pattern Anal. Mach. Intell. **30**(11), 2055–2063 (2008)
15. C.R. Jung, Efficient background subtraction and shadow removal for monochromatic video sequences. IEEE Trans. Multimedia **11**(3), 571–577 (2009)
16. J.-F. Lalonde, A.A. Efros, S.G. Narasimhan, Detecting ground shadows in outdoor consumer photographs, in *European Conference on Computer Vision* (Springer, 2010), pp. 322–335
17. J. Liu, T. Fang, D. Li, Shadow detection in remotely sensed images based on self-adaptive feature selection. IEEE Trans. Geosci. Remote Sens. **49**(12), 5092–5103 (2011)
18. Z. Liu, K. Huang, T. Tan, L. Wang, Cast shadow removal combining local and global features, in *2007 IEEE Conference on Computer Vision and Pattern Recognition* (IEEE, 2007), pp. 1–8
19. D.R. Martin, C.C. Fowlkes, J. Malik, Learning to detect natural image boundaries using local brightness, color, and texture cues. IEEE Trans. Pattern Anal. Mach. Intell. **26**(5), 530–549 (2004)
20. A. Prati, I. Mikic, M.M. Trivedi, R. Cucchiara, Detecting moving shadows: algorithms and evaluation. IEEE Trans. Pattern Anal. Mach. Intell. **25**(7), 918–923 (2003)
21. D. Rüfenacht, C. Fredembach, S. Süsstrunk, Automatic and accurate shadow detection using near-infrared information. IEEE Trans. Pattern Anal. Mach. Intell. **36**(8), 1672–1678 (2013)
22. A. Sanin, C. Sanderson, B.C. Lovell, Shadow detection: a survey and comparative evaluation of recent methods. Pattern Recogn. **45**(4), 1684–1695 (2012)
23. K. Siala, M. Chakchouk, F. Chaieb, O. Besbes. Moving shadow detection with support vector domain description in the color ratios space, in *Proceedings of the 17th International Conference on Pattern Recognition, 2004. ICPR 2004*, vol. 4 (IEEE, 2004), pp. 384–387
24. D. Toth, I. Stuke, A. Wagner, T. Aach, Detection of moving shadows using mean shift clustering and a significance test, in *Proceedings of the 17th International Conference on Pattern Recognition, 2004. ICPR 2004*, vol. 4 (IEEE, 2004), pp. 260–263
25. V.J. Tsai, A comparative study on shadow compensation of color aerial images in invariant color models. IEEE Trans. Geosci. Remote Sens. **44**(6), 1661–1671 (2006)
26. T.F.Y. Vicente, C.-P. Yu, D. Samaras, Single image shadow detection using multiple cues in a supermodular MRF, in *BMVC* (2013)
27. M.-T. Yang, K.-H. Lo, C.-C. Chiang, W.-K. Tai, Moving cast shadow detection by exploiting multiple cues. IET Image Process. **2**(2), 95–104 (2008)
28. J. Zhu, K. Samuel, S. Z. Masood, Learning to recognize shadows in monochromatic natural images. *IEEE Conference on Computer Vision and Pattern Recognition* (2010), pp. 223–230

Chapter 5
Intrinsic-Image Deriving and Decomposition

Abstract Extracting or separating intrinsic information and illumination from natural images is important for better solving robot vision tasks. In this chapter, we first derive grayscale intrinsic images based on the tricolor attenuation model and the tricolor linear model. To obtain color intrinsic images, we then propose a pixel-wise orthogonal decomposition model for outdoor image illumination processing. In our model, we set up a linear equation set for each pixel value vector based on physically based shadow invariants, deduce a pixel-wise orthogonal decomposition for its solutions, and then get an illumination-invariant vector for each pixel value vector on an image. Only requiring pixel-wise calculation, our physical illumination modeling for color illumination-invariant image can be further applied to real-time applications to resist illumination variation. Finally, we present a new illumination-based color space, the intrinsic information and lighting level (IL) space. Its first two channels represent 2D intrinsic information, and the third channel is for lighting levels. The IL color space has a one-to-one correspondence with the RGB color space. One valuable benefit of the IL color space is that illumination-related processing can be realized by directly operating on the lighting channel. As an example, based on the extracted lighting channel, we propose a new algorithm to estimate the intrinsic-lighting level of an image such that the shadow-free color image and relighting series are obtained.

5.1 Introduction

The concept of intrinsic images was first proposed by Barrow and Tenenbaum in 1978 [2]. The document pointed out that any image can be decomposed into two components: the one related to the material and the other one related to illumination. The material-related component is called intrinsic images, albedo, or reflectance, and the light-related component is called illumination maps. In view of the shadow removal, researchers often seek features that are not sensitive to shadows. The images generated by these features are generally also called intrinsic images.

In this chapter, we first derive grayscale intrinsic images based on the tricolor attenuation model and the tricolor linear model, respectively. We then designed a simple and effective algorithm to obtain RGB intrinsic images without relying on feature

J. Tian, *All Weather Robot Vision*, Research on Intelligent Manufacturing,
https://doi.org/10.1007/978-981-16-6429-8_5

extraction or shadow detection. Starting from the physical imaging mechanism, this method takes the linear shadow model in Section 4.3 as the research basis. It establishes a linear equation system for each pixel of the image and performs orthogonal decomposition on the solution of the linear equation system to obtain an illumination-invariant RGB intrinsic image. Compared with the traditional grayscale illumination-invariant images, the RGB illumination-invariant image based on orthogonal decomposition not only removes the adverse effects of illumination changes, but also retains part of the color information. Based on the RGB illumination-invariant image, we then present a new illumination-based color space to get shadow-free color images.

5.2 Intrinsic-Image Deriving

5.2.1 *Intrinsic-Image Extraction Based on Tricolor Linear Model*

Based on the tricolor linear relationship inside and outside the shaded area, we propose a new method which uses an RGB image to generate a gray intrinsic image.

Previously, we have deduced that the following equation holds.

$$F_H = \kappa_H f_H + \mu_H, \tag{5.1}$$

where f_H represents the shadow pixel value in a certain channel H, F_H represents the corresponding non-shadow background pixel value, and κ_H and μ_H are the parameters of the linear model. Then we have,

$$F_H - \frac{\mu_H}{1-\kappa_H} = \kappa_H \left(f_H - \frac{\mu_H}{1-\kappa_H} \right), \tag{5.2}$$

taking logarithms of both sides, we have,

$$\log(F_H - \frac{\mu_H}{1-\kappa_H}) = \log\left(\kappa_H) + \log(f_H - \frac{\mu_H}{1-\kappa_H}\right), \tag{5.3}$$

In detail, we have,

$$\begin{cases} \log(F_R - \frac{\mu_1}{1-\kappa_1}) = \log\left(\kappa_1) + \log(f_r - \frac{\mu_1}{1-\kappa_1}\right), \\ \log(F_G - \frac{\mu_2}{1-\kappa_2}) = \log\left(\kappa_2) + \log(f_g - \frac{\mu_2}{1-\kappa_2}\right), \\ \log(F_B - \frac{\mu_3}{1-\kappa_3}) = \log\left(\kappa_3) + \log(f_b - \frac{\mu_3}{1-\kappa_3}\right). \end{cases} \tag{5.4}$$

From Eq. 5.4, we have the following equation holds,

$$\log\left(F_{\mathrm{R}}-\frac{\mu_1}{1-\kappa_1}\right)+\log\left(F_{\mathrm{G}}-\frac{\mu_2}{1-\kappa_2}\right)-\beta\cdot\log\left(F_{\mathrm{B}}-\frac{\mu_3}{1-\kappa_3}\right)$$
$$=\log\left(f_r-\frac{\mu_1}{1-\kappa_1}\right)+\log\left(f_g-\frac{\mu_2}{1-\kappa_2}\right)-\beta\cdot\log\left(f_b-\frac{\mu_3}{1-\kappa_3}\right),\qquad(5.5)$$

where $\beta=\frac{\log(\kappa_1)}{\log(\kappa_3)}+\frac{\log(\kappa_2)}{\log(\kappa_3)}$. Denoting $\psi=[11-\beta]$, Eq. 5.5 can be rewritten as,

$$\psi\left(\left[\log\left(F_{\mathrm{R}}-\frac{\mu_1}{1-\kappa_1}\right)\ \log\left(F_{\mathrm{G}}-\frac{\mu_2}{1-\kappa_2}\right)\ \log\left(F_{\mathrm{B}}-\frac{\mu_3}{1-\kappa_3}\right)\right]\right)$$
$$=\psi\left(\left[\log\left(f_r-\frac{\mu_1}{1-\kappa_1}\right)\ \log\left(f_g-\frac{\mu_2}{1-\kappa_2}\right)\ \log\left(f_b-\frac{\mu_3}{1-\kappa_3}\right)\right]\right).\qquad(5.6)$$

Equation 5.6 shows that if we subtract $\frac{\mu_1}{1-\kappa_1},\frac{\mu_2}{1-\kappa_2},\frac{\mu_3}{1-\kappa_3}$ in R, G, B channel and take logarithm for them, respectively, we find that after ψ operation, the shadow pixels and the non-shadow pixels are strictly equal. Apparently, we obtain an intrinsic image. In Fig. 5.1, we give two intrinsic-image results calculated according to the above formula. We can see that shadows are disappeared or are much attenuated. It can

Fig. 5.1 Intrinsic-image results. © [2011] IEEE. Reprinted, with permission, from [Jiandong Tian and Yandong Tang, "Linearity of Each Channel Pixel Values from a Surface in and out of Shadows and Its Applications," IEEE Conference on Computer Vision and Pattern Recognition, pp. 985–992, 2011.]

be seen from Eq.4.26 that the parameter of the model is K_H. Because we have no information about test images, we just use the mean SPD of daylight and skylight at 20–70° sun angles to approximate the linear parameters.

5.2.2 Intrinsic-Image Extraction Based on Tricolor Attenuation Model

In this section, we propose a simple T operator to transform a color image into the gray image without shadow. Assuming $[F_{\mathrm{S_R}}\ F_{\mathrm{S_G}}\ F_{\mathrm{S_B}}]$ is the shadow we have detected, and $[F_{\mathrm{NS_R}}\ F_{\mathrm{NS_G}}\ F_{\mathrm{NS_B}}]$ is the corresponding non-shadow background, then the following formula is established

$$
\begin{aligned}
&(F_{\mathrm{NS_R}} - F_{\mathrm{S_R}}) + (F_{\mathrm{NS_G}} - F_{\mathrm{S_G}}) \\
&= (F_{\mathrm{NS_B}} - F_{\mathrm{S_B}}) \cdot \frac{(F_{\mathrm{NS_R}} - F_{\mathrm{S_R}}) + (F_{\mathrm{NS_G}} - F_{\mathrm{S_G}})}{(F_{\mathrm{NS_B}} - F_{\mathrm{S_B}})} \\
&\Rightarrow (F_{\mathrm{NS_R}} - F_{\mathrm{S_R}}) + (F_{\mathrm{NS_G}} - F_{\mathrm{S_G}}) = (F_{\mathrm{NS_B}} - F_{\mathrm{S_B}}) \cdot \left(\frac{\Delta R}{\Delta B} + \frac{\Delta G}{\Delta B}\right).
\end{aligned}
\tag{5.7}
$$

So

$$
F_{\mathrm{NS_R}} + F_{\mathrm{NS_G}} - \left(\frac{\Delta R}{\Delta B} + \frac{\Delta G}{\Delta B}\right) \cdot F_{\mathrm{NS_B}} = F_{\mathrm{S_V}} + F_{\mathrm{S_G}} - \left(\frac{\Delta R}{\Delta B} + \frac{\Delta G}{\Delta B}\right) \cdot F_{\mathrm{S_B}}.
\tag{5.8}
$$

Denoting T operator as $T = [1\ 1\ \ -(\Delta R/\Delta B + \Delta G/\Delta B)]$, Eq. (5.8) can be rewritten as

$$
T(F_{\mathrm{NS_R}}, F_{\mathrm{NS_G}}, F_{\mathrm{NS_B}}) = T(F_{\mathrm{S_R}}, F_{\mathrm{S_G}}, F_{\mathrm{S_B}}).
\tag{5.9}
$$

We find that after T operation, the shadow pixels and the non-shadow pixels are strictly equal. Apparently, we obtain a shadow-invariant image. Two results are given in Fig. 5.2. $(\Delta R/\Delta B + \Delta G/\Delta B)$ is calculated by taking the region within 10-pixel distance around the shadow as the non-shadow background based on shadow detection. It can be seen from the results that after the transformation of T operator, the shadow disappears completely, which shows the effectiveness of our shadow-invariant operator.

5.2.3 Pixel-Wise Orthogonal Decomposition

The tricolor linear model shows that the sRGB tristimulus values of a surface illuminated by daylight are proportional to those of the same surface illuminated by skylight in each of the three color channels, i.e.,

Fig. 5.2 Using T operator to obtain intrinsic images. **a**, **c** Original image; **b**, **d** shadow-invariant image after T operator. © [2009] IEEE. Reprinted, with permission, from [Jiandong Tian, Jing Sun, and Yandong Tang, "Tricolor Attenuation Model for Shadow Detection," IEEE Transactions on Image Processing, Vol. 18, No. 10, pp. 2355–2363, 2009]

$$\log(F_H + 14) = \frac{\log(K_H)}{2.4} + \log(f_H + 14), \tag{5.10}$$

where F_H denotes the RGB values of a surface in non-shadow area and f_H denotes the RGB values for the same surface in shadow area. The proportional coefficients K_H are independent of reflectance and are approximately equal to constants determined by Eq. 5.11.

$$K_H = \arg\min_{K_H} \sum_{\lambda=400}^{700} \left| E_{\text{day}}(\lambda) \cdot Q_H(\lambda) - K_H \cdot E_{\text{sky}}(\lambda) \cdot Q_H(\lambda) \right|, \tag{5.11}$$

where E_{day} and E_{sky} represent the SPD of daylight and skylight, respectively.

Fig. 5.3 Grayscale illumination-invariant image results based on the linear model. Reprinted with permission from Liangqiong Qu, Jiandong Tian, Zhi Han, and Yandong Tang, "Pixel-wise Orthogonal Decomposition for Color Illumination Invariant and Shadow-free Image," Optics Express, Vol. 23, Issue 3, pp. 2220–2239, 2015. © The Optical Society

From Eq. 4.26 and Eq. 5.6, we have

$$\begin{aligned}&\log(F_R + 14) + \log(F_G + 14) - \beta_1 \cdot \log(F_B + 14)\\ &= \log(f_R + 14) + \log(f_G + 14) - \beta_1 \cdot \log(f_B + 14),\end{aligned} \tag{5.12}$$

where

$$\beta_1 = \frac{\log(K_R) + \log(K_G)}{\log(K_B)}. \tag{5.13}$$

For a pixel in an image, Eq. 5.12 represents a shadow invariant. For an arbitrary pixel and its RGB value vector, $(v_R, v_G, v_B)^T$, $\log(v_R + 14) + \log(v_G + 14) - \beta_1 \cdot \log(v_B + 14)$ keeps the same no matter whether the pixel is on shadow region or not. Apparently, a grayscale illumination-invariant image is obtained. Two results of this type of illumination-invariant image are shown in Fig. 5.3.

The grayscale illumination-invariant image has some limitations in some applications due to the loss of color and contrast information. We expect to get an illumination (including shadow)-invariant image, which can retain the color information of the original image and appear the same as the original image without shadow. For this purpose, based on the tricolor linear model, we deduce a pixel-wise orthogonal decomposition and obtain an illumination-invariant image which retains the basic color information of the original image.

For an RGB value vector $(v_R, v_G, v_B)^T$, we first define the following grayscale illumination-invariant values I_1 according to Eq. 5.12:

$$\begin{aligned}I_1 &= \log(F_R + 14) + \log(F_G + 14) - \beta_1 \cdot \log(F_B + 14)\\ &= \log(f_R + 14) + \log(f_G + 14) - \beta_1 \cdot \log(f_B + 14)\\ &= \log(v_R + 14) + \log(v_G + 14) - \beta_1 \cdot \log(v_B + 14)\end{aligned}. \tag{5.14}$$

Similar to Eq. 5.14, we can obtain two other grayscale illumination-invariant values I_2 and I_3,

$$\begin{aligned}I_2 &= \log(F_R + 14) - \beta_2 \cdot \log(F_G + 14) + \log(F_B + 14)\\ &= \log(f_R + 14) - \beta_2 \cdot \log(f_G + 14) + \log(f_B + 14)\\ &= \log(v_R + 14) - \beta_2 \cdot \log(v_G + 14) + \log(v_B + 14)\end{aligned}, \tag{5.15}$$

$$
\begin{aligned}
I_3 &= -\beta_3 \cdot \log(F_R + 14) + \log(F_G + 14) + \log(F_B + 14) \\
&= -\beta_3 \cdot \log(f_R + 14) + \log(f_G + 14) + \log(f_B + 14) \quad , \\
&= -\beta_3 \cdot \log(v_R + 14) + \log(v_G + 14) + \log(v_B + 14)
\end{aligned}
\tag{5.16}
$$

where

$$
\beta_2 = \frac{\log(K_R) + \log(K_B)}{\log(K_G)}, \ \beta_3 = \frac{\log(K_G) + \log(K_B)}{\log(K_R)}. \tag{5.17}
$$

For convenience, let $\boldsymbol{u} = (u_R, u_G, u_B)^T$ defines a Log-RGB value vector of a pixel, where $u_H = log(v_H + 14)$, $H = \{R, G, B\}$. From Eqs. 5.12, 5.15, and 5.16, we have the following linear equations,

$$
\begin{cases}
u_R + u_G - \beta_1 \cdot u_B = I_1 \\
u_R - \beta_2 \cdot u_G + u_B = I_2 \\
-\beta_3 \cdot u_R + u_G + u_B = I_3
\end{cases} . \tag{5.18}
$$

where I_i, $i = 1, 2, 3$, correspond to the invariant values calculated by Eqs. 5.14–5.16, respectively. The matrix format of Eq. 5.18 is as follows:

$$
A\boldsymbol{u} = \boldsymbol{I}, \tag{5.19}
$$

where $A = \begin{bmatrix} 1 & 1 & -\beta_1 \\ 1 & -\beta_2 & 1 \\ -\beta_3 & 1 & 1 \end{bmatrix}$, $\boldsymbol{u} = (u_R, u_G, u_B)^T$, and $\boldsymbol{I} = (I_1, I_2, I_3)^T$. According to the definitions and calculations of β_1, β_2, and β_3, we have

$$
2 + \beta_1 + \beta_2 + \beta_3 - \beta_1\beta_2\beta_3 = 0. \tag{5.20}
$$

which leads to $\text{rank}(A) = 2$. For a given image, from algebraic theory, we know that Eq. 5.19 has infinite number of solutions. The solution space can be decomposed into a particular solution plus one-dimensional nullspace solutions. For a pixel on an image, we only know one of the solutions, i.e., pixel Log-RGB value vector. We aim to find one solution of Eq. 5.19 which is illumination invariant. In the following, we prove that this solution exists and is unique.

From algebraic theory, any solution $\boldsymbol{u}$ of Eq. 5.19 can be expressed as follows:

$$
\boldsymbol{u} = \boldsymbol{u}_s + \alpha \boldsymbol{u}_0, \tag{5.21}
$$

where $\boldsymbol{u}_s$ is an arbitrary particular solution and $\boldsymbol{u}_0$ is the normalized free solution of Eq. 5.19, such that $A\boldsymbol{u}_0 = 0$ and $\|\boldsymbol{u}_0\| = 1$, $\alpha \in R$. The symbol $\|\cdot\|$ denotes L^2 norm. One of the free solutions is solved in Formula 5.22,

$$
\begin{aligned}
\boldsymbol{u}_0' &= (\beta_1\beta_2 - 1, 1 + \beta_1, 1 + \beta_2) \\
&= \log(K_R K_G K_B) \cdot (\log(K_R), \log(K_G), \log(K_B)).
\end{aligned}
\tag{5.22}
$$

From Formula 5.22, the normalized free solution $\boldsymbol{u_0}$ can be calculated as follows:

$$\boldsymbol{u_0} = \frac{1}{\left\|\boldsymbol{u}'_0\right\|}\boldsymbol{u}'_0. \tag{5.23}$$

Here the free solution has no relationship with the image itself but is determined by matrix A, i.e., illumination condition. Formula 5.22 further reveals that this free solution is a representation of the ratio of illuminants in imaging environments.

Given a particular solution $\boldsymbol{u_s}$ and the normalized free solution $\boldsymbol{u_0}$ of Eq. 5.19, according to Eq. 5.21 and algebraic theory, we define $\boldsymbol{u_p}$ as

$$\begin{cases} \boldsymbol{u_p} = \boldsymbol{u_s} + \alpha_p \boldsymbol{u_0} \\ \alpha_p = -\langle \boldsymbol{u_s}, \boldsymbol{u_0} \rangle \end{cases}. \tag{5.24}$$

It is a particular solution of Eq. 5.19, where $\langle\cdot,\cdot\rangle$ denotes vector inner product. It can be proved that $\boldsymbol{u_p}$, satisfying $\boldsymbol{u_p} \perp \boldsymbol{u_0}$, is the unique particular solution of Eq. 5.19. For a Log-RGB value vector $\boldsymbol{u}$, whether it is on shadow or not, it is the solution of Eq. 5.19 and can be decomposed as

$$\boldsymbol{u} = \boldsymbol{u_p} + \alpha \boldsymbol{u_0}, \tag{5.25}$$

where $\alpha = \langle \boldsymbol{u}, \boldsymbol{u_0} \rangle$ and $\boldsymbol{u_p}$, $\boldsymbol{u_p} = \boldsymbol{u} - \alpha \boldsymbol{u_0}$, is the unique particular solution of Eq. 5.19. Due to $\boldsymbol{u_p} \perp \boldsymbol{u_0}$, we call this process as pixel-wise orthogonal decomposition. For a given pixel, no matter how different the values of the pixel are with different illumination conditions, this pixel-wise orthogonal decomposition will yield a unique particular solution $\boldsymbol{u_p}$.

From the deducing procedure above and the orthogonal decomposition, we know that $\boldsymbol{u_0}$, representation of illuminants ratio, is only related with $(\beta_1, \beta_2, \beta_3)^T$, and $\boldsymbol{u_p}$, perpendicular to $\boldsymbol{u_0}$, is illumination invariant. It means that for a pixel with Log-RGB value vector $\boldsymbol{u}$, no matter whether it is on shadow or not, $\boldsymbol{u_p}$ is invariant and only α reflects the variation of pixel RGB values caused by shadow or different illuminants. The different solutions caused by different α component in Eq. 5.25 can be regarded as Log-RGB values of a pixel on different illumination conditions. Therefore, given an image under multiple illumination conditions, for each pixel with Log-RGB value vector $\boldsymbol{u}$ of this image, we can directly calculate its illumination-invariant vector $\boldsymbol{u_p}$ as

$$\begin{cases} \boldsymbol{u_p} = \boldsymbol{u} + \alpha_p \boldsymbol{u_0} \\ \alpha_p = -\langle \boldsymbol{u}, \boldsymbol{u_0} \rangle \end{cases}. \tag{5.26}$$

By exponential transform on $\boldsymbol{u_p}$ for each pixel, we can get an illumination-invariant image which retains basic color information of the original image in RGB color space.

This pixel-wise orthogonal decomposition process can be shown more clearly from the illustration in Fig. 5.4. For a pixel, $\boldsymbol{u}_s^1$ and $\boldsymbol{u}_s^2$ denote its log-RGB vector within shadow and log-RGB vector without shadow, respectively, which are the solutions of Eq. 5.19. Here we denote any solution of Eq. 5.19 by vector $\boldsymbol{u}$. According to our orthogonal decomposition Eqs. 5.22, 5.23, and 5.25, $\boldsymbol{u}_s^1$, $\boldsymbol{u}_s^2$, and $\boldsymbol{u}$ can be

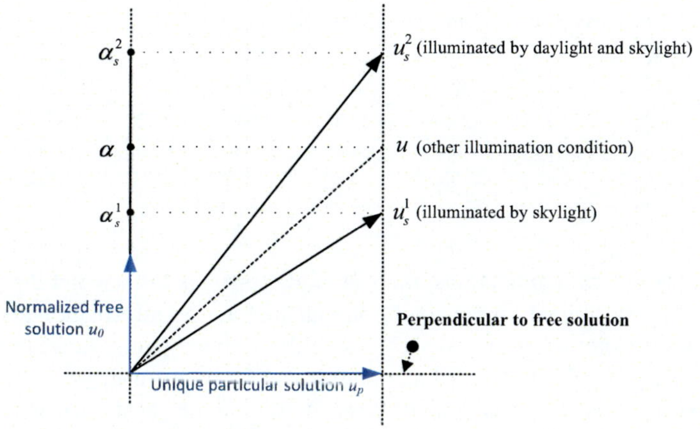

Fig. 5.4 An illustration of pixel-wise orthogonal decomposition. Reprinted with permission from Liangqiong Qu, Jiandong Tian, Zhi Han, and Yandong Tang, "Pixel-wise Orthogonal Decomposition for Color Illumination Invariant and Shadow-free Image," Optics Express, Vol. 23, Issue 3, pp. 2220–2239, 2015. © The Optical Society

projected along the vector $\boldsymbol{u_0}$ into a specific solution of Eq. 5.21, $\boldsymbol{u_p}$, which is perpendicular to $\boldsymbol{u_0}$. Therefore, for a pixel on an image, it means that, no matter how different the values of the pixel are with different illumination conditions (within shadow, without shadow or other illuminated), we can get an unique invariant $\boldsymbol{u_p}$ by our orthogonal decomposition.

5.2.3.1 Analysis of Parameters $\beta_1, \beta_2, \beta_3$ and Color Illumination-Invariant Image

The estimation of parameters $\beta_1, \beta_2, \beta_3$ is important in determining whether our pixel-wise orthogonal decomposition works well or not. Luckily, from Eqs. 5.13, 5.17, and 5.11, we know that these parameters are reflectance irrelevant and only determined by the illumination condition (i.e, the SPD of daylight and skylight), such as zenith angle (time) and aerosol (weather). In clear weather that shadows mostly take place, these parameters are mainly determined by zenith angle. The SPD can be measured by spectrometer or calculated in atmosphere optics (software like MODTRAN and SMARTS2) with zenith angle and aerosol (for weathers) in advance. In our experiment, this SPD of daylight and skylight was obtained by calculating the mean values of the SPD of daylight and skylight measured by an Avantes USB 2.0 spectrometer under different sun angles on ten different days in sunny weather. This optimization problem of K_H is solved by setting the sampling stepsize of in Eq. 5.11 as 5nm and the stepsize of K_H as 0.01. We have calculated parameters $\beta_1, \beta_2, \beta_3$ in sunny weather at different sun angles from 9:00 AM till 4:00 PM in our city and some of the values are shown in Table 5.1. The mean values in Table 5.1 are calculated by the mean SPD of daylight and skylight at 20°–70°. From Table 5.1, we notice that the values of parameters $\beta_1, \beta_2, \beta_3$ from sun angle 20° to sun angle 80° are

Table 5.1 Parameters (PRMs) from representative sun angles

PRMs	Angle							
	20°	30°	40°	50°	60°	70°	80°	Mean
β_1	2.353	2.321	2.299	2.371	2.648	2.520	2.473	2.557
β_2	1.963	1.963	1.977	1.982	1.925	1.996	1.985	1.889
β_3	1.745	1.767	1.770	1.716	1.604	1.617	1.652	1.682

quite stable. In real applications, the sun angle (time) is known and the parameters can then be determined. In the condition that we have no information about at what sun angles the pictures were taken, the stability of parameters shows that the mean values for $\beta_1, \beta_2, \beta_3$ at 20°–70° sun angles can be used for most situations in our decomposition, except some images taken at twilight, i.e., at sunrise or sunset, when sun angle is less than 10°. Also, for more precise parameter values, an automatic entropy minimization method similar to Finlayson et al. [3] is employed. In addition to the factors (zenith angle and aerosol) mentioned above that may affect parameters $\beta_1, \beta_2, \beta_3$, some other aspects are also worthy of attention. For example, cloud and fog will affect the SPD of daylight and skylight and further affect the parameters $\beta_1, \beta_2, \beta_3$ in our algorithm. From clear weather to cloudy, as the clouds increase (aerosol increases), shadows vary from strong to weak. When shadows are weak, the error of the color illumination-invariant results caused by parameters $(\beta_1, \beta_2, \beta_3)$ deviation from cloud is also weaker. Therefore, the influence caused by the cloud distribution would be small and can be corrected by automatic entropy minimization [3]. This similar situation also goes for foggy weather.

The above analysis shows that our pixel-wise orthogonal decomposition can work well in most situations. Our experimental results on many images in different scenes and reflectance also support this analysis. Two original images, their color illumination-invariant images and α information, are presented in Fig. 5.5. The shadows disappear and the main color information is maintained in our color illumination-invariant images.

5.2.3.2 Color Restoration and Shadow-Free Image

By the orthogonal decomposition on an image, some image color information gets lost (Fig. 5.5). As shown in Table 5.1 and Eqs. 5.22 and 5.23, the directions of free solution in different sun angles (from sun angle 20° to sun angle 80°) are all near to $(1, 1, 1)^T$. From the orthogonal decomposition equation, we have, if $\boldsymbol{u} \approx \alpha \boldsymbol{u_0}$, the illumination invariant $\boldsymbol{u_p} \approx (0, 0, 0)^T$. It implies that the closer to the neutral surface on RGB space a pixel RGB value vector is, the more color information is got lost. To restore this loss, we present a color restoration method to generate a shadow-free image by correcting color vectors on the illumination-invariant image which are near RGB color neutral surface.

Let $\boldsymbol{v}(x, y) = (v_R(x, y), v_G(x, y), v_B(x, y))^T$ be an RGB image, where $x = 1, 2, ..., M$; $y = 1, 2, ..., N$; M and N denote the image width and height, respectively. We define

Fig. 5.5 Pixel-wise orthogonal decomposition on images. Left: Original image. Middle: color illumination-invariant image. Right: α information. Reprinted with permission from Liangqiong Qu, Jiandong Tian, Zhi Han, and Yandong Tang, "Pixel-wise Orthogonal Decomposition for Color Illumination Invariant and Shadow-free Image," Optics Express, Vol. 23, Issue 3, pp. 2220–2239, 2015. © The Optical Society

$$\boldsymbol{u}(x, y) = (\log(v_{\mathrm{R}}(x, y) + 14), \log(v_{\mathrm{G}}(x, y) + 14),$$
$$\log(v_{\mathrm{B}}(x, y) + 14))^T. \tag{5.27}$$

Algorithm 5.1 Algorithm pipeline: Pixel-wise orthogonal decomposition for shadow-free image

Input: Original image I, Coefficient K_H;
Output: Shadow-free image O;
Some notations:
$\boldsymbol{u}, \boldsymbol{u}_{\mathbf{c}}, \boldsymbol{u}_{\mathbf{p}}$: Image in the Log-RGB color space;
$\boldsymbol{u}_{\mathbf{p}}^{\mathrm{RGB}}, \boldsymbol{u}_{\mathbf{c}}^{\mathrm{RGB}}$: Image in the RGB color space, the superscript *RGB* indicates image in the RGB color space;
$\boldsymbol{u}_{\mathbf{p}}^{\mathrm{Lab}}, \boldsymbol{u}_{\mathbf{c}}^{\mathrm{Lab}}, \boldsymbol{u}_{\mathbf{f}}^{\mathrm{Lab}}$: Image in the Lab color space, the subscript *Lab* indicates images in the Lab color space;
Algorithm:
1: Perform a logarithmic transformation on I to get the Log-RGB image $\boldsymbol{u}$;
2: Make a pixel-wise orthogonal decomposition on $\boldsymbol{u}$ according to Eq. 5.26 to get the color illumination-invariant values $\boldsymbol{u}_{\mathbf{p}}$, where we have $\boldsymbol{u}_{\mathbf{p}} = \boldsymbol{u} - < \boldsymbol{u}, \boldsymbol{u}_0 > \boldsymbol{u}_0$;
3: Calculate the parameter vector T according to Eqs. 5.29 and 5.30. This parameter vector is used to correct the colors in the original image that lies close to the neutral surface;
4: Perform color correction on $\boldsymbol{u}_{\mathbf{p}}$ according to Eq. 5.31 to get $\boldsymbol{u}_{\mathbf{c}}$;
5: Transfer $\boldsymbol{u}_{\mathbf{p}}$ and $\boldsymbol{u}_{\mathbf{c}}$ back to the RGB color space by an exponentiation transformation and get $\boldsymbol{u}_{\mathbf{p}}^{\mathrm{RGB}}$ and $\boldsymbol{u}_{\mathbf{c}}^{\mathrm{RGB}}$, respectively. This image $\boldsymbol{u}_{\mathbf{p}}^{\mathrm{RGB}}$ has almost the same chrominance as original image and image $\boldsymbol{u}_{\mathbf{c}}^{\mathrm{RGB}}$ has almost the same luminance as original image;
6: Convert $\boldsymbol{u}_{\mathbf{p}}^{\mathrm{RGB}}$ and $\boldsymbol{u}_{\mathbf{c}}^{\mathrm{RGB}}$ from RGB color space into Lab color space and get $\boldsymbol{u}_{\mathbf{p}}^{\mathrm{Lab}}$ and $\boldsymbol{u}_{\mathbf{c}}^{\mathrm{Lab}}$;
7: Combine the luminance component of $\boldsymbol{u}_{\mathbf{c}}^{\mathrm{Lab}}$ and chrominance components of $\boldsymbol{u}_{\mathbf{p}}^{\mathrm{Lab}}$ to obtain the shadow-free image $\boldsymbol{u}_{\mathbf{f}}^{\mathrm{Lab}}$ in Lab space according to Eqs. 5.32 and 5.33;
8: Convert $\boldsymbol{u}_{\mathbf{f}}^{\mathrm{Lab}}$ to the RGB color space to get the final shadow-free image O;

With the orthogonal decomposition on $\boldsymbol{u}$ (to see Eq. 5.25), we obtain

$$\boldsymbol{u}(x, y) = \alpha(x, y)\boldsymbol{u_0} + \boldsymbol{u_p}(x, y), \tag{5.28}$$

where $\boldsymbol{u_p}(x, y)$ is illumination invariant on pixel (x, y), $\alpha(x, y) \in R$. For $\boldsymbol{u}$, we define the following pixel set, which indicates pixels that are near RGB color neutral surface,

$$\begin{aligned} S = \{(x, y) | \left\| \tfrac{\boldsymbol{u}(x,y)}{\|\boldsymbol{u}(x,y)\|} - \boldsymbol{u_0} \right\| \le \varepsilon; \\ x = 1, 2, .., M; y = 1, 2, ..., N\}, \end{aligned} \tag{5.29}$$

where parameter ε is set to 0.15 empirically. We then calculate the following parameter vector,

$$\boldsymbol{T} = \frac{1}{G} \sum_{(x,y)\in S} \left(\boldsymbol{u_0} - \frac{\boldsymbol{u}(x, y)}{\|\boldsymbol{u}(x, y)\|} \right), \tag{5.30}$$

where G is the number of the pixels in set S. This parameter vector is used to measure the average deviation of the pixel vectors of image $\boldsymbol{u}$ in set S from the vector $\boldsymbol{u_0}$. Then the color correction on $\boldsymbol{u_p}$ is calculated as follows:

$$\begin{aligned} \boldsymbol{u_c}(x, y) = \| \boldsymbol{u_p}(x, y) \| \cdot \\ \left(\frac{\boldsymbol{u_p}(x,y)}{\|\boldsymbol{u_p}(x,y)\|} + \frac{1}{\kappa \left\| \frac{\boldsymbol{u}(x,y)}{\|\boldsymbol{u}(x,y)\|} - \boldsymbol{u_0} \right\|^3 + 1} \boldsymbol{T} \right). \end{aligned} \tag{5.31}$$

From the formula above, if $\boldsymbol{u}(x, y) = \alpha(x, y)\boldsymbol{u_0}$, we have $\boldsymbol{u_c}(x, y) = \boldsymbol{u_p}(x, y) + \boldsymbol{T}$. This color correction aims to correct the color for pixels in set S. The role of function $\frac{1}{\kappa \left\| \frac{\boldsymbol{u}(x,y)}{\|\boldsymbol{u}(x,y)\|} - \boldsymbol{u_0} \right\|^3 + 1}$ in Eq. 5.31 is used to minimize the impact of this correction on pixels not in set S and keep the correction smooth. The parameter κ is set to 0.02 empirically. With exponentiation, we transform $\boldsymbol{u_c}$ and $\boldsymbol{u_p}$ into RGB space and get $\boldsymbol{u_c}^{\mathrm{RGB}}$ and $\boldsymbol{u_p}^{\mathrm{RGB}}$, respectively, where image $\boldsymbol{u_p}^{\mathrm{RGB}}$ has almost the same chrominance as original image and image $\boldsymbol{u_c}^{\mathrm{RGB}}$ has almost the same luminance as original image. This is verified by experimental results on many images in different scenes and reflectance.

Then a shadow-free image is obtained by transferring the right luminance component of $\boldsymbol{u_c}^{\mathrm{RGB}}$ to the color illumination-invariant image. We adopt Lab color space (L, the luminance component; a, b, two chrominance components) for this transfer. First we convert $\boldsymbol{u_c}^{\mathrm{RGB}}$ and $\boldsymbol{u_p}^{\mathrm{RGB}}$ from RGB space into Lab space and get $\boldsymbol{u_c}^{\mathrm{Lab}}$ and $\boldsymbol{u_p}^{\mathrm{Lab}}$, respectively,

$$\begin{cases} \boldsymbol{u_c}^{\mathrm{Lab}}(x, y) = (L_c(x, y), a_c(x, y), b_c(x, y))^T \\ \boldsymbol{u_p}^{\mathrm{Lab}}(x, y) = (L_p(x, y), a_p(x, y), b_p(x, y))^T \end{cases}. \tag{5.32}$$

Then the final shadow-free image $\boldsymbol{u}_{\mathbf{f}}^{\mathrm{Lab}}$ in Lab space is obtained as follows,

$$\boldsymbol{u}_{\mathbf{f}}^{\mathrm{Lab}}(x, y) = (L_c(x, y), a_p(x, y), b_p(x, y))^T. \tag{5.33}$$

An overview of our shadow-free image generation pipeline is presented in Algorithm 5.1. Figure 5.6 shows the stages of the algorithm implemented on a shadow image. From the results, we can see that our method for shadow-free image effectively removes shadows from images while retaining the color and texture information of the original image. It is worth noting that, unlike shadow removal method that only recovers pixels in the shadow region, our shadow-free image, without the process of shadow detection, is operating on the entire image. In our algorithm, all the brighter pixels and darker pixels are pulled to the same illumination intensity and it may reduce the contrast of the image. Therefore, our shadow-free image may sometimes looks a bit dull compared with the original image.

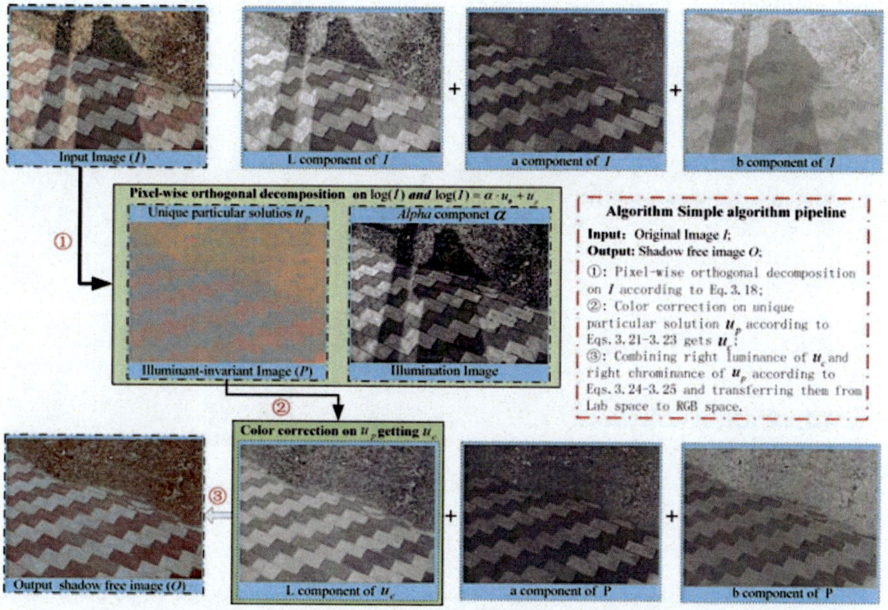

Fig. 5.6 Stages of our algorithm implemented on a shadow image. Note that the arrow in the first row represents the color space conversion, which is converting from RGB space to Lab space. These three components have no actual use in our algorithm. They are only used to act as the reference for the Lab component of our shadow-free image O. Reprinted with permission from Liangqiong Qu, Jiandong Tian, Zhi Han, and Yandong Tang, "Pixel-wise Orthogonal Decomposition for Color Illumination Invariant and Shadow-free Image," Optics Express, Vol. 23, Issue 3, pp. 2220–2239, 2015. © The Optical Society

5.2.3.3 Experiments

In our experiment, we evaluate the results of color illumination-invariant image and the relevant shadow-free image on more outdoor real images. These images consist of common scenes or objects under a variety of illumination conditions without shadow or with different types of shadows, such as soft shadows, cast shadows, and self shadows. We first compare our method with two state-of-the-art full-automatic methods (a statistical learning method [4] and a physical-based method [6] (Figs. 5.7 and 5.8)) and a shadow removal method [1] with human intervention (Fig. 5.9), respectively. Then the results obtained by our method and Shen's method [5] on images under different illumination conditions show that our color illumination-invariant image is much closer to the intrinsic feature than Shen's from the view of yielding an identical reflectivity result regardless of illumination condition (Fig. 5.10).

Comparison with full-automatic methods. In our experiment, we compare our method with two state-of-the-art methods ([4, 6]) in terms of shadow-free results and running time. These two methods are based on shadow feature with statistical learning and physical model, respectively. These methods are executed with the codes published by the authors of [4, 6].

Fig. 5.7 Comparisons with two state-of-the-art methods (Guo et al. [4] and Yang et al. [6]). (b1), (b2) are close-ups of the red rectangles in (a1), (a2). Reprinted with permission from Liangqiong Qu, Jiandong Tian, Zhi Han, and Yandong Tang, "Pixel-wise Orthogonal Decomposition for Color Illumination Invariant and Shadow-free Image," Optics Express, Vol. 23, Issue 3, pp. 2220–2239, 2015. © The Optical Society

Fig. 5.8 Comparisons with two state-of-the-art methods (Guo et al. [4] and Yang et al. [6]) on more images. Reprinted with permission from Liangqiong Qu, Jiandong Tian, Zhi Han, and Yandong Tang, "Pixel-wise Orthogonal Decomposition for Color Illumination Invariant and Shadow-free Image," Optics Express, Vol. 23, Issue 3, pp. 2220–2239, 2015. © The Optical Society

In Fig. 5.7, we show two shadow-free image results. For the first image (Fig. 5.7(a1)), both our method and Yang's method [6] remove the soft shadow of leaves successfully. However, Guo's method [4] fails to recover a shadow-free image due to the wrong detection of shadows. Due to the lack of robust shadow features, this shadow feature and learning-based method [9] are not always effective, especially in soft shadow region or some complex scenarios (e.g., Figs. 5.8d, e). Close-ups of the first image (Fig. 5.7(b1)) show that, compared with our result, Yang's method [6] fails to retain the white arrow on the ground. This is because that Yang's method is not solid for neutral regions when its neighbor regions are also neutral [6]. For the second image (Fig. 5.7(a2)), our method gets a shadow-free image with texture preserved, while Guo's method [4] fails to retain the original texture even when it detects the shadows successfully due to the inappropriate relighting on pixels. Specially, some pixels are relighted serious wrongly, such that it looks like that some objects that do not exist in the original image appear (e.g., the tiny yellow flowers shown in Guo's shadow-free image in Fig. 5.7(b2)). Close-ups of the red rectangle (Fig. 5.7(b2)) show that some of the grasses are blurred in Yang's method [6] due to bilateral filtering, while our method, without any filtering operation, retains the texture well.

Some other comparison results are shown in Fig. 5.8. The images in Fig. 5.8 show that our method removes shadows effectively from a single image with texture well-preserved. However, Guo's method [4] fails to recover a shadow-free image either due to the wrong detection on shadow regions (Fig. 5.8b–e) or the inappropriate relighting on pixels (Fig. 5.8a, c). Due to the complexity of image reflectance and the irregular distribution of shadows in Fig. 5.8e, none of the shadow regions are correctly detected in Guo's shadow-free image. For images with dark shadows or in which shadows pervade in a scenario (Fig. 5.8a–e), Yang's method [4] blurs the shadow-free images. For example, the clear water in the original image in Fig. 5.8e is seriously blurred in Yang's method. Some small and tiny objects such as leaves are also removed in Yang's results.

Figures 5.7 and 5.8 verify that our method removes shadows more effectively from a single image and performs steadily on images with complex outdoor scenes. Our shadow-free image also better preserves the original texture compared to the other two methods. In running time, our method also outperforms Guo's [4] and Yang's [6]. In Table 5.2, we compare the running time of these three methods on the images shown in Fig. 5.8. All experiments were conducted on a computer with Intel (R) Core (TM) i3-2120 3.3GHz CPU and 2G RAM memory. The source codes that Guo and Yang provided are implemented in MATLAB and C++, respectively. For better comparison, we implement our method both in MATLAB and C++. Without any statistical learning or complex filtering operation, our method is faster than Guo's [4] and Yang's [6]. Note that, when the reflectance of image becomes complex and shadows on the image get intricate, it will be extremely hard for Guo's method to detect shadow regions and then remove them, and the running time rose up to 6.4827e + 003 s, almost 2 h in Fig. 5.8e. All these tested examples and quantitative results show that our method performs better in terms of accuracy, consistency, and efficiency than the other two methods on shadow images.

Comparison with shadow removal method with human intervention. Apart from previous comparisons with full-automatic methods for shadow-free image, in this section, we also compare our method with a shadow removal method [1]. In Arbel's method, the shadow mask is firstly extracted based on region growing and SVM learning with the user inputs (Figs. 5.9b, c). Then the shadow-free image is constructed by multiplying a constant scalar to cancel the effect of shadows in umbra regions, while on penumbra regions, the shadows are removed based on the intensity surface construction.

Figure 5.9 gives the comparison of our method with Arbel's method on shadow-free image. Figure 5.9(r1,e) shows that even though Arbel's method effectively removes shadows from image with dark shadows, the texture in the original shadow

Table 5.2 Comparison of running time of the three methods on images shown in Fig. 5.8 (measured in seconds)

Images (Resolution)	a (287 * 163)	b (682 * 453)	c (683 * 450)	d (495 * 326)	e (496 * 334)	f (498 * 332)	g (500 * 475)
Guo's (MATLAB)	36.36	138.63	41.38	49.22	158.2786	286.59	6.5e+003
Ours (MATLAB)	0.13	0.78	0.77	0.375	0.384	0.386	0.546
Yang's (C++)	0.171	0.94	1.73	0.92	0.602	0.614	0.75
Ours (C++)	0.041	0.187	0.34	0.1829	0.119	0.092	0.157

area is not consistent with that in the non-shadow area. Figure 5.9(r2,d) and (r3,d) further show that Arbel's method fails to reconstruct a shadow-free image from images with complex shadows. In these complex scenarios, the constant scalar in the umbra area in Arbel's method no longer works. The wrong constant scalar relights the shadow regions either excessively or insufficiently. As a contrast, our method can not only remove the shadows, but also keep the texture consistency between the shadow and non-shadow area as shown in Fig. 5.9(r1,d), (r2,e), and (r3,e).

The "intrinsic" of Color Illumination-Invariant Image. Both Guo's [4] method, Yang's [6] method, and Arbel's method [1] are designed for removing shadows from an image. They cannot yield an identical reflectivity result regardless of illumination condition. In this section, we will show that our color illumination-invariant image has intrinsic feature. We apply our algorithm to a series of images taken in outdoor scenes at different times on a sunny day to testify the intrinsic feature of our color illumination-invariant image.

Shown in Fig. 5.10(r1), these tested images were taken under four different illumination conditions, one under total daylight (Fig. 5.10(r1,a)), two under partly daylight and partly skylight (Fig. 5.10(r1,b) and (r1,c)), and one under total skylight (Fig. 5.10(r1,d)). Taking the image under total daylight as reference image, we compare the differences between the corresponding processed results by different methods. For an intrinsic-image derivation algorithm, the difference between the reference intrinsic image and other three intrinsic images should be approximately the same or at least much closer to each other than the original image sequences.

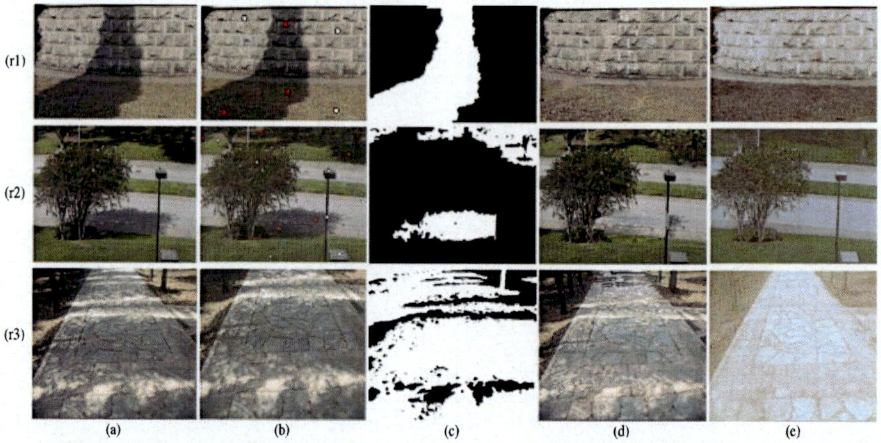

Fig. 5.9 Comparison with Arbel's shadow removal method [1]. **a** Original image. **b** Examples of user-guided shadow and non-shadow observations in Arbel's method denoted by red circles and white circles, respectively. **c** The extracted shadow masks in Arbel's method. **d** Arbel's shadow-free image. **b** Our shadow-free image. Reprinted with permission from Liangqiong Qu, Jiandong Tian, Zhi Han, and Yandong Tang, "Pixel-wise Orthogonal Decomposition for Color Illumination Invariant and Shadow-free Image," Optics Express, Vol. 23, Issue 3, pp. 2220–2239, 2015. © The Optical Society

Fig. 5.10 Images under different illumination conditions. For rows: (r1) original image, (r2) intrinsic image by Shen et al. [5], (r3) our color illumination-invariant image, (r4) our shadow-free image. For columns: **a** (reference image) in daylight, **b**, **c** partly in daylight and partly in skylight, **d** in skylight. Reprinted with permission from Liangqiong Qu, Jiandong Tian, Zhi Han, and Yandong Tang, "Pixel-wise Orthogonal Decomposition for Color Illumination Invariant and Shadow-free Image," Optics Express, Vol. 23, Issue 3, pp. 2220–2239, 2015. © The Optical Society

In our experiment, we calculated the differences both for our color illumination-invariant images and shadow-free images. As a comparison, we also present the intrinsic images obtained by Shen et al. [5] and the corresponding differences. The root mean square error (RMSE) is used as the measurement. Since being compressed, the range of our color illumination is much lower compared with original images and shadow-free images. Therefore, besides RMSE, we also adopt relative error ($\frac{\text{RMSE}}{\text{The Range of Measured Image}}$) as our measurement. All these measured images are 24-bit RGB images.

As shown in Fig. 5.10(r3), the four color illumination-invariant images generated by our method are basically the same. It verifies that our orthogonal decomposition on an image yields an identical reflectivity result regardless of illumination condition. The quantitative measurement is given in Table 5.3. Taken the images under total daylight and total skylight as an example, the RMSE difference of the original images is decreased from 101.85 to 4.67 in our color illumination-invariant images. Also, when considering the relative error, this value is decreased from 39.94% to 5.7%. Even after a color restoration operation on illumination-invariant images, the RMSE difference (24.88) between our shadow-free images is still much smaller than original images. However, the intrinsic images obtained by Shen et al. [5] still have a big difference, 86.36. The quantitative measurements demonstrate that our color

Table 5.3 MSE and relative error between images

Images	RMSE			Relative error (%)		
	a and b	a and c	a and d	a and b	a and c	a and d
Original img.	45.96	80.03	101.85	18.02	31.39	39.94
Intrinsic img. by Shen et al. [5]	45.17	72.67	86.36	17.71	28.50	33.87
Our invariant img.	4.98	4.49	4.67	5.79	5.22	5.70
Our shadow-free img.	15.12	17.87	24.88	6.00	7.45	9.99

illumination-invariant images are considerably closer to generate an intrinsic representation than Shen et al. [5]. The corresponding shadow-free images also maintain this "intrinsic" to some extent.

In this section, we propose a novel, effective, and fast method to obtain a color illumination-invariant image and shadow-free image from a single outdoor image. Different from state-of-the-art methods that either need shadow detection or statistical learning, our shadow-free image is obtained by a pixel-wise orthogonal decomposition and color restoration. Also, our shadow-free image better preserves the texture information compared to the state-of-the-art method. Intrinsic to the illumination condition, our color illumination-invariant image is useful for relighting by adjusting α information. Besides, only requiring pixel-wise calculation, our method for color illumination-invariant image and shadow-free image can be further applied to the real-time applications to resist illumination variation.

5.3 A New Intrinsic-Lighting Color Space for Daytime Outdoor Images

The observed colors of natural scenes are mostly influenced by illumination. It may generate uneven lightness or shadows, which greatly affect the quality of various vision tasks. Therefore, increasingly more attention has been focused on extracting and representing illumination invariance from natural images, which is generally defined as an "intrinsic" image. An illumination-based color space should be constructed via the intrinsic and lighting information and be suitable for "direct" lighting processing, as shown by the example presented in Fig. 5.11.

Generally, a color space is designed for a certain purpose. The RGB and CMYK color models are the most well-known color spaces for digital display or printing, and they are device-dependent color models. For approximating human vision, the Lab color space was designed. The YUV color space was developed when engineers wanted color television in a black-and-white infrastructure. Based on different concepts, the HSV/HSL color spaces represent colors by hue, saturation, and value (brightness) / lightness (luminance). They are based more on how colors are organized and conceptualized in human vision in terms of color-making attributes. However, the color attributes here do not correspond to the optical spectra in physics.

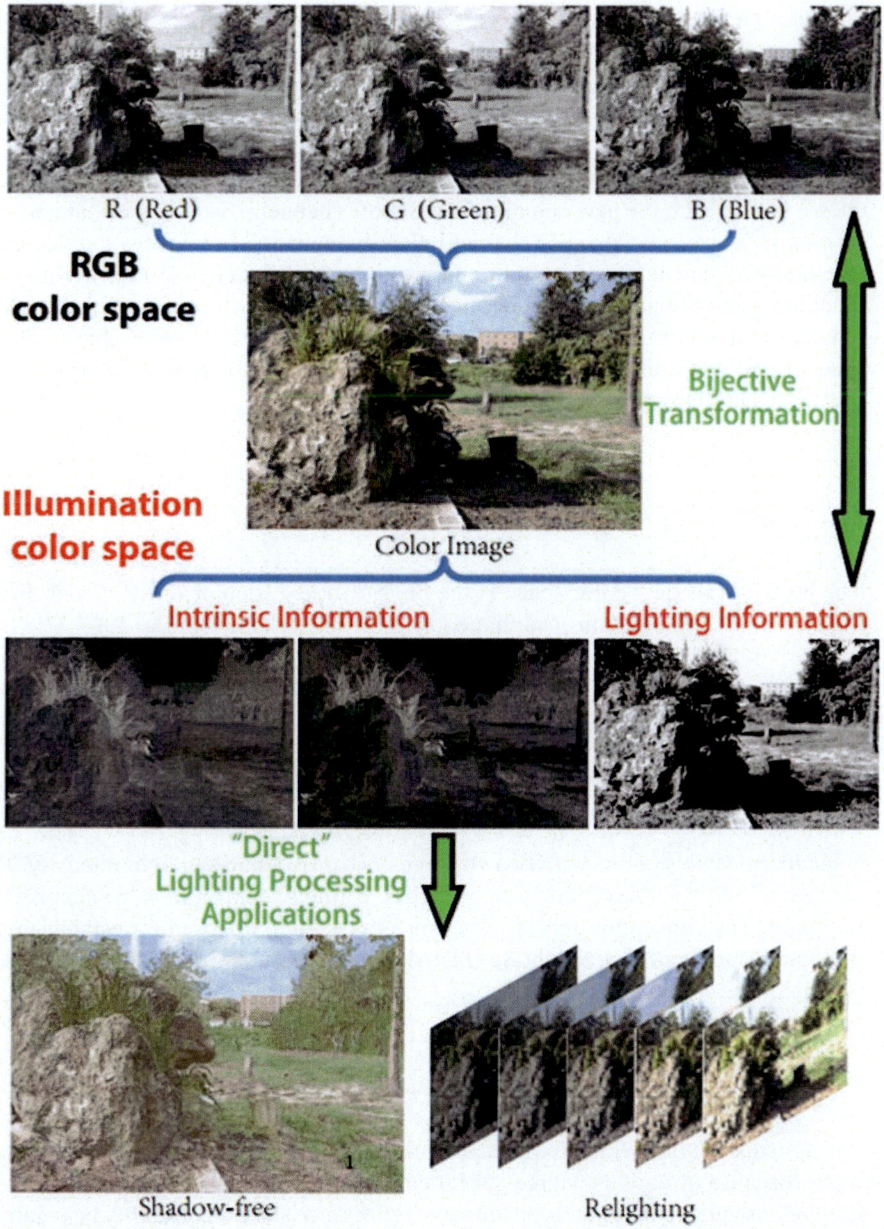

Fig. 5.11 Illumination color space separates colors into intrinsic information and lighting information. It is bijective with the RGB color representation and is more convenient for dealing with illumination-related processings. © [2017] IEEE. Reprinted, with permission, from [Zhi Han, Jiandong Tian, Liangqiong Qu, and Yandong Tang, "A New Intrinsic-Lighting Color Space for Daytime Outdoor Images," IEEE Transactions on Image Processing, Vol. 26, No. 2, pp. 1031–1039, 2017]

The aforementioned color spaces are suitable for color representation, display, painting, and so forth. However, they generally lack the ability to intuitively show the information of reflectance (material) or illumination separately. It may impede related image processing or perception, such as shadow removal, relighting, and scene understanding.

In this section, we aim at providing a new color space for image representation. As shown in Fig. 5.11, the new color space has three channels. Each of the channels has its own physical meaning such that an image is separated into intrinsic reflectance (the first two channels) and lighting information (the third channel). Consequently, a couple of illumination-related applications can be directly reached, e.g., shadow removal and extracting and relighting. Furthermore, the new color space can be transformed with other color spaces (such as RGB) directly. It guarantees a gamut for color representation.

5.3.1 *New Color Space: IL Space*

The basis of this new color space is the physics-based tricolor linear model. From the model, the illumination-invariant equations hold as shown in Eq. 5.34,

$$\begin{cases} \log(F_R) + \log(F_G) - \beta_1 \log(F_B) = \log(f_R) + \log(f_G) - \beta_1 \log(f_B) \\ \log(F_R) - \beta_2 \log(F_G) + \log(F_B) = \log(f_R) - \beta_2 \log(f_G) + \log(f_B) \\ -\beta_3 \log(F_R) + \log(F_G) + \log(F_B) = -\beta_3 \log(f_R) + \log(f_G) + \log(f_B), \end{cases} \tag{5.34}$$

where $\beta_1 = \frac{\log(K_R)+\log(K_G)}{\log(K_B)}$, $\beta_2 = \frac{\log(K_R)+\log(K_B)}{\log(K_G)}$, *and* $\beta_3 = \frac{\log(K_G)+\log(K_B)}{\log(K_R)}$. The new color space depicts colors with two attributes: intrinsic value and light intensity. The concept of intrinsic lines is first proposed for creating two intrinsic value channels of the space. The concept of lighting level surfaces is then proposed for establishing a metric measurement for the lighting channel. Finally, the new tri-channel color space is constructed.

5.3.1.1 Intrinsic Lines

The light distribution within the same illumination environment is not uniform everywhere because of various degrees of reflections and occlusions by other objects or itself. Consequently, the pixels on the same reflectance should not simply be regarded as either inside (f_H) or outside (F_H) shadows but rather under a continuous variation of light intensities. On this basis, by defining log-RGB value $L_H = \log(\mathscr{F}_H)$ for short, we can derive Eq. 5.35 from Eq. 5.34.

$$\begin{cases} L_R + L_G - \beta_1 L_B = I_1 \\ L_R - \beta_2 L_G + L_B = I_2 \\ -\beta_3 L_R + L_G + L_B = I_3. \end{cases} \tag{5.35}$$

Here, $I = (I_1, I_2, I_3)$ is defined as the *intrinsic value*. It provides an invariant for a set of log-RGB values. These log-RGB values belong to the same reflectance but under different lighting conditions.

Since Eq. 5.35 is not a full-rank equation set (the rank is 2), its solutions can be written in the form of

$$\begin{cases} L_R = \frac{\log(K_R)}{\log(K_B)} L_B + \frac{\log(K_R)}{\log(K_R)+\log(K_G)+\log(K_B)} \Delta I_1 \\ L_G = \frac{\log(K_G)}{\log(K_B)} L_B + \frac{\log(K_G)}{\log(K_R)+\log(K_G)+\log(K_B)} \Delta I_2, \end{cases} \tag{5.36}$$

where $\Delta I_1 = I_1 - I_3$, $\Delta I_2 = I_1 - I_2$, and L_B are a free variable.

Equation 5.36 defines the relationship among the three channels of log-RGB values for one given I, which constitutes a straight line in the 3D space of (L_R, L_G, L_B). We define this line as the *intrinsic line* because it identifies the intrinsic characteristic of the set of log-RGB values. As I changes, we obtain infinite numbers of intrinsic lines. They share the same slope, i.e., $\frac{\log(K_R)}{\log(K_B)}$ and $\frac{\log(K_G)}{\log(K_B)}$. Therefore, the intrinsic lines are parallel. They are distinguished by the intercepts, which are dominated by the values of ΔI_1 and ΔI_2. Theoretically, each intrinsic line corresponds to one distinct reflectance. Moreover, different (L_R, L_G, L_B) values residing on the same intrinsic line correspond to the colors from the same reflectance but under various light intensities. Larger values correspond to higher light intensities.

In the new color space, the values of ΔI_1 and ΔI_2 constitute the two channels for representing the attribute of intrinsic information.

5.3.1.2 Lighting Level Surfaces

Considering one specific reflectance under uneven lighting conditions, the log-RGB values of the pixels on the reflectance may be various because of different light intensities. However, they are distributed on the same intrinsic line. By setting the log-RGB values of all the pixels to one specific value, the entire reflectance is under the same lighting intensity. This procedure can be regarded as obtaining one specific solution of Eq. 5.36 by adding one more equation to make the equation set full rank. For simplicity, the equation can be designed as follows,

$$\log(K_R)L_R + \log(K_G)L_G + \log(K_B)L_B = T. \tag{5.37}$$

Here, T is a specific value. Geometrically, it defines a flat plane that is cutting through and vertical to the intrinsic lines. Therefore, by projecting all the pixels onto this plane, each reflectance is under the same light intensity. As the value of T changes, we have a series of parallel flat planes. Figure 5.12a presents a visual example. Ideally, each

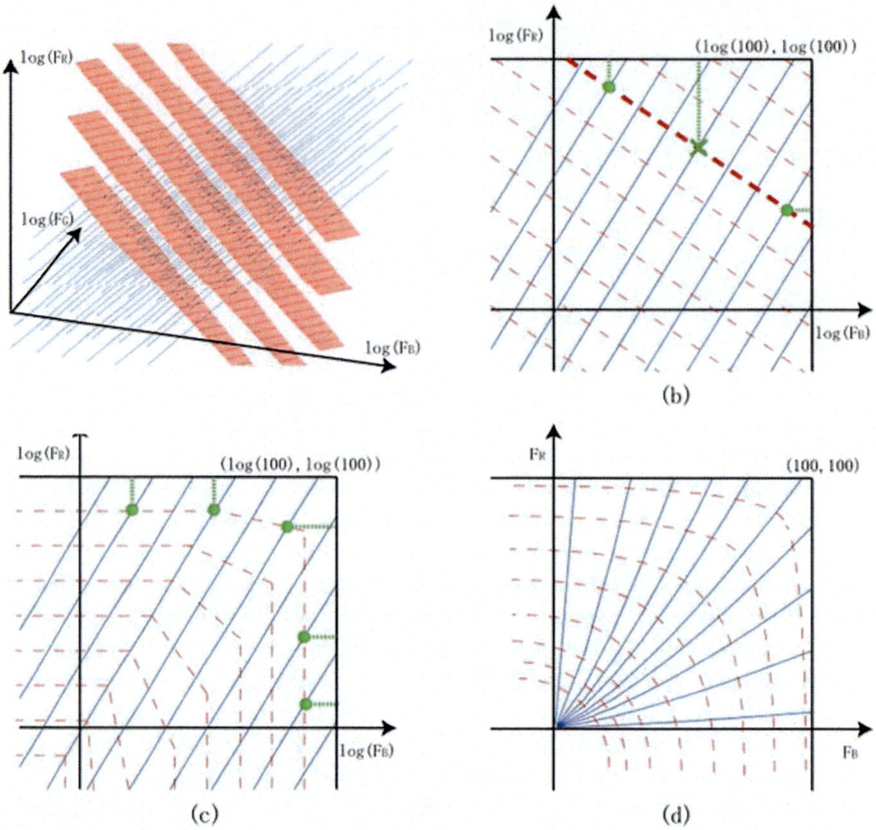

Fig. 5.12 Intrinsic lines (solid blue) and lighting level surfaces (dashed red). **a** 3D view for vertical flat surfaces in log-RGB space; **b** Simplified 2D view for (**a**); **c** 2D view for nonlinear surfaces in log-RGB space; and **d** 2D view for nonlinear surfaces in linear sRGB space. © [2017] IEEE. Reprinted, with permission, from [Zhi Han, Jiandong Tian, Liangqiong Qu, and Yandong Tang, "A New Intrinsic-Lighting Color Space for Daytime Outdoor Images," IEEE Transactions on Image Processing, Vol. 26, No. 2, pp. 1031–1039, 2017]

plane defines a unified level of relative light intensity for all reflectances, which is similar to the scale of a ruler for measuring light intensities on every intrinsic line. Therefore, we name these planes *lighting level surfaces*. T is taken as the "scale of the ruler."

However, the vertical flat planes are not rational enough for the lighting level surface system because T cannot appropriately reflect the relationship of lighting levels among different reflectances. We select a random level surface as an illustration (the thicker one in the simplified 2D view in Fig. 5.12b). The colors obtained via the intersections of the surface with every intrinsic line are supposed to be on the same relative lighting level (green circles and cross). However, for the intrinsic lines with larger absolute values of the intercept (green circles), the brightness of the colors may approximate the maximum because $\max(L_R, L_G, L_B)$ is close to the defined upper bound of the log-RGB values[1] (measured by the green dotted lines in the figure), whereas for the intrinsic lines with smaller absolute values of the intercept (green cross), the brightness may only reach the middle of the range. In this case, the two colors on the same level surface have a large difference in brightness, which deviates from the purpose of the lighting level surface.

A more rational lighting level surface system is set up in the form shown in Fig. 5.12c. On the same lighting level, the colors from different intrinsic lines (green circles) have very similar brightnesses. The buildup of this new surface system begins with introducing a new variable ζ. It is defined as *lighting level index* since it is used for indexing the relative lighting levels of the surfaces. For each intrinsic line identified by $\Delta I = (\Delta I_1, \Delta I_2)$, the ζ values and the (L_R, L_G, L_B) values have a one-to-one correspondence. We set $\zeta = 1$ while $\max(L_R, L_G, L_B)$ reaches the upper bound of the log-RGB values, and we set $\zeta = 0$ while $\max(L_R, L_G, L_B)$ reaches 0. The corresponding L_H ($H = \{R, G, B\}$) values at $\zeta = 1$ and $\zeta = 0$ are denoted by $M_{L_H}^{\Delta I}$ and $m_{L_H}^{\Delta I}$, respectively. Then, the relationship between L_H and ζ is given by

$$L_H(\zeta) = m_{L_H}^{\Delta I} + \zeta(M_{L_H}^{\Delta I} - m_{L_H}^{\Delta I}). \tag{5.38}$$

To better interpret the correspondence between the lighting level and the sRGB values, in Fig. 5.12d, the intrinsic lines and lighting level surfaces are transformed from the log-RGB space to the linear sRGB space. Note that the infinite interval of $\zeta \in (-\infty, 0)$ corresponds to the region of $\mathscr{F}_H \in (0, 1)$, which is the very tiny and insignificant area of the sRGB space from the illumination aspect.

This family of lighting level surfaces is nonlinear. By regarding T as the function of ζ and parameterized by ΔI,

$$\begin{aligned} T(\zeta) = &\tfrac{\Sigma_{K2}}{\log(K_B)}(m_{L_B}^{\Delta I} + \zeta(M_{L_B}^{\Delta I} - m_{L_B}^{\Delta I})) \\ &+\tfrac{1}{\Sigma_K}(\log^2(K_R)\Delta I_1 + \log^2(K_G)\Delta I_2), \end{aligned} \tag{5.39}$$

where $\Sigma_K = \log(K_R) + \log(K_G) + \log(K_B)$, $\Sigma_{K^2} = \log^2(K_R) + \log^2(K_G) + \log^2(K_B)$, and Eq. 5.37 is still rewritten in a linear form as

$$\log(K_R)L_R + \log(K_G)L_G + \log(K_B)L_B = T(\zeta). \tag{5.40}$$

ζ provides the metric measurement for lighting levels such that it is defined as the third channel for depicting the attribute of lighting.

[1] log(100) for linear sRGB or log(255) for gamma-corrected sRGB.

5.3.1.3 Intrinsic and Lighting Color Space

By solving the union of Eqs. 5.36 and 5.40, the relationships between $(L_{\mathrm{R}}, L_{\mathrm{G}}, L_{\mathrm{B}})$ and $(\Delta I_1, \Delta I_2, \zeta)$ are obtained as shown in Eqs. 5.41 and 5.13.

$$\begin{pmatrix} \Delta I_1 \\ \Delta I_2 \\ T(\zeta) \end{pmatrix} = \begin{pmatrix} \frac{\Sigma_K}{\log(K_{\mathrm{R}})} & 0 & -\frac{\Sigma_K}{\log(K_{\mathrm{B}})} \\ 0 & \frac{\Sigma_K}{\log(K_{\mathrm{G}})} & -\frac{\Sigma_K}{\log(K_{\mathrm{B}})} \\ \log(K_{\mathrm{R}}) & \log(K_{\mathrm{G}}) & \log(K_{\mathrm{B}}) \end{pmatrix} \begin{pmatrix} L_{\mathrm{R}} \\ L_{\mathrm{G}} \\ L_{\mathrm{B}} \end{pmatrix}. \tag{5.41}$$

If we define the domain of the sRGB space as $\{\mathscr{F}_H > 0, H = \{R, G, B\}\}$ (optically meaningful), in Eqs. 5.41 and 5.13, the $\log(\cdot)$ function is monotone and differentiable, and the lighting level surfaces $T(\zeta)$ are dense in the space without any intersections. Therefore, these two relationships build the objective mapping between the sRGB value and $(\Delta I_1, \Delta I_2, \zeta)$ value (visualized in Fig. 5.13). When transforming a single sRGB image to the new representation, the meaning of each channel is clear. ΔI_1 and ΔI_2 provide the intrinsic values, which reflect the intrinsic information (free from illumination) of every pixel. ζ records the lighting distribution of the scene.

When the environment coefficients $(K_{\mathrm{R}}, K_{\mathrm{G}}, K_{\mathrm{B}})$ are given, $(\Delta I_1, \Delta I_2, \zeta)$ constitute a new tri-channel color space based on the **intrinsic** information and **lighting** level. Consequently, we name the new space *IL color space*.

$$\begin{pmatrix} L_{\mathrm{R}} \\ L_{\mathrm{G}} \\ L_{\mathrm{B}} \end{pmatrix} = \begin{pmatrix} \frac{\log(K_{\mathrm{R}})(\log^2(K_{\mathrm{G}})+\log^2(K_{\mathrm{B}}))}{\Sigma_K \Sigma_{K^2}} & -\frac{\log(K_{\mathrm{R}})\log^2(K_{\mathrm{G}})}{\Sigma_K \Sigma_{K^2}} & \frac{\log(K_{\mathrm{R}})}{\Sigma_{K^2}} \\ -\frac{\log(K_{\mathrm{G}})\log^2(K_{\mathrm{R}})}{\Sigma_K \Sigma_{K^2}} & \frac{\log(K_{\mathrm{G}})(\log^2(K_{\mathrm{R}})+\log^2(K_{\mathrm{B}}))}{\Sigma_K \Sigma_{K^2}} & \frac{\log(K_{\mathrm{G}})}{\Sigma_{K^2}} \\ -\frac{\log(K_{\mathrm{B}})\log^2(K_{\mathrm{R}})}{\Sigma_K \Sigma_{K^2}} & -\frac{\log(K_{\mathrm{B}})\log^2(K_{\mathrm{G}})}{\Sigma_K \Sigma_{K^2}} & \frac{\log(K_{\mathrm{B}})}{\Sigma_{K^2}} \end{pmatrix} \begin{pmatrix} \Delta I_1 \\ \Delta I_2 \\ T(\zeta) \end{pmatrix}. \tag{5.13}$$

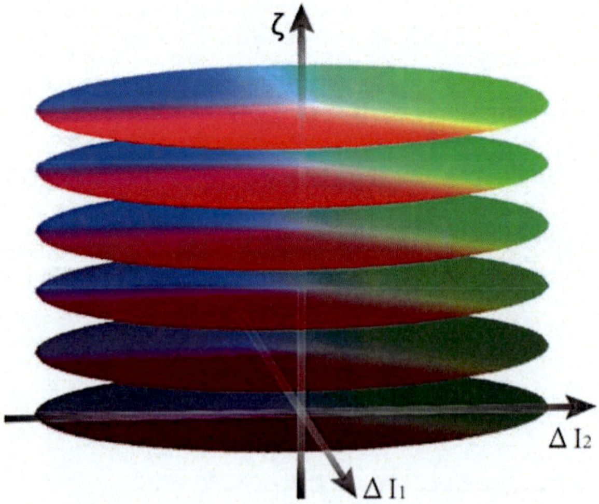

Fig. 5.13 Visualized color distribution in the IL color space. © [2017] IEEE. Reprinted, with permission, from [Zhi Han, Jiandong Tian, Liangqiong Qu, and Yandong Tang, "A New Intrinsic-Lighting Color Space for Daytime Outdoor Images," IEEE Transactions on Image Processing, Vol. 26, No. 2, pp. 1031–1039, 2017]

5.3.2 *Shadow/Lighting Processing via the IL Space*

Shadow removal and relighting are two typical examples of illumination-related processing. Shadows are the area under less illumination generated by the uneven lighting distribution. From the aspect of illumination, shadow removal consists of placing the entire scene under a uniform light intensity. Relighting consists of placing different lighting distributions on the same scene. The fundamental purpose of these two tasks is to estimate the true lighting levels of the entire scene under even illumination.

5.3.2.1 Intrinsic-Lighting Levels

Lighting level surfaces are the scales for measuring relative light intensities on every intrinsic line. It means that even though all the reflectance in one specific scene is ideally under the same light intensity, the colors from different reflectances are probably on different lighting level surfaces (different ζ values). It is mainly caused by the metamerism. To unify the light intensities of the scene, we need to estimate the *intrinsic-lighting levels* ζ^* of all the corresponding intrinsic lines. ζ^* reflects the lighting levels of all the image pixels under the true illumination environment without lighting variations. The surface $S(\zeta^*)$ spanned by this set of optimal values is defined as *intrinsic-lighting contour surface*. It predicts the lighting levels of all possible reflectances in the illumination environment provided by the given scene.

Figure 5.14 presents a synthetic example of 24-color ColorChecker. (b) is the partly shadowed image of (a) with the same shadow intensity on every color patch. In (i), the IL spatial coordinates of the 24 colors in and out of shadows are displayed by the points in corresponding colors. As shown, each pair of the in and out of

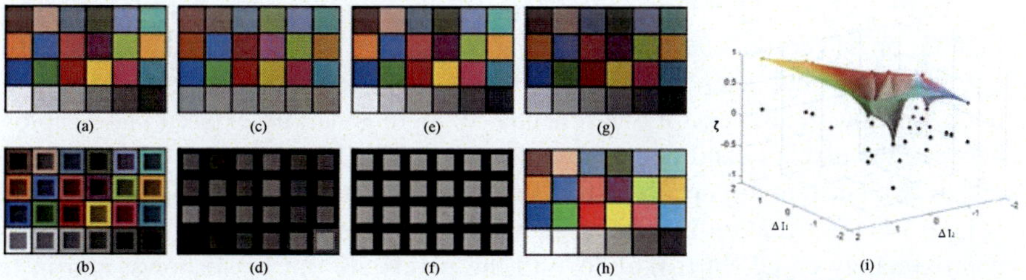

Fig. 5.14 **a** 24-color ColorChecker; **b** ColorChecker with shadows of the same shadow intensity; **c** RGB result by setting uniform lighting level $\zeta = 0.816$; **d** removed shadow intensities by (**c**); **e** RGB result with intrinsic-lighting level ζ^*; **f** removed shadow intensities by (**e**); **g** RGB result with $\zeta^* - 0.2$; **h** RGB result with $\zeta^* + 0.2$; and **i** colors represented in IL space and the intrinsic-lighting contour surface $S(\zeta^*)$ of this example. © [2017] IEEE. Reprinted, with permission, from [Zhi Han, Jiandong Tian, Liangqiong Qu, and Yandong Tang, "A New Intrinsic-Lighting Color Space for Daytime Outdoor Images," IEEE Transactions on Image Processing, Vol. 26, No. 2, pp. 1031–1039, 2017]

shadow colors in one color patch have the same $(\Delta I_1, \Delta I_2)$ value. Moreover, all the pairs have similar differences on ζ because of the same added shadow intensity. The intrinsic-lighting contour surface $S(\zeta^*)$ is spanned by the true colors (with ζ^*), or equally here by the out-of-shadow colors (with larger ζ) in this synthetic example. (c) is the RGB image by setting all ζ to the same specific value, while (e) is by setting ζ to ζ^*. (d) and (f) are the extracted shadow intensities. Comparing (c) and (e) with (a), the advantage of ζ^* is quite obvious. Although the shadows are all removed, the majority of the colors become distorted in (c), particularly the monochrome colors in the 4th row, while (e) preserves all the colors perfectly. (f) correctly reflects the fact that all the color patches are shadowed with the same shadow intensity. Moreover, by tuning the intrinsic-lighting contour surface up and down, we obtain the shadow-free image under different illumination environments, as shown in (g) and (h).

5.3.2.2 Pursuing Intrinsic-Lighting Contour Surface

Natural images are considerably more complicated than the synthetic example because of noises, textures, uneven lighting conditions, and so forth. The major issues are manifested in two aspects: the deviation of intrinsic values and the dispersion of light intensities.

The deviation of intrinsic values is caused by various perturbations. Theoretically, the pixels of the same reflectance should share the same intrinsic value I such that their log-RGB values lie exactly on the same intrinsic line. However, in the real world, they are actually distributed continuously in a small area around the true intrinsic line. Therefore, to group pixels of the same reflectance for estimating ζ^*, they should not be assigned to lines but rather divided into N bundles $\{I\}_n$, $n = 1, ..., N$, and the corresponding estimated intrinsic levels are denoted by ζ_n^*. This goal is achieved by applying a clustering method (e.g., K-means) on the $(\Delta I_1, \Delta I_2)$ values. The number of bundles N depends on the parameter of the clustering algorithm.

The dispersion of light intensities is primarily caused by the slight fluctuation of lighting or reflecting conditions. Consequently, the ζ values from the same bundle $\{\zeta\}_n$ are distributed in a relatively wider range rather than only two values for in and out of shadow. However, for one reflectance, there generally exists a large gap of ζ between the pixels in and out of shadows, and thus, $\{\zeta\}_n$ should be distributed into one or two clusters. Figure 5.15 shows four typical situations of the distributions of $\{\zeta\}_n$. (c) is the situation in which the number of pixels in and out of shadows are approximately equal. In (d), almost all the pixels are out of shadows, and in (e), the out-of-shadow pixels constitute the majority. Conversely, (f) shows the situation in which most of the pixels are in the shadows. As expected, the distributions of $\{\zeta\}_n$ can be represented by a mixture of two Gaussians. Consequently, we adopt the mixture of Gaussians (MoG) model for analyzing $\{\zeta\}_n$. The lower expectation of the two Gaussians is for ζ inside shadows, while the higher one is for out of shadows and chosen as ζ_n^*.

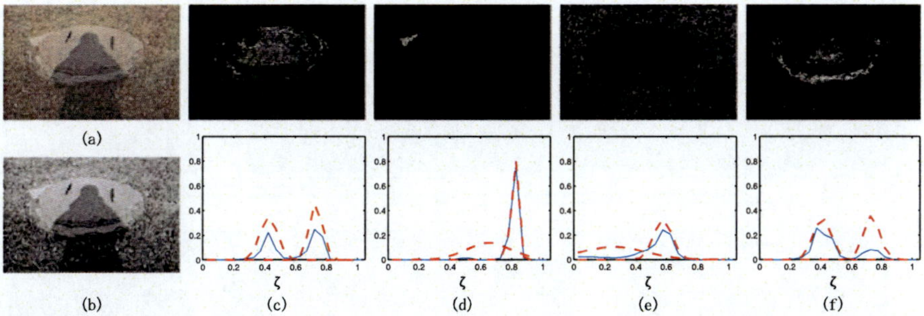

Fig. 5.15 Distributions of $\{\zeta\}_n$. **a** Input image; **b** ζ image; and **c–f** four typical cases of distribution: the upper row are the positions in the image of the pixels with similar I; the lower row are the distributions of $\{\zeta\}_n$ (blue solid lines) and the fitted Gaussian curves (red-dashed lines) © [2017] IEEE. Reprinted, with permission, from [Zhi Han, Jiandong Tian, Liangqiong Qu, and Yandong Tang, "A New Intrinsic-Lighting Color Space for Daytime Outdoor Images," IEE Transactions on Image Processing, Vol. 26, No. 2, pp. 1031–1039, 2017]

In Algorithm 5.2, we provide a brief statement of the algorithm for pursuing the intrinsic-lighting contour surface.

Algorithm 5.2 Pursuing intrinsic-lighting contour surface

Initial: Derived intrinsic values of a single image, $\{I\}$;
Extracted lighting level indices of the image, $\{\zeta\}$;
Environment coefficients, (K_R, K_G, K_B);
Classify and Update: Intrinsic-lighting contour surface, $S(\zeta^*)$;
1: Divide $\{I\}$ into $\{I\}_1, \{I\}_2, ...\{I\}_N$ by clustering on $(\Delta I_1, \Delta I_2)$ values;
2: **for** n=1 to N **do**
3: Select $\{\zeta\}_n$ corresponding to $\{I\}_n$;
4: Calculate cluster centers $\{\zeta_c\}$ by EM algorithm;
5: $\zeta_n^* \leftarrow \max(\zeta_c)$;
6: **end for**
7: Get $S(\zeta^*)$ from the interpolations on $\{\zeta_n^*\}_{n=1}^N$ by Delaunay triangulation.

In applications, ζ^* can be regarded as an alternative channel for synthesizing shadow-free images. As shown in Fig. 5.16, the IL representation $(\Delta I_1, \Delta I_2, \zeta)$ can be transformed back to the RGB representation. By replacing ζ by ζ^*, $(\Delta I_1, \Delta I_2, \zeta^*)$ can provide a shadow-free color image. Via the differences between ζ and ζ^*, the shadow intensities are extracted, by which the shadow regions are "detected." Simultaneously, it realizes the decomposition of one shadowed image into an intrinsic color representation plus shadow information.

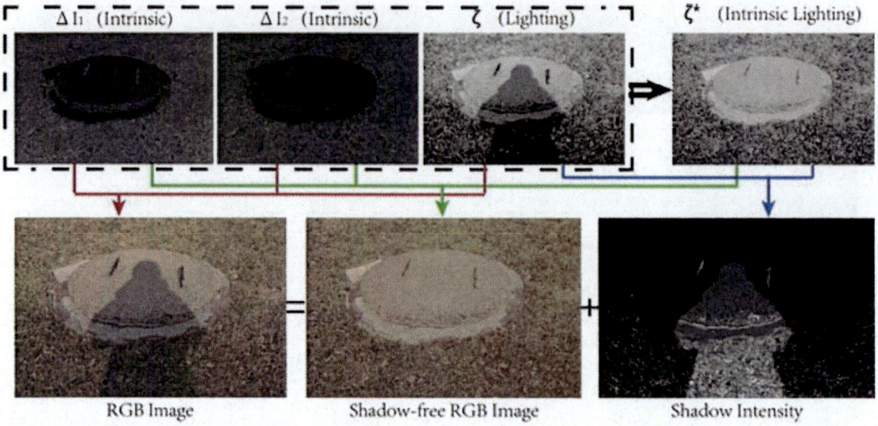

Fig. 5.16 Transformation from IL to RGB representation © [2017] IEEE. Reprinted, with permission, from [Zhi Han, Jiandong Tian, Liangqiong Qu, and Yandong Tang, "A New Intrinsic-Lighting Color Space for Daytime Outdoor Images," IEEE Transactions on Image Processing, Vol. 26, No. 2, pp. 1031–1039, 2017]

5.3.3 *Experiments for Shadow-Free Color Images*

For obtaining shadow-free color images using the IL color space, our goals are threefold: (1) removing shadows from the given images; (2) recovering the "true" colors of the scenes; and (3) showing the shadow intensities of the original images. In the following experiments, all the images are in common formats, e.g., BMP, JPG, and PNG; hence, an inverse gamma correction is needed before processing. Because we do not know the environment information of each test image, the parameters K_H are chosen (6.79, 5.92, 4.24) via the average of the measured SPD. The comparisons are with the state-of-the-art paired regions strategy [4] and bilateral filtering strategy [6]. Note that the basis of [4] is quite different from [6] and IL. [4] is based on "shadow removal," i.e., a shadow detection and removal algorithm via a mathematical method. [6] and IL are based on "intrinsic image" from the aspect of physical optics. Furthermore, the goal of IL is to provide a "color representation," and shadow-free image is a valuable application of the color representation. The comparisons aim to verify the ability of shadow removal by IL.

We first compare IL with [4] on the outdoor images from the dataset provided by [4]. The comparison is on two measurements, as shown in Table 5.4: (1) the recovery deviation of the shadow removal region, for which the average by IL is 17.86 compared to 41.60 by [4] (24-bit RGB); (2) the shadow region detection accuracy, for which the average by IL is 93.0% compared to 86.7% by [4]. IL outperforms [4] on both the measurements. In addition, the detection accuracy by IL is even underestimated because IL detects more tiny shadow areas, such as in leaves, grass, or sand, which are often not detected by [4] and also not labeled in the ground truth.

Figure 5.17 presents more visual comparisons of the three models. IL and [6] provide better shadow removal results than [4]. The difference can also be realized

Table 5.4 Comparisons on outdoor images in Guo's dataset: (1) the average recovery deviation of the shadow removal regions (RGB range: 0–255); (2) the average accuracy of shadow region detection. © [2017] IEEE. Reprinted, with permission, from [Zhi Han, Jiandong Tian, Liangqiong Qu, and Yandong Tang, "A New Intrinsic-Lighting Color Space for Daytime Outdoor Images," IEEE Transactions on Image Processing, Vol. 26, No. 2, pp. 1031–1039, 2017]

Model	Recovery deviation	Detection accuracy (%)
IL	17.86	93.0
[4]	41.60	86.7

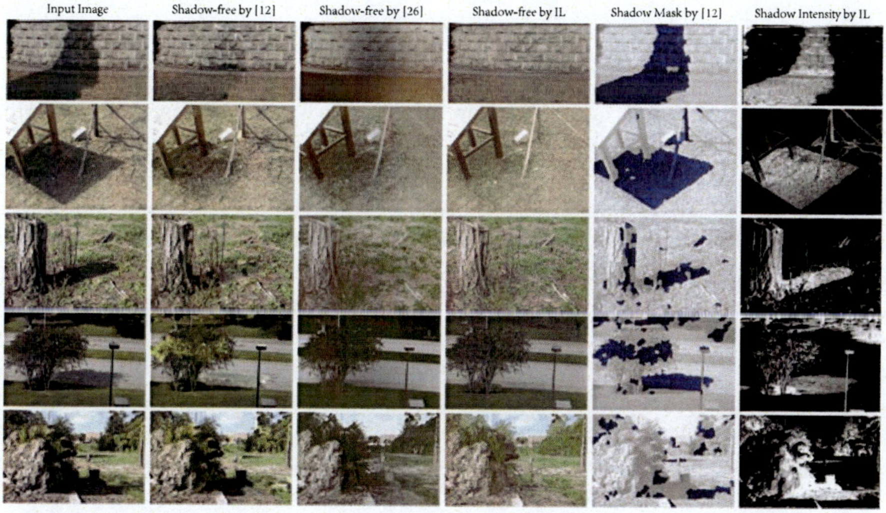

Fig. 5.17 Comparisons of shadow-free color images by [4, 6] and IL (our method); shadow mask by [6] and shadow intensity by IL © [2017] IEEE. Reprinted, with permission, from [Zhi Han, Jiandong Tian, Liangqiong Qu, and Yandong Tang, "A New Intrinsic-Lighting Color Space for Daytime Outdoor Images," IEEE Transactions on Image Processing, Vol. 26, No. 2, pp. 1031–1039, 2017]

by comparing shadow intensity images by IL with shadow mask images by [4]. IL provides more precise shadow areas than [4], particularly for some tiny shadow areas. Quite different from traditional approaches such as [4], in which shadow removal depends on the result of shadow detection, our algorithm obtains the shadow area simultaneously with the shadow removal results. Therefore, it is more robust, particularly on the cases with multiple unconnected shadow regions. The shadow intensity image can be directly applied to shadow processing. The refinement on the relighting series in the next subsection is one immediate utilization.

Then, we compare IL and [6] in more detail. Visually, IL is superior on both the effects of shadow removal and color recovery. To provide a more objective measurement for color recovery, we perform a quantitative comparison of the average RGB deviations of the out-of-shadow area, as shown in Table 5.5. Our results suffer smaller color deviations, and the deviation ratios are only approximately 5%.

Table 5.5 Average RGB deviations of out-of-shadow area (RGB range: 0–255)

Ex. in Fig. 5.17	IL transformation			Bilateral filtering [6]		
	R	G	B	R	G	B
1.	7.7	7.0	6.8	13.4	14.7	20.8
2.	13.6	13.3	12.8	25.9	21.7	15.5
3.	14.9	14.5	13.5	28.2	26.1	23.0
4.	9.5	9.2	7.4	21.4	20.7	28.3
5.	13.4	12.9	11.5	33.4	30.4	22.1

Although the proposed algorithm is capable of recovering the "true" colors of the scene, when comparing the shadow-free image with its corresponding original image, the percepts by human vision may suffer minor differences, e.g., feeling whiten or dull. The major reasons are two-fold. First, the intrinsic-lighting contour surface is estimated from the given scene. The deviation may occur for various reasons, such as improper clustering results or irregular lighting level distribution of one reflectance. Second, the perception differences come from the decrease of image contrast or entropy. Our shadow-free image is obtained by setting the lighting levels of all the pixels to the intrinsic-lighting contour surface. By such processing, not only the shadows but also the over-reflecting phenomenon are removed. All the darker and brighter areas are pulled to the same brightness, which reduces the contrast of the image. Since a large number of textures are essentially generated by the variation of light intensities (not the cases by the mixtures of colors or materials), the disappearance of lighting variations makes the texture area more smooth and lowers the entropy of the image, which makes the image more cartoon like. Consequently, the visual percept for the entire scene is affected.

5.3.4 *Intrinsic-Lighting Contour Surface Embedding and Relighting*

In the concept of the intrinsic-lighting contour surface, the meaning of "contour" is that the colors distributed on this surface are under the same light intensity in a given scene. Therefore, one estimated intrinsic-lighting contour surface can be shared by other images to make the scenes of the new images under the same lighting condition. Figure 5.18 presents a simple example. The shadow-free image of the first input (a) is calculated via estimating the intrinsic contour surface, whereas the following three shadow-free images are obtained by directly embedding the original images into the surface. In detail, for each of the images, the values of channels ΔI_1 and ΔI_2 are kept unchanged, whereas the ζ^* values are calculated via the interpolation from the intrinsic contour surface provided by (a). Not only can similar scenes with different shadows (b) or with different illumination conditions (c) be processed to be similar

Fig. 5.18 Embedding images with different illumination conditions or scenes into an estimated intrinsic-lighting contour surface © [2017] IEEE. Reprinted, with permission, from [Zhi Han, Jiandong Tian, Liangqiong Qu, and Yandong Tang, "A New Intrinsic-Lighting Color Space for Daytime Outdoor Images," IEEE Transactions on Image Processing, Vol. 26, No. 2, pp. 1031–1039, 2017]

Fig. 5.19 Original image with ζ_0 is relighted to the lighting levels of $\zeta_0 - 0.4$, $\zeta_0 - 0.2$, $\zeta_0 - 0.1$, and $\zeta_0 + 0.2$, which simulates illumination conditions ranging from heavy clouds to bright sun © [2017] IEEE. Reprinted, with permission, from [Zhi Han, Jiandong Tian, Liangqiong Qu, and Yandong Tang, "A New Intrinsic-Lighting Color Space for Daytime Outdoor Images," IEEE Transactions on Image Processing, Vol. 26, No. 2, pp. 1031–1039, 2017]

under the same illumination environment as (a) but it is also effective for completely different scenes only with similar reflectance combination (d). This experiment also demonstrates that ΔI_1 and ΔI_2 preserve the intrinsic information of lighting invariant.

Relighting consists of placing various lighting distributions upon the same scene to simulate different illuminations. In the IL representation, the relighting effects can be achieved by directly tuning the lighting channel ζ. Figure 5.19 shows two relighting series for simulating illumination conditions ranging from heavy clouds to bright sun. Each relighting image is obtained by adding or subtracting a uniform lighting variation $d\zeta$ from the original lighting condition ζ_0. For the scene with slight shadows (upper row), the light intensities of all the pixels vary from low to high altogether. However, for the scene with heavy shadows (lower row), when the illumination environment becomes brighter, the shadow areas should become even darker. It is caused by the stronger direct sunlight and the weaker scattered skylight in such a situation. Therefore, the intensities of pixels in and out of the shadows areas change in opposite directions. Note that more than a simple contrast enhancement, our relighting processing is based on physical optics. It is consistent with real-world situations and makes the effect more real.

In this section, we propose a novel intrinsic information and illumination-based color space named the IL space. A simple bijective mapping exists between the IL and RGB spaces. To verify the merit of the new space, we further present an algorithm for extracting the intrinsic-lighting information. It can be applied to realize the goal of decomposing a single image into the intrinsic color representation and lighting information, which leads to applications of shadow removal, shadow detection, relighting, and so forth.

References

1. E. Arbel, H. Hel-Or, Shadow removal using intensity surfaces and texture anchor points. IEEE Trans. Pattern Anal. Mach. Intell. **33**(6), 1202–1216 (2010)
2. H. Barrow, J. Tenenbaum, A. Hanson, E. Riseman, Recovering intrinsic scene characteristics. Comput. Vis. Syst **2**(3–26), 2 (1978)
3. G.D. Finlayson, M.S. Drew, C. Lu, Entropy minimization for shadow removal. Int. J. Comput. Vision **85**(1), 35–57 (2009)
4. R. Guo, Q. Dai, D. Hoiem, Paired regions for shadow detection and removal. IEEE Trans. Pattern Anal. Mach. Intell. **35**(12), 2956–2967 (2013)
5. J. Shen, X. Yang, Y. Jia, X. Li, Intrinsic images using optimization, in *CVPR 2011* (IEEE, 2011), pp. 3481–3487
6. Q. Yang, K.-H. Tan, N. Ahuja, Shadow removal using bilateral filtering. IEEE Trans. Image Process. **21**(10), 4361–4368 (2012)

Chapter 6
Shadow and Highlight Removal

Abstract Shadow removal is a challenging task as it usually requires the detection or annotation of shadows as well as semantic understanding of the scene. In this chapter, we introduce a shadow removal algorithm based on the shadow linear model and then introduce an end-to-end shadow removal algorithm based on deep learning without relying on shadow detection. Our deep neural network (DeshadowNet) is designed with a multi-context architecture, where the output shadow matte is predicted by embedding information from three different perspectives, i.e., high-level semantic context, mid-level appearance information, and local image details. Experimental results show that DeshadowNet is adaptive to shadows with widely varying penumbra widths and works well for shadows cast on different semantic regions. For highlight removal, in this chapter, we derive a global color-lines constraint from dichromatic reflection model to robustly recover specular and diffuse reflection. Our key observation is that each image pixel lies along a color line in normalized RGB space and the different color lines representing distinct diffuse chromaticities intersect at one point, namely the illumination chromaticity. For pixels along the same color line, they spread over the entire image and their distances to the illumination chromaticity reflect the amount of specular reflection components. With global (non-local) information from these color lines, our method can effectively separate specular and diffuse reflection components in a pixel-wise manner for a single image, and it is suitable for real-time applications. Experimental results on synthetic and real images show that our method performs better than the existing methods in terms of both specular reflection removal accuracy and computational efficiency.

6.1 Introduction

In the task of image illumination processing, local illumination processing mainly focuses on two aspects: shadow processing and highlight processing. The existing shadow removal algorithm is generally divided into two steps: shadow detection and shadow removal. These algorithms firstly use shadow detection or human–computer interaction to locate the shadow area and then use gradient-domain reconstruction to achieve shadow removal [7, 18] or directly perform shadow removal in the image

J. Tian, *All Weather Robot Vision*, Research on Intelligent Manufacturing,
https://doi.org/10.1007/978-981-16-6429-8_6

pixel domain [1, 9]. In this chapter, we first introduce an image pixel domain shadow removal algorithm based on the shadow linear model and then introduce an end-to-end shadow removal algorithm based on deep learning without relying on shadow detection. Unlike the previous chapter, which uses the intrinsic-image decomposition to directly restore a non-shadow image, the shadow removal algorithm in this chapter does not change the pixel values of the non-shadow area.

The highlight region generated by the specular component of the reflected light will mask the texture details of the surface. It is still a challenging subject to eliminate the highlight region and restore the covered texture details in an image. This chapter proposes a simple and effective method to determine and eliminate highlight regions under natural lighting.

6.2 Shadow Removal Based on Tricolor Linear Model

This section will introduce how to apply the tricolor linear model to remove shadows in an image. According to the shadow linear model, we already know that the following equation holds:

$$F_H = \kappa_H f_H + \mu_H, \tag{6.1}$$

where f_H represents the shadow pixel value in a certain channel, F_H represents the corresponding non-shadow background pixel value, and κ_H and μ_H are the parameters of the linear model. Obviously, based on the known shadow area, if the linear model parameters are obtained, then the shadow pixel value can be restored to the non-shadow pixel value, so as to achieve the purpose of shadow removal. We firstly use pixels on the line segments perpendicular to the shadow boundary, as shown in Fig. 6.1, to estimate linear parameters in Eq. (6.1). We assume a blue line in shadow and its corresponding red line out of shadow have the same reflectance.

$$\frac{1}{M}\sum_{i=1}^{M} F_{H_i} = \kappa_H\left(\frac{1}{M}\sum_{i=1}^{M} f_{H_i}\right) + u_H, \tag{6.2}$$

where M is the number of sampling counterparts. Equation (6.2) actually uses the mean value of shadow pixel and that of non-shadow pixel to estimate the linear parameters. First, averaging the sampling points does not change its linear characteristics. Secondly, in the actual image, some corresponding points may not be on a reflective surface, and the averaging operation can smooth these noise points.

Because we have two parameters and only one equation, we fix the constant term just as the mean value calculated at sun angle from 20° to 70°, and then the proportional term can be determined by Eq. (6.2). Because umbra is only illuminated by skylight while penumbra is also illuminated by part of sunlight, the parameters in penumbra are a little different to those in umbra, which further leads to the results near the edge may be over-recovered. Therefore, we finally need to perform a simple

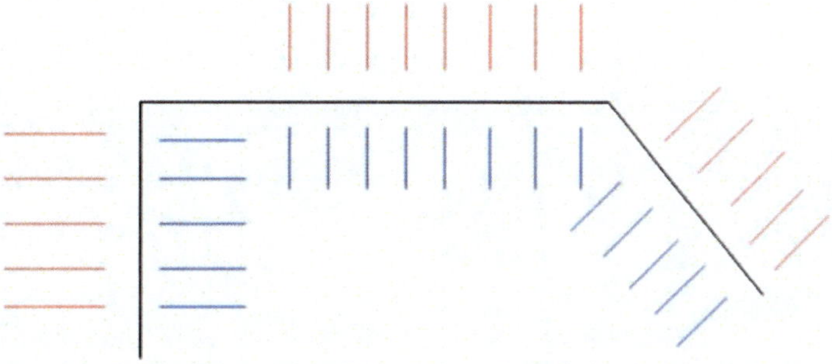

Fig. 6.1 Line segments used to estimate linear parameters. © [2011] IEEE. Reprinted, with permission, from [Tian and Tang, "Linearity of Each Channel Pixel Values from a Surface in and out of Shadows and Its Applications," IEEE Conference on Computer Vision and Pattern Recognition, pp. 985–992, 2011]

Fig. 6.2 One our shadow removal result.© [2011] IEEE. Reprinted, with permission, from [Tian and Tang, "Linearity of Each Channel Pixel Values from a Surface in and out of Shadows and Its Applications," IEEE Conference on Computer Vision and Pattern Recognition, pp. 985–992, 2011]

median filter in the perpendicular direction of shadow boundaries to smooth the edge. One of our shadow removal result is shown in Fig. 6.2. From the result, we find that the shadow is disappeared.

In Fig. 6.3, we compared the research results published by Lalonde [17]. As far as we know, this method is the first method that can detect and remove shadows in a complex single image. This method detects shadows by machine learning and then removes shadows by mathematical methods for solving partial differential equations. In general, our results are better than those of the Lalonde method. Our method can better preserve the texture and details of the shadow area while removing the shadow. After removing the shadow, the Lalonde method blurs the shadow area, and even causes the non-shadow area to blur. For example, in the second image of second row, the person under the house is seriously smoothed but is preserved in our result.

Fig. 6.3 Comparisons of our shadow removal results with that in [17]. Left: Original images; Middle: Results in [17]; Right: Our results. © [2011] IEEE. Reprinted, with permission, from [Tian and Tang, "Linearity of Each Channel Pixel Values from a Surface in and out of Shadows and Its Applications," IEEE Conference on Computer Vision and Pattern Recognition, pp. 985–992, 2011]

6.3 Deep Learning-Based Shadow Removal

The physically based illumination model presented in the last section is designed specifically for natural illumination, which is not valid for indoor environment. In this chapter, we then propose a more universal illumination model (DeShadowNet) to process both the indoor and outdoor illumination variations.

DeshadowNet is designed with a multi-context architecture, where the output illumination attenuation effects are predicted by embedding information from three different perspectives. The first global network extracts shadow features from a global view. Two levels of features are derived from the global network and transferred to two parallel networks. While one extracts the appearance of the input image, the other one involves semantic understanding for final prediction. These two complementary networks generate multi-context features to obtain the shadow matte with fine local details. To evaluate the performance of the proposed model, in this chapter, we also construct the first large-scale benchmark with 3088 image pairs. Extensive experiments on two publicly available benchmarks and our large-scale benchmark show that the proposed method performs favorably against several state-of-the-art methods.

A shadow-free image I_{ns} can also be considered as a pixel-wise product of a shadow image I_s and a shadow matte S_m.

$$I_{ns} = S_m \cdot I_s, \tag{6.3}$$

where the shadow matte S_m represents the illumination that need to be compensated when recovering a shadow-free image.With Eq. 6.3, the shadow removal process is transformed to estimating a shadow matte for an input shadow image. Most existing methods follow this formulation to address the shadow removal problem.

In this section, we aim to explore shadow removal in an end-to-end and fully automatic framework, to address the above-mentioned problems. In contrast to the conventional pipelines that detect shadows, classify umbra/penumbra regions, and then remove shadows, we unify these steps into one and directly learn the mapping function between the shadow image and its shadow matte, which can then be used to recover a shadow-free image with Eq. 6.3.

To this end, we propose a new deep neural network for shadow removal, called *DeshadowNet*. It involves a multi-context embedding mechanism, which integrates high-level semantic information, mid-level appearance information, and local image details in the final prediction. The multi-context embedding is implemented by jointly training three networks, global localization network (G-Net), appearance modeling network (A-Net), and semantic modeling network (S-Net). The G-Net extracts shadow feature representation to describe the global structure and high-level semantic context of the scene. The A-Net and S-Net acquire the appearance information from the shallower layer and semantic information from the deeper layer of G-Net, respectively, allowing the prediction of fine shadow matte using multi-context information. The structure of the proposed DeshadowNet with three subnetworks is shown in Fig. 6.4. To evaluate different shadow removal algorithms, we further construct a new challenging and large-scale shadow removal dataset (SRD). It contains 3088 shadow and shadow-free image pairs.

6.3.1 A New Dataset for Shadow Removal—SRD

Although the shadow removal problem has been studied for decades, publicly available datasets for this purpose are still limited. Among them, the most widely adopted shadow removal dataset is [12], which contains only 76 shadow/shadow-free image pairs. To facilitate the evaluation of shadow removal methods, we have constructed a large-scale dataset called *SRD*, which contains 3088 shadow and shadow-free image pairs. To the best of our knowledge, SRD is the first large-scale benchmark for shadow removal.

To construct our dataset, we use a Canon 5D camera with a tripod and a wireless remote controller for image capturing. We set the manual capture mode with a fixed exposure parameter to capture a shadow image, where the shadow is cast by different objects. We then remove the shadow source to capture the corresponding

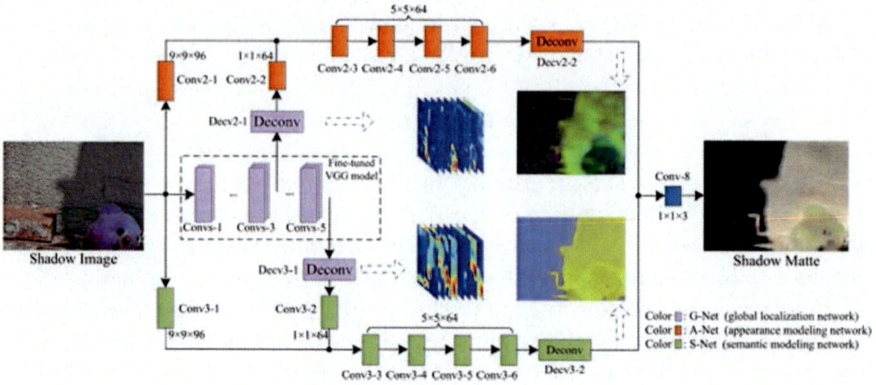

Fig. 6.4 Architecture of DeshadowNet. DeshadowNet consists of three cooperative subnetworks: a global localization network (G-Net), an appearance modeling network (A-Net), and a semantic modeling network (S-Net). These three subnetworks are marked in different colors. © [2017] IEEE. Reprinted, with permission, from [Qu et al. "DeshadowNet: A Multi-context Embedding Deep Network for Shadow Removal," IEEE Conference on Computer Vision and Pattern Recognition, pp. 2308–2316, 2017]

shadow-free image. These arrangements minimize the illumination difference between two captured images.

We enrich the diversity of the proposed dataset in the following four aspects:

- **Illumination:** We take the shadow images at different illumination conditions to include hard and soft shadows in the dataset. Specifically, we capture shadows on cloudy and sunny days, and at different time of the day (e.g., dawn, morning, midday, afternoon, dusk). For example, in Fig. 6.5, the first two are hard shadow images and the third to fifth are soft shadow images.
- **Scene**: We capture shadow images from a variety of scenes, e.g., parks, campuses, buildings, streets, mountains, and beaches.
- **Reflectance**: We cast shadows on different semantic objects to obtain different reflectance phenomena. The sixth and seventh images in Fig. 6.5 show two examples.
- **Silhouette**: We use occluders of various shapes and geometries to cast shadows of different silhouettes and penumbra widths. The fourth and fifth images in Fig. 6.5 show two examples.

6.3.2 *Proposed Model*

The proposed multi-context embedding deep network, called DeshadowNet, is shown in Fig. 6.4. It aims to learn a mapping function between the shadow image and its shadow matte. In this section, we first discuss the motivation and architecture of DeshadowNet and then present the details of the training procedure.

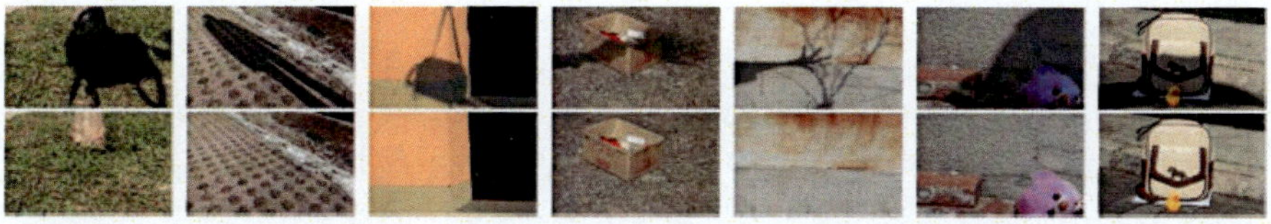

Fig. 6.5 An illustration of several captured shadow and shadow-free image pairs in SRD. © [2017] IEEE. Reprinted, with permission, from [Qu et al. "DeshadowNet: A Multi-context Embedding Deep Network for Shadow Removal," IEEE Conference on Computer Vision and Pattern Recognition, pp. 2308–2316, 2017]

6.3.2.1 Multi-context Convolutional Architecture

Our ideas are that an accurate shadow matte estimation method needs to understand the image content from a global perspective and model the precise illumination compensation with local image details. Hence, in DeshadowNet, we implement these two ideas by designing three cooperative networks. The first network, G-Net, takes a shadow image as input and extracts shadow feature representation which describes the global structure and high-level semantic information of the scene. The other two networks, A-Net and S-Net, acquire the appearance information from the shallower layer and semantic information from the deeper layer of G-Net, respectively, facilitating the prediction of fine shadow matte using multi-context information.

G-Net: Global localization network. G-Net is constructed on the basis of VGG16 network [22], which is originally designed for object recognition. Recent works suggest that CNNs trained with a large amount of data on image classification task can be well-generalized across datasets and tasks such as semantic segmentation and depth prediction [5, 8, 19]. Thus, we adopt the convolutional layers of a pretrained VGG16 model [22] and transfer their feature representation to the shadow matte prediction task with fine-tuning.

The VGG16 network contains thirteen 3×3 convolutional layers (five convolution blocks) and three fully connected layers, along with five max-pooling layers and subsampling layers. These five convolutional groups and spatial poolings substantially increase the receptive field of the network and are thus able to extract the global context and semantic information of the scene. However, these five max-pooling layers introduce a stride of 32 pixels in the network, making the final prediction map coarse. Hence, instead of directly applying the original VGG16 architecture, we set the pooling stride to 1 in the last two max-pooling layers to get a denser prediction. Except this modification, we further replace the fully connected layers in VGG16 network with 1×1 convolutional layers [19], followed by a deconvolution layer (see Fig. 6.4). These 1×1 convolutional layers enable our network to run in a fully convolutional fashion.

A-Net: Appearance modeling network/S-Net: Semantic modeling network. After extracting the global shadow features with G-Net, we then design two parallel and complementary networks (A-Net and S-Net) to predict a fine shadow matte with multi-context features.

Table 6.1 The model architecture

	Layer	convs1	convs2	convs3	convs4	convs5	Decv2-1	Decv3-1
G-Net	# of convs	2	2	3	3	3	1	1
	# of channels	64	128	256	512	512	256	256
	Filter size	3 × 3	3 × 3	3 × 3	3 × 3	3 × 3	8 × 8	8 × 8
	Conv. stride	1	1	1	1	1	4	4
	Zero-Padding	1	1	1	1	1	2	2
	Pool-size	2 × 2	2 × 2	2 × 2	2 × 2	2 × 2	–	–
	Pool-stride	2	2	2	**1**	**1**	–	–
	Output size	112 × 112	56 × 56	28 × 28	28 × 28	28 × 28	112 × 112	112 × 112
	Layer	conv2-1 (conv3-1)	conv2-2 (conv3-2)	conv2-3 (conv3-3)	conv2-4 (conv3-4)	conv2-5 (conv3-5)	conv2-6 (conv3-6)	Decv2-2 (Decv3-2)
A-Net (S-Net)	# of channels	96	64	64	64	64	64	3
	Filter size	9 × 9	1 × 1	5 × 5	5 × 5	5 × 5	5 × 5	4 × 4
	Conv. stride	1	1	1	1	1	1	2
	Zero-Padding	4	–	2	2	2	2	1
	Pool-size	3 × 3	–	–	–	–	–	–
	Pool-stride	2	–	–	–	–	–	–
	Output size	112 × 112	112 × 112	112 × 112	112 × 112	112 × 112	112 × 112	224 × 224

It takes a shadow image of resolution $8n \times 8n$ as input and outputs a shadow matte of the same size, where n is an arbitrary natural number. In this table, we set $n = 56$ (i.e., input size of 224×224) for illustration

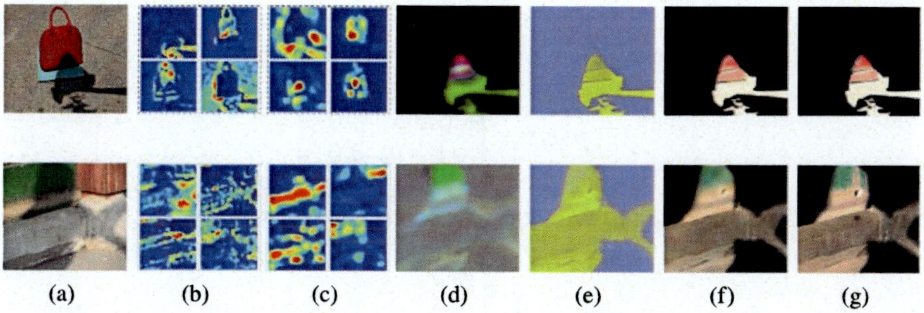

Fig. 6.6 Visualization of the intermediate results of the proposed network. **a** Shows the original shadow images and (**g**) shows the shadow mattes obtained from Eq. 6.3 and the original image pairs. **b** Shows some example feature maps of the Conv3 layer of G-Net that produce (**d**) via A-Net, which encode appearance information of the shadow regions. **c** Show some example feature maps of the Conv5 layer of G-Net that produce **e** via S-Net, which encode semantic information of shadow regions. The final predicted shadow mattes **f** embed multi-context information.© [2017] IEEE. Reprinted, with permission, from [Qu et al. "DeshadowNet: A Multi-context Embedding Deep Network for Shadow Removal," IEEE Conference on Computer Vision and Pattern Recognition, pp. 2308–2316, 2017]

In G-Net, each convolution block is followed by a max-pooling layer, thus each of them has a progressively larger receptive field. The deeper layers of G-Net are good at capturing high-level semantic context but poor for accurate localization due to the resulted coarse features. While the shallower layers, which capture more local appearance information, cannot inject contextual information into final prediction. To better localize the shadow regions and predict fine details for the shadow matte, we further design a multi-context mechanism for local detail refinement. With this mechanism, two levels of features are derived from the G-Net and transferred to the two parallel networks (i.e., A-Net and S-Net). Specifically, while the A-Net acquires the appearance information from the shallower layer of G-Net to help model the appearance of the shadow image with local image details, the S-Net extracts the semantic information from the deeper layer of G-Net to provide semantic understanding in the final prediction. These two networks are then integrated with a convolution layer.

Figure 6.6 shows some intermediate results of the proposed network. By feeding with the mid-level appearance information in Fig. 6.6b, A-Net predicts shadow matte in coarse scale but helps model the appearance of shadow matte (e.g., the color value of the bag and wall). On the other hand, S-Net predicts shadow matte with the guidance of high-level semantic context (e.g., semantic objects and shadows in Fig. 6.6c). It can predict shadow matte in fine scale (e.g., fine object boundary in Fig. 6.6e compared with Fig. 6.6d). These predicted intermediate shadow mattes demonstrate that the convolutional features in the shallower layer and deeper layer are complementary in predicting the final shadow matte. We will further analyze the effectiveness of different subnetworks in the experiment section.

To avoid the overfitting problem and achieve an optimal local minimum, drop-out is applied after each convolutional layer, and all the rectified linear units (ReLUs) in the networks are replaced with parametric rectified linear units (PReLUs) [13]. In contrast to ReLU, the coefficient of PReLU is adaptively learned and defined as:

$$p(x_i) = \begin{cases} x_i, & x_i \geq 0 \\ ax_i, & x_i < 0 \end{cases}, \tag{6.4}$$

where x_i is the input of the activation function p at channel i, and a is the learned parameter.

6.3.2.2 Training

The relationship of a shadow image I_s and its shadow matte S_m is given by Eq. 6.3. During training, we transform them into log space as:

$$\log(I_{ns}) = \log(S_m) + \log(I_s). \tag{6.5}$$

Given a shadow and shadow-free image pair, we first calculate the corresponding ground-truth shadow matte S_m according to Eq. 6.5. Then, our goal is to learn a mapping function that infers the relationship between a shadow image and its shadow matte as:

$$\log S_m = \log F(I_s, \Theta), \tag{6.6}$$

where Θ represents the learned parameters of the deep network. We adopt the mean squared error (MSE) as the loss function in the log space to train our model:

$$L(\Theta) = \frac{1}{K} \sum_{i=1}^{K} \left\| \log(F(I_s^i, \Theta)) - \log(S_m^i) \right\|, \tag{6.7}$$

where K is the total number of training samples in a batch. We minimize the loss using stochastic gradient descent (SGD) with back-propagation.

Training strategy. Although the performance improves significantly with the increase of the network depth, training a very deep network is a non-trivial task due to the instability of the gradient vanishing/exploiding problems [3]). In this chapter, we adopt the following four strategies for fast convergence and to prevent overfitting:

1. *Multi-stage training strategy.* We train DeshadowNet with two stages. The appearance and semantic streams (G-Net+A-Net and G-Net+S-Net) are first trained separately. These two streams are then connected with a convolution layer, and all three networks are jointly optimized.
2. *Multi-size training strategy.* The fully convolutional fashion of DeshadowNet enables our model to train on images of resolution $8n \times 8n$. To inject scale-invariance to the network [14], we adopt a multi-size training strategy by feeding

images of three sizes: coarse scale 64 × 64, medium scale 128 × 128, and fine scale 224 × 224.

3. *Data synthesis.* To prevent overfitting and improve the robustness of the network, we pretrain the proposed method on a large-scale synthetic shadow removal dataset. Similar to [11], we apply computer graphics techniques to synthesize shadow and shadow-free image pairs. We configure Maya with realistic light sources to project light on occluder objects, thus casting shadows on a projection plane. We have rendered 60,000 640 × 480 shadow/shadow-free image pairs by changing the light sources, occluder objects, and projection planes. We randomly change the shape and the position of the light source. There are 256 segmented objects in [10] that are used as the occluder objects. Finally, we collect more than 1000 real images (without shadows) from the Internet as the projection plane.
4. *Data augmentation.* We augment the training data with three different operations: image translations, flipping, and cropping.

Implementation. We have implemented DeshadowNet using Caffe [15]. All the networks described in this chapter are trained and tested on a single NVIDIA Tesla K40m. The proposed network takes 3–5 weeks of training to converge. The detailed configuration of DeshadowNet is shown in Table 6.1. The filter weights in A-Net and S-Net are initialized with random Gaussian variables (with mean value $\mu = 0$ and standard deviation $\sigma = 0.001$). We set the momentum to 0.9 and the weight decay to 0.0005 for training. The learning rate for G-Net is set to 10^{-5}. The learning rate for the rest of the network is set to 10^{-4}, and it is progressively decreased during training. In general, the proposed DeshadowNet is fast, and it takes only 0.3s to recover a shadow-free image of resolution 640 × 480.

6.3.3 Experiments

In this section, we extensively compare DeshadowNet with several state-of-the-art shadow removal methods on the publicly available UIUC dataset [12] and LRSS dataset [11], as well as on our proposed SRD dataset.

UIUC dataset [12]. It contains 76 shadow and shadow-free image pairs. We use all these images for testing.

LRSS dataset [11]. It contains 37 image pairs. It is specially designed to evaluate the performance on soft shadow removal. LRSS mainly contains shadows with diverse penumbra widths. We use all these images for testing.

SRD dataset. Our new dataset contains 3088 image pairs. We divide them into two parts randomly: 2680 for training and the remaining 408 images for testing.

Compared methods. We compare the proposed DeshadowNet with five state-of-the-arts methods: three automatic methods [12, 16, 27] and two interactive methods [9, 11] (requiring annotations of shadow and non-shadow regions). For fair comparison,

Table 6.2 Quantitative results using RMSE (smaller is better)

Dataset	Different reg.	Original	Guo 2013	Yang 2012	Gong 2014	Gryka 2015	Khan 2016	Ours
UIUC	Shadow	42	13.9	21.6	11.8	13.9	12.1	**9.6**
	Non-shadow	**4.6**	5.4	20.3	4.9	7.6	5.1	4.8
	All	13.7	7.4	20.6	6.6	9.1	6.8	**5.9**
LRSS	Shadow	44.45	31.58	23.35	22.27	–	–	**14.21**
	Non-shadow	**4.1**	4.87	19.35	4.39	–	–	4.17
	All	17.73	13.89	20.70	10.43	–	–	**7.56**
SRD	Shadow	42.38	29.89	23.43	19.58	–	–	**11.78**
	Non-shadow	**4.56**	6.47	22.26	4.92	–	–	4.84
	All	14.41	12.60	22.57	8.73	-	-	**6.64**

The original difference between the shadow and shadow-free images is reported in the third column. The best and second best results are marked in **red** and *blue* colors, respectively

we use either the publicly available source codes provided by the authors [9, 12, 27] or the quantitative/qualitative results reported in the papers [11, 16].[1] Note that [16] and [11] only report their performances on the UIUC dataset [12].

6.3.3.1 Performance Comparisons

Following the setting in [12] and [25], we adopt the root mean square error (RMSE) and structure similarity index (SSIM) in the LAB color space as evaluation metrics. While RMSE directly measures the per-pixel error between the recovered images and the ground-truth images, SSIM considers structural information, which is more consistent with human visual perception.

Tables 6.2 and 6.3 report the RMSE and SSIM values, respectively, of different shadow removal methods on the UIUC dataset [12], LRSS dataset [11], and the proposed SRD test dataset. We evaluate the performance of different methods on the shadow regions, non-shadow regions, and the whole image. We can see that the proposed DeshadowNet achieves the best performance among all the compared methods and datasets.

In Fig. 6.7, we show some qualitative shadow removal results from different methods, marked with their RMSE errors with respect to different regions (shadow, non-shadow, and whole image). The images in the first three rows (Fig. 6.7a, b, c) contain shadows cast on different semantic regions. Both Guo [12] and Gong [9] may perform well on a specific semantic region, but fail to remove the shadows on others. For example, in Fig. 6.7b, Guo [12] successfully removes the shadow on the ground, but fails on the box surface. As a contrast, the proposed DeshadowNet works well on these situations by integrating multi-context information and local image details.

[1] The shadow-free results of [11] are obtained from their project website: http://visual.cs.ucl.ac.uk/pubs/softshadows/.

Table 6.3 Quantitative results using SSIM (larger is better)

Dataset	Different reg.	Original	Guo 2013	Yang 2012	Gong 2014	Gryka 2015	Ours
UIUC	Shadow	0.6227	0.9228	0.8757	0.9551	0.9418	**0.9751**
	Non-shadow	**0.9861**	0.9811	0.9230	0.9839	0.9695	0.9859
	All	0.8975	0.9669	0.9114	0.9769	0.9627	**0.9832**
LRSS	Shadow	0.6194	0.7905	0.8814	0.8723	–	**0.9518**
	Non-shadow	0.9882	0.9813	0.9226	0.9863	–	**0.9888**
	All	0.8637	0.9169	0.9087	0.9478	–	**0.9763**
SRD	Shadow	0.5403	0.7381	0.8601	0.8695	–	**0.9487**
	Non-shadow	**0.9843**	0.9685	0.8735	0.9790	–	0.9823
	All	0.8687	0.9087	0.8700	0.9509	–	**0.9735**

The original difference between the shadow and shadow-free image is reported in the third column. The best and second best results are marked in **red** and *blue* colors, respectively

The images in Fig. 6.7d, e contain shadows of widely varying penumbra widths. It is difficult for existing methods [9, 12] to detect these shadows accurately. As shown in Fig. 6.7d, Guo [12] fails to detect the soft shadow. Even though the shadow is detected perfectly in [9] (through user annotation), the recovery of the shadow-free image is still unsatisfactory, due to the difficulty in automatically identifying umbra and penumbra regions from this image. Yang [27] can directly obtain a shadow-free image without shadow detection, but it also alters the colors of the non-shadow regions. Figure 6.7f, g shows results of different methods on more complicated situations, i.e., shadow cast on multiple semantic regions (e.g., brick, wall, and teddy bear) in Fig. 6.7f and the highly complex shadow in Fig. 6.7g. These quantitative and qualitative comparison results demonstrate that the proposed DeshadowNet can effectively recover high-quality shadow-free images from shadow images, even though with shadows cast on different semantic regions.

6.3.3.2 Component Analysis

Our DeshadowNet consists of three subnetworks, i.e., G-Net, A-Net, and S-Net. To further analyze the effectiveness and necessity of different subnetworks, we have trained three variant models of DeshadowNet and conducted a series of experiments on the UIUC shadow dataset [12]. These three variant models are: a model using S-Net only (or A-Net, since without the feature map of G-Net, A-Net is identical to S-Net in architecture), a model using G-Net and A-Net, and a model using G-Net and S-Net.

Table 6.4 shows the shadow removal performances of these three models on the UIUC dataset [12], measured in RMSE. We can see that none of the three models perform better than DeshadowNet. When removing G-Net, S-Net alone has relative poor performance and the RMSE error on the shadow regions reaches 14.2, compared

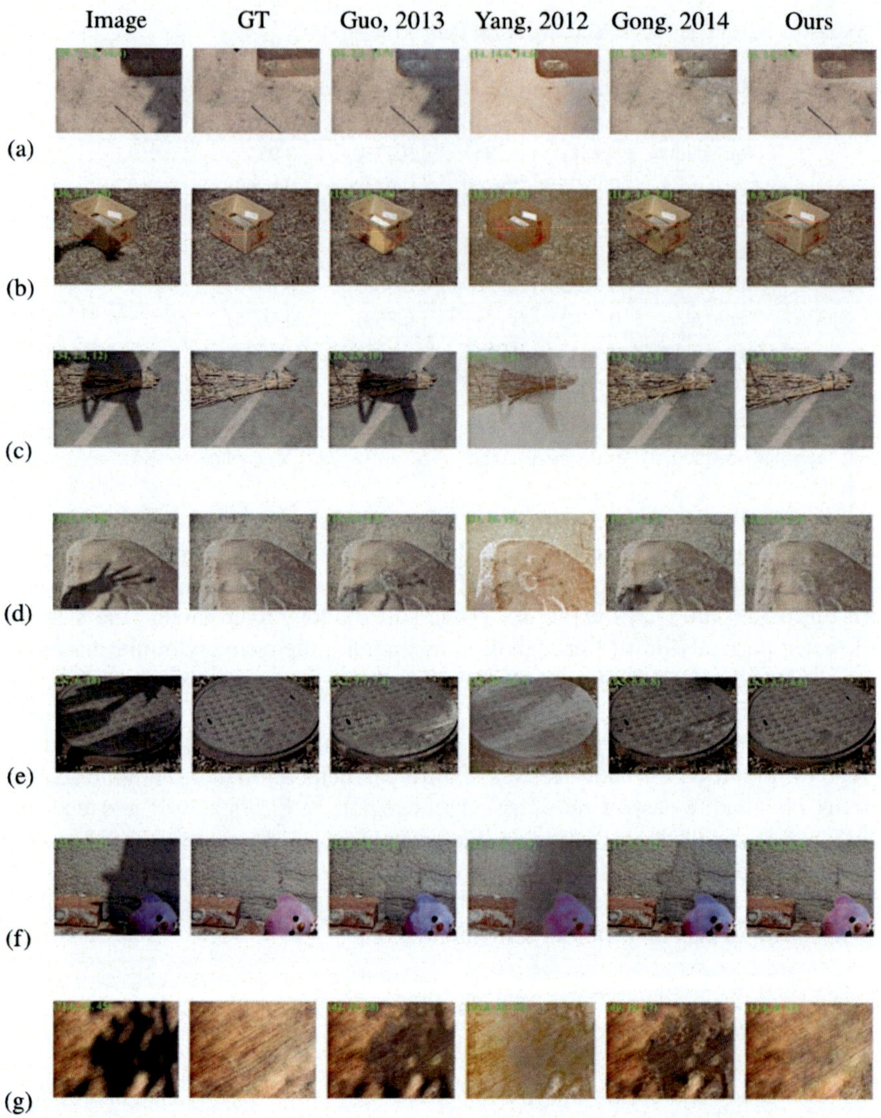

Fig. 6.7 Shadow removal results of different methods on images with different types of shadows. The RMSE errors with respect to different regions (shadow, non-shadow, and the entire image) are marked at the top-left hand corner of each image. © [2017] IEEE. Reprinted, with permission, from [Qu et al. "DeshadowNet: a Multi-context Embedding Deep Network for Shadow Removal," IEEE Conference on Computer Vision and Pattern Recognition, pp. 2308–2316, 2017]

Table 6.4 Effectiveness of different subnetworks in DeshadowNet (measured by RMSE on the UIUC dataset [12])

Regions	S-Net	G-Net+A-Net	G-Net+S-Net	DeshadowNet
Shadow	14.2	11.85	10.3	**9.6**
Non-shadow	5.85	5.04	4.82	**4.8**
All	7.90	6.7	6.2	**5.9**

(a) Shadow (b) S-Net (c) G+A (d) G+S (e) Final (f) Result

Fig. 6.8 Effectiveness of different subnetworks in DeshadowNet. From left to right: **a** shadow images, shadow mattes predicted by **b** S-Net, **c** G-Net+A-Net, **d** G-Net+S-Net, and **f** DeshadowNet; **g** the final shadow removal results by DeshadowNet. © [2017] IEEE. Reprinted, with permission, from [Qu et al. "DeshadowNet: A Multi-context Embedding Deep Network for Shadow Removal," IEEE Conference on Computer Vision and Pattern Recognition, pp. 2308–2316, 2017]

with 10.3 by G-Net+S-Net. This demonstrates the effectiveness of the embedding mechanism in our network, where G-Net provides the appearance information and semantic context for the A-Net and S-Net, respectively.

Figure 6.8 qualitatively compares these three models. Feeding with high-level semantic context and local image details, the G-Net+S-Net predicts more accurate shadow matte, as shown in Fig. 6.8d. On the other hand, G-Net+A-Net combines local image details with the mid-level appearance information from G-Net and predicts shadow matte in coarse scale but helps model the appearance of shadow matte (i.e., Fig. 6.8c obtains more accurate values in shadow matte than Fig. 6.8d but coarse segmentation). Thus in DeshadowNet, this multiple contextual information is incorporated for fine and accurate shadow matte prediction.

In this section, we have proposed an end-to-end model (DeshadowNet) to recover a shadow-free image from a single shadow image. Unlike the conventional pipeline that requires shadow detection or user annotations and removes shadows with handcrafted features, DeshadowNet unifies these steps into one and directly learns the mapping function between the shadow image and its shadow matte. It does not require a separate shadow detection step nor any post-processing refinement step. Thus, DeshadowNet is adaptive to shadows with widely varying penumbra widths and works well for shadows cast on different semantic regions.

The proposed multi-context embedding network, which integrates both high-level semantic context, mid-level appearance information and local image details, provides new insight for the research on low-level computer vision tasks.

6.4 Highlight Removal with Color-Lines Constraint

In this section, we derive a global color-lines constraint from dichromatic reflection model to effectively recover specular and diffuse reflection for textured images. We find that each image pixel lies along a color line in normalized RGB space and different color lines representing distinct diffuse chromaticities intersect at one point, namely the illumination chromaticity. By using two highlight invariant variables, we propose an efficient algorithm to cluster image pixels for obtaining color lines. We find that the pixels along the same color line spread over the entire image and their distances to the illumination chromaticity reflect the amount of specular reflection. With global (non-local) information from these color lines, we propose a method to completely separate specular reflection for textured images. The proposed method separates specular and diffuse reflection of an image in a pixel-wise manner, and it is suitable for real-time applications.

6.4.1 Global Color-Lines Constraint and Pixel Clusters

6.4.1.1 Dichromatic Reflection Model

According to dichromatic reflection model [20], a color image $\mathbf{I}(x)$ can be computed by the integration over the light spectrum:

$$\begin{aligned} \mathbf{I}(x) = \omega_d(x) \int \mathbf{q}(\lambda)\, S_d(\lambda, x)\, E(\lambda)\, \mathrm{d}\lambda \\ + \omega_s(x) \int \mathbf{q}(\lambda)\, S_s(\lambda, x)\, E(\lambda)\, \mathrm{d}\lambda, \end{aligned} \tag{6.8}$$

where $\mathbf{I} = [I_r, I_g, I_b]$ is the color vector of camera response or image intensity of a pixel, $x = [x, y]$ are the two dimensional image coordinates of a pixel, and $\mathbf{q} = [\mathrm{q}_r, \mathrm{q}_g, \mathrm{q}_b]$ is the three-element-vector of camera sensor sensitivity; $\omega_d(x)$ and $\omega_s(x)$ are the geometric scale factors of diffuse reflection and specular reflection, respectively; S_d and S_s are the spectral distribution function of diffuse reflection and specular reflection, respectively; $E(\lambda)$ is the spectral power distribution of illumination light.

Similar to [24], we define the diffuse chromaticity $\mathbf{\Lambda}(x)$ and the specular chromaticity $\mathbf{\Gamma}(x)$ as:

$$\mathbf{\Lambda}(x) = \frac{\int \mathbf{q}(\lambda) S_d(\lambda, x) E(\lambda) \, d\lambda}{\sum_{c\in\{r,g,b\}} \int q_c(\lambda) S_d(\lambda, x) E(\lambda) \, d\lambda}, \tag{6.9}$$

$$\mathbf{\Gamma}(x) = \frac{\int \mathbf{q}(\lambda) S_s(\lambda, x) E(\lambda) \, d\lambda}{\sum_{c\in\{r,g,b\}} \int q_c(\lambda) S_s(\lambda, x) E(\lambda) \, d\lambda}. \tag{6.10}$$

Then, the color intensity of a pixel becomes:

$$\mathbf{I}(x) = m_d(x)\mathbf{\Lambda}(x) + m_s(x)\mathbf{\Gamma}(x), \tag{6.11}$$

where

$$m_d(x) = \omega_d(x) \left(\sum_{c\in\{r,g,b\}} \int q_c(\lambda) S_d(\lambda, x) E(\lambda) \, d\lambda \right), \tag{6.12}$$

$$m_s(x) = \omega_s(x) \left(\sum_{c\in\{r,g,b\}} \int q_c(\lambda) S_s(\lambda, x) E(\lambda) \, d\lambda \right). \tag{6.13}$$

Here, $m_{\mathrm{d}}(x)$ and $m_s(x)$ are the diffuse and specular reflection coefficients, respectively, which depend on scene geometry at image pixels. The diffuse chromaticity $\mathbf{\Lambda} = [\Lambda_r, \Lambda_g, \Lambda_b]$ implies the intrinsic color of the surface, and it retains the same for uniform color surface. The specular reflection chromaticity $\mathbf{\Gamma} = [\Gamma_r, \Gamma_g, \Gamma_b]$ is usually considered as illumination chromaticity. In most cases, the illumination chromaticity can be assumed to be uniform for a given image such that $\Gamma_r = \Gamma_g = \Gamma_b = \frac{1}{3}$. In the real world, illumination chromaticity is not pure white, and this can be achieved by normalizing the real illumination chromaticity estimated by Tan et al. [24] as a preprocessing.

In our model, chromaticity of a color is defined as:

$$\widetilde{\mathbf{I}}(x) = \frac{\mathbf{I}(x)}{\sum_{c\in\{r,g,b\}} \mathrm{I}_c(x)}, \tag{6.14}$$

where $\mathbf{I}(x)$ is normalized by dividing the summation of its three channels. Substituting (6.11) into (6.14), we obtain

$$\widetilde{\mathbf{I}}(x) = \frac{m_d(x)}{\sum_{c\in\{r,g,b\}} \mathrm{I}_c(x)} \mathbf{\Lambda}(x) + \frac{m_s(x)}{\sum_{c\in\{r,g,b\}} \mathrm{I}_c(x)} \mathbf{\Gamma}(x). \tag{6.15}$$

In general, chromaticities $\mathbf{\Gamma}(x)$ and $\mathbf{\Lambda}(x)$ are normalized, i.e., $\sum_{c\in\{r,g,b\}} \Lambda_c = 1$ and $\sum_{c\in\{r,g,b\}} \Gamma_c = 1$. As a result, the sum of color intensities for a pixel can be written as $\sum_{c\in\{r,g,b\}} \mathrm{I}_c(x) = m_d(x) + m_s(x)$. Thus, (6.15) can be rewritten as:

$$\widetilde{\mathbf{I}}(x) = \frac{m_d(x)}{m_d(x) + m_s(x)}\mathbf{\Lambda}(x) + \frac{m_s(x)}{m_d(x) + m_s(x)}\mathbf{\Gamma}(x). \tag{6.16}$$

Based on (6.16), most of the specular reflection removal methods first seek for an appropriate diffuse chromaticity $\mathbf{\Lambda}(x)$ within a local patch and then propagate it within the patch to recover $m_s(x)$ or $m_d(x)$. Due to a lack of global information, these methods are accurate in separating the reflection components when only the exact diffuse chromaticities are given .

6.4.1.2 Global Color-Lines Constraint

For simplicity, we set $\frac{m_d(x)}{m_d(x)+m_s(x)} = \alpha(x)$, and rewrite (7.17) as:

$$\begin{aligned}\widetilde{\mathbf{I}}(x) &= \alpha(x)\mathbf{\Lambda}(x) + (1 - \alpha(x))\mathbf{\Gamma}(x) \\ &= \alpha(x)(\mathbf{\Lambda}(x) - \mathbf{\Gamma}(x)) + \mathbf{\Gamma}(x).\end{aligned} \tag{6.17}$$

In (6.17), illumination chromaticity $\mathbf{\Gamma}(x)$ is fixed for a color image, and $\alpha(x)$ can be considered as a matting coefficient between diffuse chromaticity $\mathbf{\Lambda}(x)$ and illumination chromaticity $\mathbf{\Gamma}(x)$. The pixels with the same diffuse chromaticity $\mathbf{\Lambda}(x)$ make (6.17) a three-dimensional line, and (6.17) is a line set for an image with various diffuse chromaticities. In recent literature, the authors of [4, 6] also use color lines for haze removal, but their color lines are different from ours in highlight removal model.

Fig. 6.9a, d is color distributions of synthetic image (b) and natural image (e), respectively, in the normalized color space, and (c) is the ground truth of synthetic image (b). As we discussed above, the pixels with the same $\mathbf{\Lambda}(x)$ fit a line, which is well-represented in (a). The highlight image (b) has two chromaticities: blue one and red one, and they distribute along two lines shown in (a). We extend the two lines and calculate their intersection, namely the illumination chromaticity. The estimated illumination for (b) is [0.342, 0.342, 0.316], and the real one is $\left[\frac{1}{3}, \frac{1}{3}, \frac{1}{3}\right]$. The pixels of natural image (e) cluster together in (d), and each cluster in (d) represents a chromaticity of (e). We fit a line for each cluster and calculate the intersection of these lines. The red five-pointed star is the intersection, and its value is [0.370, 0.335, 0.295]. The illumination chromaticity estimated by Tan et al. [24] is [0.370, 0.340, 0.290], and it is almost the same as ours. By calculating the intersection of color lines, we provide an accurate approach to estimate the illumination chromaticity of a highlight image, and this illumination chromaticity is helpful for separating specular reflection.

Pixels in (a) and (d) which are far away from illumination chromaticity are diffuse chromaticities of images, whereas the ones near the illumination chromaticity are highlight pixels. According to highlight pixels' distances to the illumination chromaticity, we can calculate their specular reflection components. For (e), pixels belonging to one cluster marked red triangles are depicted in (d), and they spread over the entire image in RGB space. Therefore, a color line can capture the global information that is not limited to small patches.

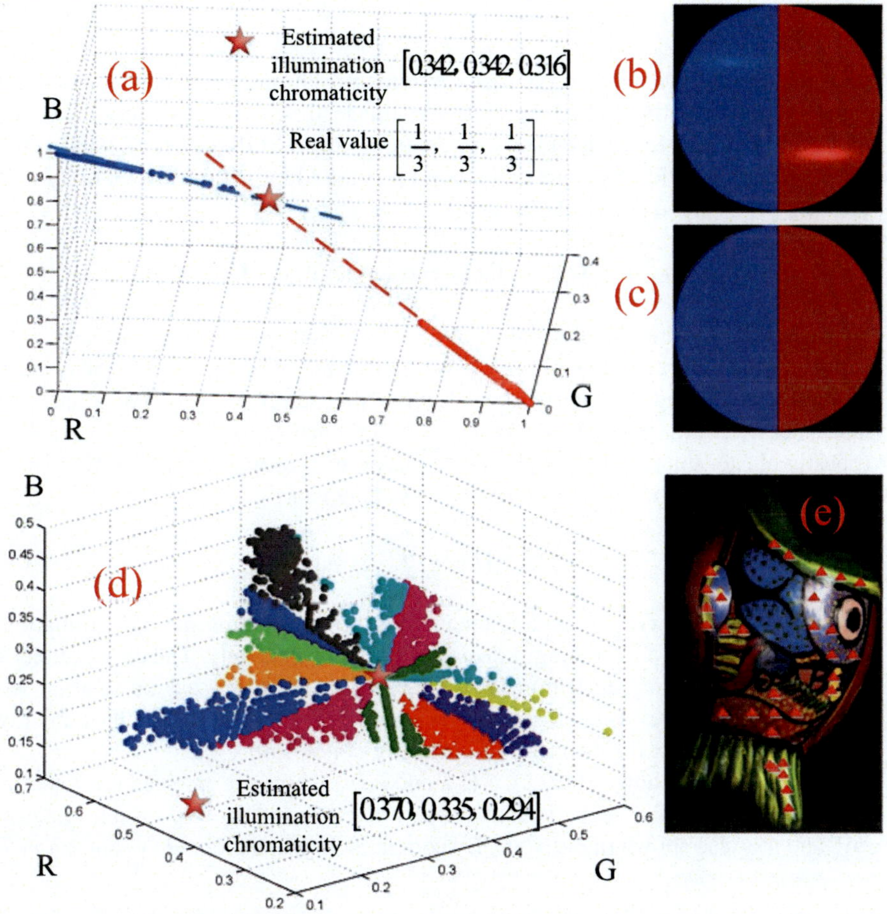

Fig. 6.9 Global color-lines constraint for images. **a** and **d** are color distributions of synthetic image **b** and natural image **e**, respectively, in the normalized color space, and **c** is the ground truth of synthetic image (**b**). Highlight image **b** has two chromaticities: blue one and red one, and they distribute along two lines shown in (**a**). The pixels of natural image (**e**) cluster together in (**d**), and each cluster in (**d**) represents a chromaticity of (**e**). We thus can fit a line for each cluster. By calculating the intersection of color lines, we provide an approach to estimate the illumination chromaticity for a highlight image. For (**e**), pixels belonging to one cluster marked red triangles are depicted in (**d**), and they are spread over the entire image in RGB space. © [2017] IEEE. Reprinted, with permission, from [Ren et al. "Specular Reflection Separation with Color-Lines Constraint," IEEE Transactions on Image Processing, Vol. 26, Issue: 5, pp: 2327–2337, 2017]

6.4.1.3 Clusters of Image Pixels

In 6.4.1.2, we have introduced that illumination chromaticity can be estimated through the color lines fitted by clustered image pixels. In this section, we will explain how to cluster image pixels. In the normalized RGB space, all the pixels of an image locate inside a sphere, and its center is the illumination chromaticity $\mathbf{\Gamma}(x)$ (see Eq. 6.17). Let us define $\widetilde{\mathbf{I}}_d(\mathbf{x})$ as:

$$\widetilde{\mathbf{I}}_d(\mathbf{x}) = \widetilde{\mathbf{I}}(x) - \mathbf{\Gamma}(x) = \alpha(x)(\mathbf{\Lambda}(x) - \mathbf{\Gamma}(x)), \tag{6.18}$$

where $\widetilde{\mathbf{I}}_d(\mathbf{x})$ is a directional vector to the illumination chromaticity. Initially, $\mathbf{\Gamma}(x)$ is unknown, and we set it as $\left[\frac{1}{3}, \frac{1}{3}, \frac{1}{3}\right]$. We then refine the illumination chromaticity $\mathbf{\Gamma}(x)$ by calculating the intersection of color lines (Sect. 6.4.1.2). For a particular chromaticity $\mathbf{\Lambda}(x)$, the distance $\left\|\widetilde{\mathbf{I}}_d(x)\right\|_2$ is only determined by $\alpha(x)$. In order to robustly cluster image pixels, we transfer Cartesian coordinate system (RGB space) into spherical coordinate system, and express $\widetilde{\mathbf{I}}_d(\mathbf{x})$ as:

$$\widetilde{\mathbf{I}}_d(x) = [r(x), \theta(x), \varphi(x)], \tag{6.19}$$

where $r(x) = \left\|\widetilde{\mathbf{I}}_d(x)\right\|_2$ is the Euclidean distance to the illumination chromaticity, and $\theta(x)$ and $\varphi(x)$ are the longitude and latitude, respectively. The chromaticity of an image pixel is only determined by $\theta(x)$ and $\varphi(x)$, and it has no relation with distance $r(x)$. However, the distance $r(x)$ determines a pixel's specular reflection. Please see Appendix for the proof of the above conclusion.

In Fig. 6.10, we show $1 - r(x)$, $\theta(x)$, and $\varphi(x)$ which are rescaled for visualization. Hue in HSI color space is usually considered as highlight invariant image [2, 26], and it is also shown in Fig. 6.10. By comparison, we find that $\theta(x)$ and $\varphi(x)$ are

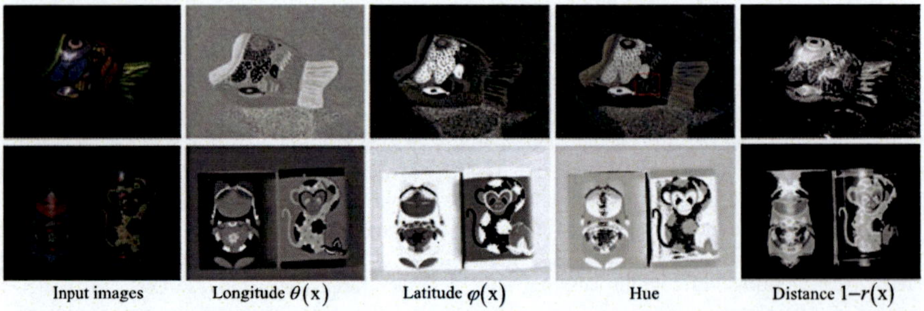

Fig. 6.10 Invariant variables for specular reflection. Longitude $\theta(x)$ and latitude $\varphi(x)$ are less affected by specular reflection than hue. The hue is sensitive to image noise and sometimes degrades the textures of an image. Distance $r(x)$ is a metric to measure how far the distance is between the position x and illumination chromaticity, and it can reflect the specular component of an image.© [2017] IEEE. Reprinted, with permission, from [Ren et al. "Specular Reflection Separation with Color-Lines Constraint," IEEE Transactions on Image Processing, Vol. 26, Issue: 5, pp: 2327–2337, 2017]

less affected by specular reflection than Hue. The Hue is sensitive to image noise and sometimes degrades the textures of an image. Distance $r(x)$ is a metric to measure how far the distance is between the position x and illumination chromaticity, and it can reflect the specular component of an image. The angles $\theta(x)$ and $\varphi(x)$ are invariant to specular reflection, and we thus use $\mathbf{\Lambda}^{P}(x) = [\theta(x), \varphi(x)]^{\mathrm{T}}$ to cluster image pixels. The chromaticity distance between two pixels x and y is defined with l_1 norm:

$$d_{\Lambda}(\mathrm{x,y}) = \left\| \mathbf{\Lambda}^{P}(x) - \mathbf{\Lambda}^{P}(\mathrm{y}) \right\|_1. \tag{6.20}$$

If the distance between two pixels is less than a threshold T, they belong to the same cluster; otherwise, they are classified into different clusters. After each pixel is assigned a label, we calculate mean values of all the clusters. Using the mean values as initial values, we employ the K-nearest neighbor (KNN) search rule to re-cluster image pixels. Algorithm 6.3 shows the process of pixel clustering, and the only parameter in it is T. Since $\theta(x)$ and $\varphi(x)$ are in the range $[0, 2\pi] \times [0, \pi]$, we set $T \in \left[\frac{\pi}{6}, \frac{\pi}{3}\right]$ for the experiments in this paper. Figures 6.11 and 6.12 show the pixel clustering results for two highlight images in chromaticity space. Yang's method [26] employs a region-growing algorithm to locate adjacent pixels with similar chromaticities, and their clustering results can capture the local details. However, the smooth regions are inclined to be impaired (see b). Shen's method [21] uses maximum and minimum chromaticity to globally cluster image pixels, but textured regions are often degraded (see c). With global (non-local) information, our results shown in (d) preserve texture details well and smooth the homogeneous regions completely.

Algorithm 6.3 Cluster image pixels

Input: Highlight image $\mathbf{I}(x)$, threshold T
Output: Labels matrix $\mathbf{L}(x)$, number of labels N
Initialize: Convert $\mathbf{I}(x)$ into $\Lambda^{\mathrm{P}}(x)$ through (2.3), (2.7), (6.19), $N \leftarrow 1, \mathbf{L}(x_1) \leftarrow N$, cluster center $m_N \leftarrow \Lambda^{P}(x_1)$
1. **For** unlabelled pixel x_i
2. $d_{\Lambda}^{n}(x_i, m_n) = \left\| \mathbf{\Lambda}^{P}(x) - m_n \right\|_1, 1 \le n \le N$;
3. $n_{\min} = \arg\min_{1 \le n \le N} d_{\Lambda}^{n}$;
4. **if** $d_{\Lambda}^{n_{\min}} \le T$;
5. $\mathbf{L}(x_i) \leftarrow n_{\min}$, update $m_{n_{\min}}$;
6. **else**
7. $N \leftarrow N + 1, \mathbf{L}(x_1) \leftarrow N, m_N \leftarrow \Lambda^{P}(x_i)$;
8. **end**
9. Search nearest neighbor for each x_i in $\{m_n | 1 \le n \le N\}$;

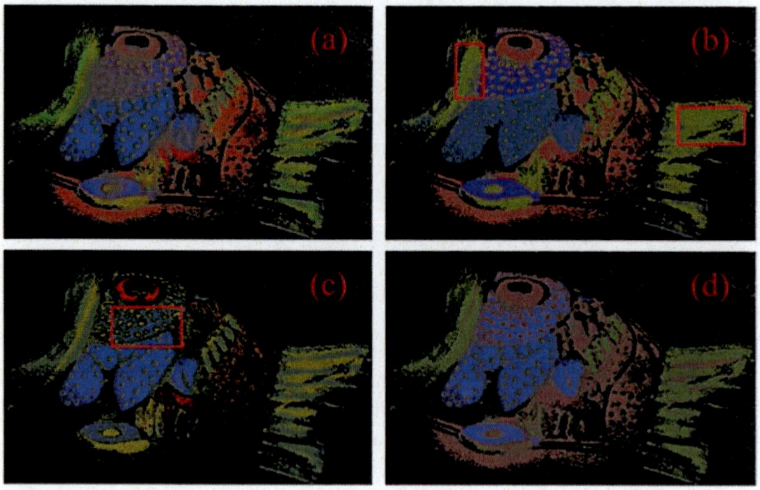

Fig. 6.11 Pixel clustering results of *Fish* image using different methods. **a** Chromaticity of input image, **b** clustering result of Yang et al. [26], **c** clustering result of shen et al.[21], **d** our result. © [2017] IEEE. Reprinted, with permission, from [Ren et al. "Specular Reflection Separation with Color-Lines Constraint," IEEE Transactions on Image Processing, Vol. 26, Issue: 5, pp: 2327–2337, 2017]

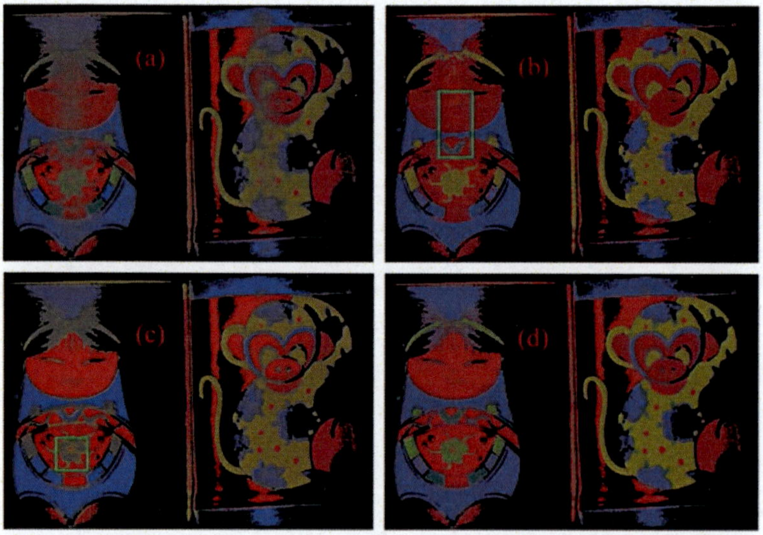

Fig. 6.12 Pixel clustering results of *Cups* image using different methods. **a** Chromaticity of input image, **b** clustering result of Yang et al. [26] **c** clustering result of shen et al. [21] **d** our result. © [2017] IEEE. Reprinted, with permission, from [Ren et al. "Specular Reflection Separation with Color-Lines Constraint," IEEE Transactions on Image Processing, Vol. 26, Issue: 5, pp: 2327–2337, 2017]

6.4.2 Separation of Specular Reflection

Our reflection separation algorithm is mainly composed of three steps. First, we transform the input image, and cluster image pixels using the highlight invariant variables $\theta(x)$ and $\varphi(x)$. Then, we fit color lines for pixels clusters and estimate the illumination chromaticity. According to each pixel's distance to illumination chromaticity, we finally separate specular reflection in a pixel-wise manner. The overall algorithm is presented in Algorithm 6.4. The first two steps are discussed in Sects. 6.4.1.2 and 6.4.1.3, and we here introduce how to separate specular reflection.

Through the first two steps of our algorithm, we obtain color lines and illumination chromaticity $\boldsymbol{\Gamma}(x)$. We express $\widetilde{\mathbf{I}}_d(\mathbf{x}) = \widetilde{\mathbf{I}}(x) - \boldsymbol{\Gamma}(x)$ and convert it into spherical coordinate system to obtain distance $r(x)$ (Eq. 6.19). For a given color line (pixels cluster), $r(x)$ depends on $\alpha(x)$:

$$r(x) = \alpha(x)\,\|\Lambda(\mathbf{x}) - \boldsymbol{\Gamma}(\mathbf{x})\|_2, 0 \le \alpha(x) \le 1. \tag{6.21}$$

Since $\alpha(x) = \frac{m_d}{m_d+m_s}$, $\alpha(x) = 1$ represents the diffuse chromaticity of the clustered image pixels. Meanwhile, $\alpha(x) = 1$ also corresponds to the pixel which has the farest distance $r_{\max}(x)$,

$$r_{\max}(x) = \max_{x\in C}\{r(x)\}, \tag{6.22}$$

where the estimation is done per pixels cluster C. Based on (6.21) (6.22), the matting coefficient $\alpha(x)$ for each pixel can be estimated in a pixel-wise manner:

$$\alpha(x) = \frac{r(x)}{r_{\max}(x)}. \tag{6.23}$$

To suppress image noise, the final diffuse coefficient $\alpha(x)$ is filtered with a 5×5 median filter. In real calculation, to avoid specular reflection being over-separated, we replace $r_{\max}(x)$ in (6.23) with

$$r_{\text{median}}(x) = \underset{x\in C}{\text{median}}\{r(x)\}. \tag{6.24}$$

According to (6.17), the diffuse component $\widetilde{\text{D}}(x)$ in chromaticity space can be obtained:

$$\widetilde{\mathbf{D}}(x) = \widetilde{\mathbf{I}}(x) - (1 - \alpha(x))\,\boldsymbol{\Gamma}(x). \tag{6.25}$$

We transform $\widetilde{\text{D}}(x)$ into RGB space based on (6.14), and obtain diffuse component:

$$\mathbf{D}(x) = \widetilde{\mathbf{D}}(x) \sum\nolimits_{c\in\{r,g,b\}} \mathbf{I}_c(x). \tag{6.26}$$

Thus, the specular reflection is obtained by subtracting diffuse reflection:

$$\mathbf{S}(x) = \mathbf{I}(x) - \mathbf{D}(x). \tag{6.27}$$

Algorithm 6.4 Algorithm for separating specular reflection

Input: Highlight image $\mathbf{I}(x)$

Output: Diffuse reflection $\mathbf{D}(x)$, specular reflection $\mathbf{S}(x)$

1. Cluster image pixels: Algorithm 3;
2. Fit color lines and estimate illumination chromaticity $\mathbf{\Gamma}(x)$;
3. Calculate distances $r(x)$ to $\mathbf{\Gamma}(x)$: (2.3) (2.7) (6.19);
4. Estimate the farest $r_{\max}(x)$ for each pixels cluster;
5. Estimate diffuse coefficient $\alpha(x)$: (2.12);
6. Separating specular reflection: (6.25) (6.26) (6.27);

6.4.3 Experiments

In the experiments, we compare our results with three recent methods proposed by Yang et al. [26], Shen et al. [21], and Yang [28]. For quantitative comparison, we first employ four test images to evaluate the reflection separation performance using peak signal-to-noise ratio (PSNR). Though human visual system is insensitive to color changes in large areas, it is sensitive to color changes at edges. Instead, the PSNR measures image difference on a pixel-wise basis and treats all pixels equally, and it is a complement for evaluating highlight removal performance.

Before we evaluate our method, we first analyze the performance of the proposed method with respect to parameter T in Algorithm 6.3. We conducted experiments on four images with ground-truth diffuse reflection, and Fig. 6.13 shows the effect of threshold T in terms of PSNR. It is observed that, for four images (*Ball*, *Cups*, *Masks*,

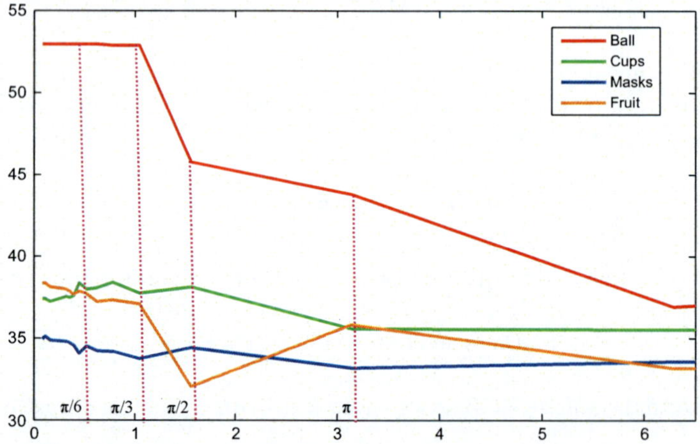

Fig. 6.13 Effect of threshold T in terms of PSNR for four images: *Ball*, *Cups*, *Masks*, and *Fruit*. It is observed that the changes of PSNR values are almost unanimous along the T direction. When $T \in [\frac{\pi}{6}, \frac{\pi}{3}]$, the PSNR values for the four images remain unchanged. When $T \geq \frac{\pi}{3}$, the PSNR values are on downward trends.© [2017] IEEE. Reprinted, with permission, from [Ren et al. "Specular Reflection Separation with Color-Lines Constraint," IEEE Transactions on Image Processing, Vol. 26, Issue: 5, pp: 2327–2337, 2017]

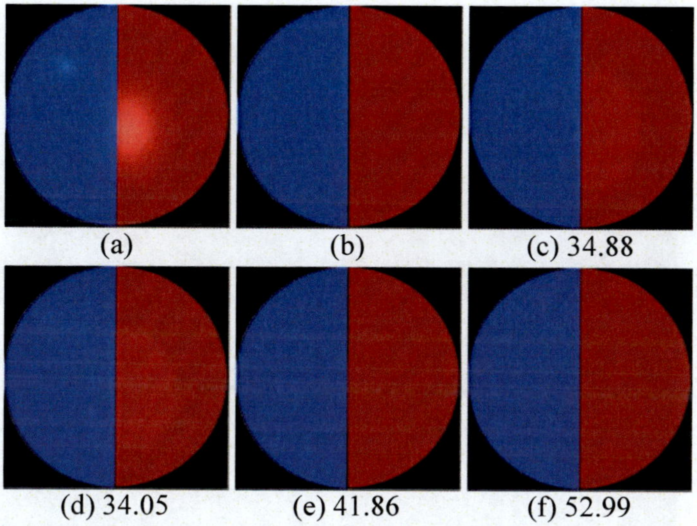

Fig. 6.14 Diffuse components of synthetic image *Ball* recovered by different methods. The PSNR value for each method is in the bottom of its corresponding separation result. **a** Input synthetic image, **b** ground truth, **c** diffuse component of Yang et al. [26], **d** diffuse component of Shen et al. [21], **e** diffuse component of Yang et al.[28], **f** ours. © [2017] IEEE. Reprinted, with permission, from [Ren et al. "Specular Reflection Separation with Color-Lines Constraint," IEEE Transactions on Image Processing, Vol. 26, Issue: 5, pp: 2327–2337, 2017]

and *Fruit*), the changes of PSNR values are almost unanimous for $T \in \left[\frac{\pi}{6}, \pi\right]$. When $T \in \left[\frac{\pi}{6}, \frac{\pi}{3}\right]$, the PSNR values for the four images remain unchanged. When $T \geq \frac{\pi}{3}$, the PSNR values are on downward trends. Intuitively, if the threshold T is small, the number of pixels clusters will increase. Image textures are thus well-segmented and preserved. However, the smaller T does not mean the better visual appearance of an image, and it is because a small T may lead to the incomplete removal of specular reflection. Similarly, if T is large, the number of pixels clusters will decrease. Though all the specular reflection in an image is separated, the chromaticities of diffuse component are inclined to be dark and distorted. In our experiments, we set $T = \frac{\pi}{6}$ for images *Ball*, *Cups*, *Masks*, and *Fruit* to quantitatively evaluate the specular reflection separation performance. For the other images, we set $T = \frac{\pi}{3}$ to completely remove the specular reflection.

Figure 6.14 shows the separation results for a synthetic image which is generated by two hemispheres rendered with two colors, respectively. The PSNR values are shown at the bottom of separation results for different methods. It is observed that the methods of Yang et al. [26] and Shen et al. [21] fail to remove all the specular reflection and thus produce low PSNR values. The diffuse image by Yang et al. [28] is relatively better, but it impairs the edge between the two hemispheres. Our method produces good visual ability and the highest PSNR value. Figure 6.15 is a real-world image with multicolored surfaces. The results of Shen et al. [21] and

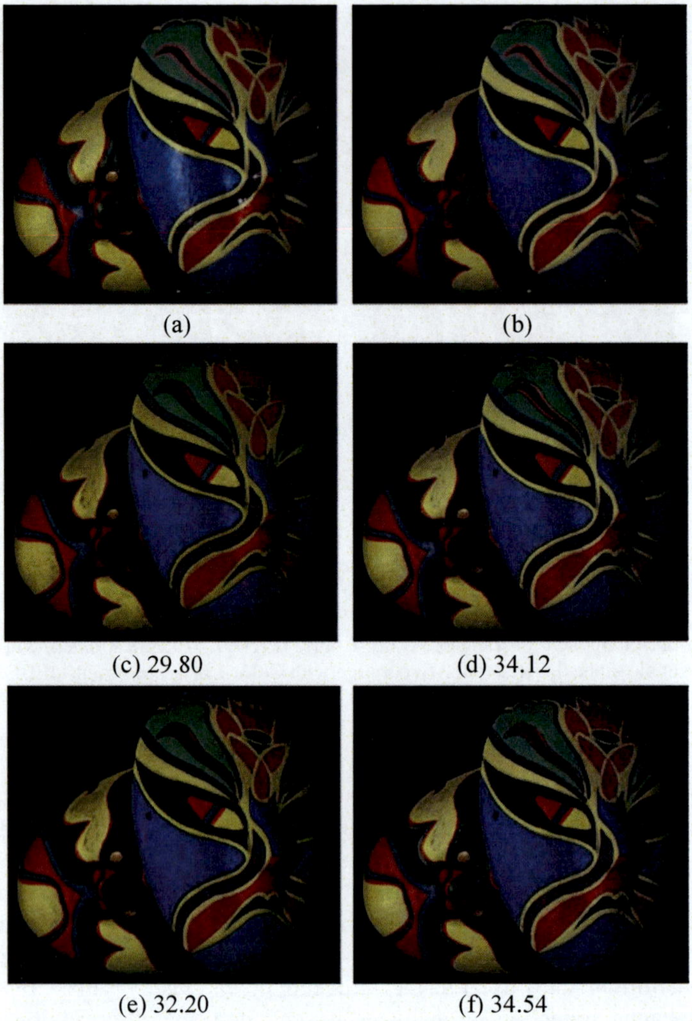

Fig. 6.15 Diffuse components of real-world image *Masks* recovered by different methods. The PSNR value for each method is in the bottom of its corresponding separation result. **a** Input synthetic image, **b** ground truth, **c** diffuse component of Yang et al. [26], **d** diffuse component of Shen et al.[21], **e** diffuse component of Yang et al. [28], **f** ours. © [2017] IEEE. Reprinted, with permission, from [Ren et al. "Specular Reflection Separation with Color-Lines Constraint," IEEE Transactions on Image Processing, Vol. 26, Issue: 5, pp: 2327–2337, 2017]

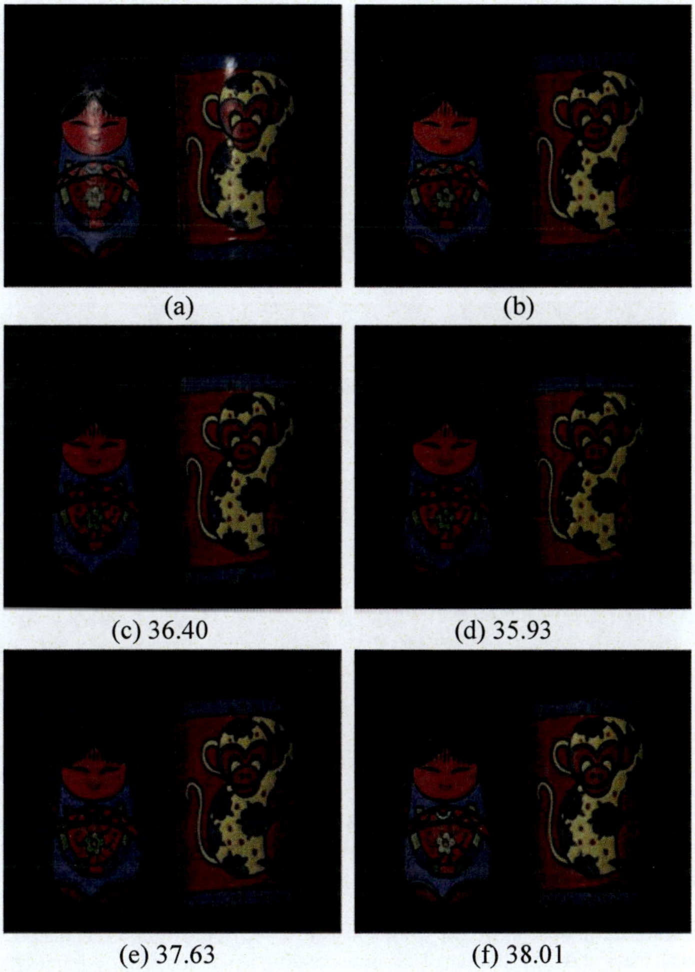

Fig. 6.16 Diffuse components of real-world image *Cups* recovered by different methods. The PSNR value for each method is in the bottom of its corresponding separation result. **a** Input synthetic image, **b** ground truth, **c** diffuse component of Yang et al. [26], **d** diffuse component of Shen et al. [21], **e** diffuse component of Yang et al.[28], **f** ours. © [2017] IEEE. Reprinted, with permission, from [Ren et al. "Specular Reflection Separation with Color-Lines Constraint," IEEE Transactions on Image Processing, Vol. 26, Issue: 5, pp: 2327–2337, 2017]

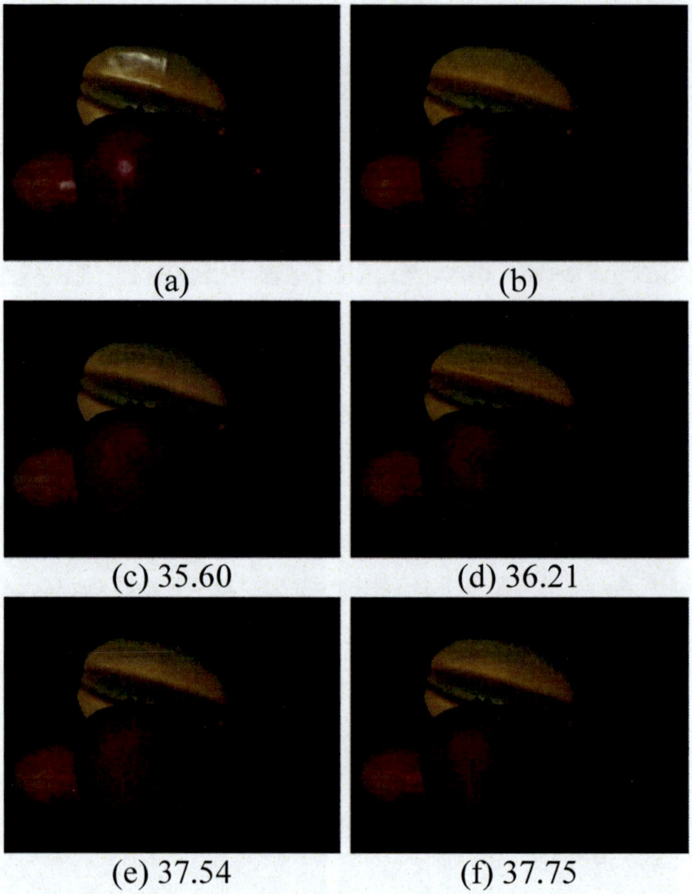

Fig. 6.17 Diffuse components of real-world image *Fish* recovered by different methods. The PSNR value for each method is in the bottom of its corresponding separation result. **a** Input synthetic image, **b** ground truth, **c** diffuse component of Yang et al. [26], **d** diffuse component of Shen et al. [21], **e** diffuse component of Yang et al. [28], **f** ours.© [2017] IEEE. Reprinted, with permission, from [Ren et al. "Specular Reflection Separation with Color-Lines Constraint," IEEE Transactions on Image Processing, Vol. 26, Issue: 5, pp: 2327–2337, 2017]

Yang et al. [28] have obvious discontinuities in the mask's face. Yang's method [26] removes too much specular reflection and thus produces a low PSNR value. Figure 6.16 shows the separated diffuse reflections of two cup surfaces that contain textures, and Fig. 6.17 shows the separation results on the *Fruit* image, which has various shapes of highlight on individual surfaces. All the methods have almost the same performance in Figs. 6.16 and 6.17 from the visual appearances, but the proposed method achieves the highest PSNR values. The above results validate that our method performs better than other state-of-the-art methods under the quantitative measure.

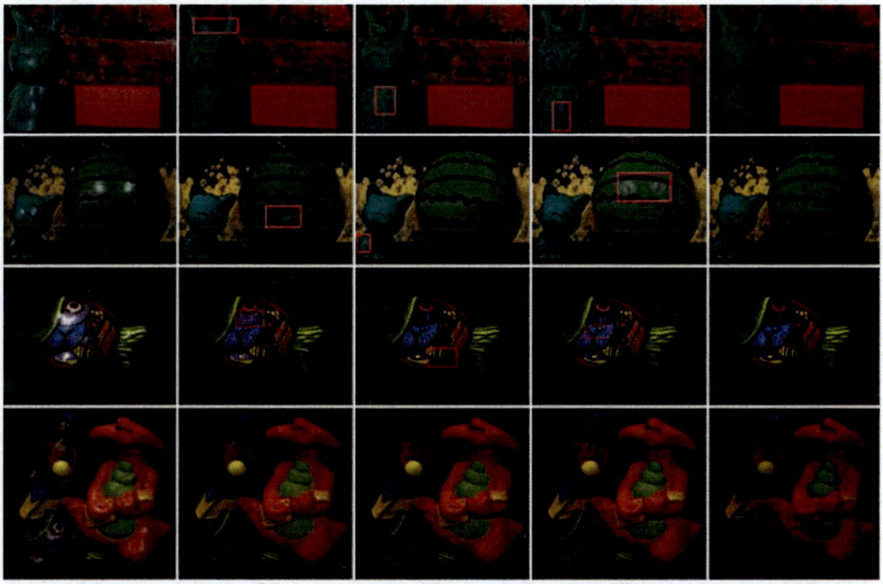

Fig. 6.18 Highlight removal results for natural scenes: *Rabbit*, *Watermelon*, *Fish* and *Toys*. **a** Input highlight images, **b** diffuse components of Yang et al. [26], **c** diffuse components of Shen et al. [21], **d** diffuse components of Yang et al. [28], **e** ours. © [2017] IEEE. Reprinted, with permission, from [Ren et al. "Specular Reflection Separation with Color-Lines Constraint," IEEE Transactions on Image Processing, Vol. 26, Issue: 5, pp: 2327–2337, 2017]

We also conduct highlight removal on four other scenes, namely *Rabbit*, *Watermelon*, *Fish*, and *Toys*. *Rabbit* and *Watermelon* in the first two rows of Fig. 6.18 were used in previous work [21], and they are illuminated by two fluorescent lamps and two LEDs. The last two rows in Fig. 6.18 are *Fish* and *Toys*, which were used in previous work [23]. As there are no ground truths, we compare the visual appearances of recovered diffuse images by different methods. Yang's method [26] employs a region-growing algorithm to locate adjacent pixels with similar diffuse chromaticities, and it often causes discontinuities in the separation results. The small highlight spots on *Rabbits* and *Fish* are not removed, and there is an obvious boundary in *Watermelon*. The method of Shen et al. [21] performs well to remove highlight for textured image *Fish*, but it cannot remove all the specular reflection for images *Rabbit* and *Watermelon* with smooth regions. Yang et al. [28] use bilateral filter to smooth the maximum fraction of the color components of the original image for removing the noise contributed by the specular pixels. Their method works effectively to remove highlight in a small region, but it fails to remove the large highlight spots on the surface of *Watermelon*. Furthermore, the highlight on the *Rabbit* and the *Fish* is not completely removed. With global chromaticity information, our method can completely remove highlight spots for smooth regions, and it outperforms other methods to preserve image textures. The specular reparation results for *Toys* in the last row of

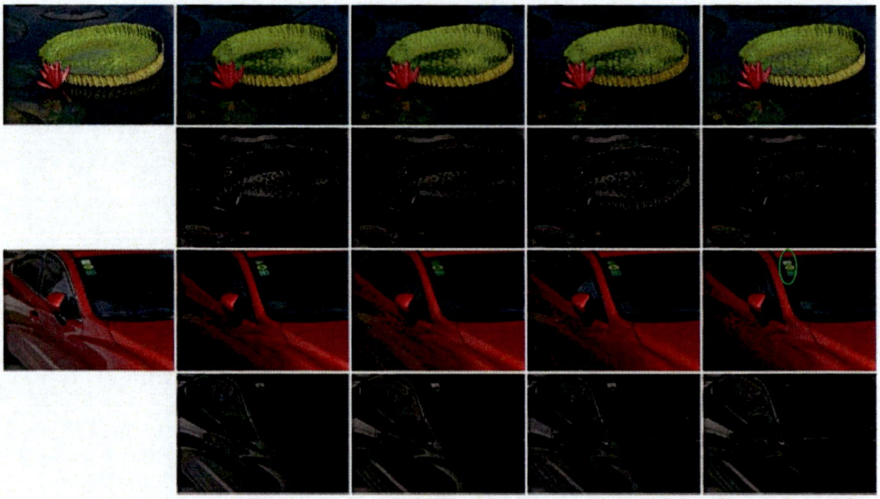

Fig. 6.19 Visual comparison for two real images: *Lotus* and *Car* taken under natural illumination. **a** Input images, **b** separation results of Yang et al. [26], **c** separation results of Shen et al.[21], **d** separation results of Yang et al.[28], **e** ours.© [2017] IEEE. Reprinted, with permission, from [Ren et al. "Specular Reflection Separation with Color-Lines Constraint," IEEE Transactions on Image Processing, Vol. 26, Issue: 5, pp: 2327–2337, 2017]

Fig. 6.18 show that all the methods can obtain visually satisfactory results for images with a small amount of specular reflection.

Figure 6.19 shows the visual comparison for two real images: *Lotus* and *Car* taken under natural illumination. The methods in [21, 26, 28] can remove all the specular reflection of the two images, but they degrade some tiny details, e.g., textures in the lotus leaf and labels in the car's window. With global (non-local) chromaticity information obtained from color lines, our method can separate specular reflection clearly and preserve tiny details well.

In Table 6.5, we compare the running time of different methods for ten highlight images. As we can see, since our method separates specular reflection in a pixel-wise manner and needs no iterative process, its computational complexity is lower than the methods in [26, 28], and analogous to [21]. Thus, it is suitable for real-time applications, e.g., stereo matching for specular surfaces.

In this section, we derived a global color-lines constraint from dichromatic reflection model to robustly recover specular and diffuse reflection. Our key observation is that each image pixel lies along a color line in the normalized RGB space and different color lines representing distinct diffuse chromaticities intersect at one point, namely the illumination chromaticity. By using two highlight invariant variables, we first proposed an algorithm to efficiently cluster image pixels for obtaining color lines. Pixels along the same color line spread over the entire image, and a color line can thus capture global chromaticity information that is not limited to small image

Table 6.5 Running time (in seconds) of different methods

Image	Size (pixels)	[26]	[21]	[28]	Ours
Ball	300×300	0.277	0.012	0.034	**0.008**
Masks	500×450	7.437	0.053	0.088	**0.048**
Cups	640×480	11.520	0.054	0.095	**0.051**
Rabbit	377×279	1.817	0.029	0.082	**0.023**
Watermelon	379×255	2.187	**0.024**	0.030	0.027
Fish	640×480	16.623	0.045	0.075	**0.037**
Toys	353×387	6.391	0.036	0.076	**0.031**
Lotus	600×349	8.225	0.069	0.226	**0.061**
Car	640×480	6.382	0.054	0.176	**0.052**
Fruit	640×480	6.673	0.068	0.081	**0.051**

patches. According to each pixel's distance to illumination chromaticity in a color line, we then separated specular reflection in a pixel-wise manner, without specular pixel identification and iterative process. Experimental results on synthetic and real images show that our method performs better than the state-of-the-art methods in terms of both specular reflection removal accuracy and computational efficiency.

Appendix

Proof of Eq. 6.19

According to (6.17), we define the chromaticity of pixels x_1 and x_2 as follows ($\mathbf{\Gamma}$ is fixed for an image):

$$\widetilde{\mathbf{I}}(x_1) = \alpha(x_1)(\mathbf{\Lambda}(x_1) - \mathbf{\Gamma}) + \mathbf{\Gamma}, \tag{6.28}$$

$$\widetilde{\mathbf{I}}(x_2) = \alpha(x_2)(\mathbf{\Lambda}(x_2) - \mathbf{\Gamma}) + \mathbf{\Gamma}. \tag{6.29}$$

Through (6.18), we first obtain $\widetilde{\mathbf{I}}_d(x_1)$ and $\widetilde{\mathbf{I}}_d(x_2)$:

$$\widetilde{\mathbf{I}}_d(x_1) = \alpha(x_1)(\mathbf{\Lambda}(x_1) - \mathbf{\Gamma}), \tag{6.30}$$

$$\widetilde{\mathbf{I}}_d(x_2) = \alpha(x_2)(\mathbf{\Lambda}(x_2) - \mathbf{\Gamma}). \tag{6.31}$$

Then, we transfer $\widetilde{\mathbf{I}}_d(x_1)$ and $\widetilde{\mathbf{I}}_d(x_2)$ into spherical coordinate system:

$$\begin{cases} \theta\left(x_1\right) = \arctan \dfrac{\Lambda_g\left(x_1\right) - \Gamma_g}{\Lambda_r\left(x_1\right) - \Gamma_r}, \\ \theta\left(x_2\right) = \arctan \dfrac{\Lambda_g\left(x_2\right) - \Gamma_g}{\Lambda_r\left(x_2\right) - \Gamma_r}, \\ \psi\left(x_1\right) = \arctan \dfrac{\Lambda_b\left(x_1\right) - \Gamma_b}{\sqrt{\left(\Lambda_r\left(x_1\right) - \Gamma_r\right)^2 + \left(\Lambda_g\left(x_1\right) - \Gamma_g\right)^2}}, \\ \psi\left(x_2\right) = \arctan \dfrac{\Lambda_b\left(x_2\right) - \Gamma_b\left(x_2\right)}{\sqrt{\left(\Lambda_r\left(x_2\right) - \Gamma_r\right)^2 + \left(\Lambda_g\left(x_2\right) - \Gamma_g\right)^2}}, \\ r\left(x_1\right) = \alpha\left(x_1\right)\sqrt{\left(\Lambda_r\left(x_1\right) - \Gamma_r\right)^2 + \left(\Lambda_g\left(x_1\right) - \Gamma_g\right)^2 + \left(\Lambda_b\left(x_1\right) - \Gamma_b\right)^2}, \\ r\left(x_2\right) = \alpha\left(x_2\right)\sqrt{\left(\Lambda_r\left(x_2\right) - \Gamma_r\right)^2 + \left(\Lambda_g\left(x_2\right) - \Gamma_g\right)^2 + \left(\Lambda_b\left(x_2\right) - \Gamma_b\right)^2}. \end{cases} \tag{6.32}$$

From (6.32), if pixels in an image have the same diffuse chromaticity $\boldsymbol{\Lambda}$, they must have equal longitude θ and latitude φ no matter how much specular components they have. In other words, θ and φ represent the chromaticity of an image. Distance r is proportional to α for a fixed chromaticity $\boldsymbol{\Lambda}$, and it determines the specular reflection of the image.

References

1. E. Arbel, H. Hel-Or, Shadow removal using intensity surfaces and texture anchor points. IEEE Trans. Pattern Anal. Mach. Intell. **33**(6), 1202–1216 (2010)
2. A. Artusi, F. Banterle, D. Chetverikov, A survey of specularity removal methods. Comput. Graph. Forum **30**(8), 2208–2230 (2011)
3. Y. Bengio, P. Simard, P. Frasconi, Learning long-term dependencies with gradient descent is difficult. IEEE Trans. Neural Netw. **5**(2), 157–166 (1994)
4. D. Berman, T. Treibitz, S. Avidan, Non-local image dehazing, in *2016 IEEE Conference on Computer Vision and Pattern Recognition (CVPR)*. (IEEE 2016), pages 1674–1682
5. D. Eigen, C. Puhrsch, R. Fergus, Depth map prediction from a single image using a multi-scale deep network, in *Advances in neural information processing systems* (2014), pages 2366–2374
6. R. Fattal, Dehazing using color-lines. ACM Trans. Graph. **34**(1), 1–14 (2014)
7. G.D. Finlayson, S.D. Hordley, C. Lu, M.S. Drew, On the removal of shadows from images. IEEE Trans. Pattern Anal. Mach. Intell. **28**(1), 59–68 (2005)
8. L.A. Gatys, A.S. Ecker, M. Bethge, Image style transfer using convolutional neural networks, in *Proceedings of the IEEE Conference on Computer Vision and Pattern Recognition*, (2016), pages 2414–2423
9. H. Gong, D. Cosker, Interactive shadow removal and ground truth for variable scene categories, in *BMVC* (2014)
10. G. Griffin, A. Holub, P. Perona, Caltech-256 object category dataset (2007)
11. M. Gryka, M. Terry, G.J. Brostow, Learning to remove soft shadows. ACM Trans. Graph. (TOG) **34**(5), 1–15 (2015)
12. R. Guo, Q. Dai, D. Hoiem, Paired regions for shadow detection and removal. IEEE Trans. Pattern Anal. Mach. Intell. **35**(12), 2956–2967 (2013)

13. K. He, X. Zhang, S. Ren, J. Sun, Delving deep into rectifiers: Surpassing human-level performance on imagenet classification, in *Proceedings of the IEEE International Conference on Computer Vision* 2015a, pages 1026–1034
14. K. He, X. Zhang, S. Ren, J. Sun, Spatial pyramid pooling in deep convolutional networks for visual recognition. IEEE Trans. Pattern Anal. Mach. Intell. **37**(9), 1904–1916 (2015)
15. Y. Jia, E. Shelhamer, J. Donahue, S. Karayev, J. Long, R. Girshick, S. Guadarrama, T. Darrell. Caffe: convolutional architecture for fast feature embedding, in *Proceedings of the 22nd ACM international conference on Multimedia* (2014), pages 675–678
16. S.H. Khan, M. Bennamoun, F. Sohel, R. Togneri, Automatic shadow detection and removal from a single image. IEEE Trans. Pattern Anal. Mach. Intell. **38**(3), 431–446 (2016)
17. J.-F. Lalonde, A.A. Efros, S.G, Narasimhan, Detecting ground shadows in outdoor consumer photographs. In *European conference on computer vision*. (Springer, 2010), pages 322–335
18. F. Liu, M. Gleicher, Texture-consistent shadow removal, in *European Conference on Computer Vision* (Springer, 2008), pages 437–450
19. J. Long, E. Shelhamer, T. Darrell, Fully convolutional networks for semantic segmentation, in *Proceedings of IEEE CVPR* (2015), pages 3431–3440
20. S.A. Shafer, Using color to separate reflection components. Color. Res. Appl. **10**(4), 210–218 (1985)
21. H.L. Shen, Z.H. Zheng, Real-time highlight removal using intensity ratio. Appl. Opt. **52**(19), 4483–4493 (2013)
22. K. Simonyan and A. Zisserman, Very deep convolutional networks for large-scale image recognition. *arXiv preprint*arXiv:1409.1556 (2014)
23. R.T. Tan, K. Ikeuchi, Separating reflection components of textured surfaces using a single image. IEEE Trans. Pattern Anal. Mach. Intell. **27**(2), 178–193 (2005)
24. R.T. Tan, K. Nishino, K. Ikeuchi, Illumination chromaticity estimation using inverse-intensity chromaticity space, in *2003 IEEE Computer Society Conference on Computer Vision and Pattern Recognition, 2003. Proceedings*, volume 1 (IEEE, 2003), pages 1–8
25. Y. Xiao, E. Tsougenis, C.-K. Tang. Shadow removal from single rgb-d images, in *Proceedings of the IEEE Conference on Computer Vision and Pattern Recognition* (2014), pages 3011–3018
26. J. Yang, L. Liu, S.Z. Li, Separating specular and diffuse reflection components in the hsi color space, in *IEEE International Conference on Computer Vision Workshops* (2013), pages 891–898
27. Q. Yang, K.-H. Tan, N. Ahuja, Shadow removal using bilateral filtering. IEEE Trans. Image Process. **21**(10), 4361–4368 (2012)
28. Q. Yang, J. Tang, N. Ahuja, Efficient and robust specular highlight removal. IEEE Trans. Pattern Anal. Mach. Intell. **37**(6), 1304–11 (2015)

Chapter 7
Rain and Snow Removal

Abstract Raining and snowing have negative effects on the visual quality and degrade the performance of various vision algorithms in outdoor environment. In this chapter, we propose three rain and snow removal algorithms including two for videos and one for single images. We first introduce an algorithm to remove snowflakes in videos by using the global and local low-rank property; i.e., assume the snowflake is sparse in snowflake-removed scenes. The proposed framework contains two steps of low-rank decomposition. The first step is to extract stationary background with the whole frames, and the second step focuses on local blocks obtained by patch-matching for removing snowflakes in front of moving foreground. Then, we introduce a rain and snow removal algorithm based on matrix decomposition no longer with the assumption that all the snowflakes/rain streaks are sparse in snow/rain scenes. We divide snowflakes and rain streaks into two categories: sparse ones and dense ones. With background fluctuations and optical flow information, the detection of moving objects and sparse snowflakes/rain streaks is formulated as a multi-label Markov Random Fields (MRFs), making the rain and snow removal model adaptable to different rain and snow scenes. Finally, we propose an encoder–decoder deraining net for single-image rain removal in heavy scenes, which is only supervised by MSE loss and adopts dually skip connections for rain removal in one stage. The dual connections, skip sum connection and skip concatenation connection, enable the deraining net to promote information flow between layers, and thus can help to localize rain streaks. In addition, the learnable pixel-wise attention can adaptively recalibrate the decoded rain features to preserve more image details.

7.1 Introduction

Raining and snowing have negative effects on the visual quality and degrade the performance of various vision algorithms such as object detection [38], video tracking [2] and segmentation [42]. In order to make the robot vision system, such as autonomous driving, drone flying, be more robust in bad weather, it is necessary to model and remove rain and snow from an image/video, but heavy rain/snow presents various properties (e.g., directions, shapes, and transparencies), making rain/snow removal

J. Tian, *All Weather Robot Vision*, Research on Intelligent Manufacturing,
https://doi.org/10.1007/978-981-16-6429-8_7

a challenging problem. In recent years, researchers have proposed many algorithms for removing rain and snow. The literature deals with rain and snow mostly from the video, and the information redundancy of adjacent frames in the video helps the detection and removal of rain and snow. A pioneering work on detecting and removing rain streaks in a video was proposed in [18], where the authors suggested a correlation model capturing the dynamics of rain and a physics-based motion blur model characterizing the photometric properties of rain. Then they proposed a method to remove rain effects based on the observation that some camera parameters, such as exposure time and depth of field, can be selected to mitigate the effects of rain without altering the appearance of the scene [19]. Early research works usually assumed that rain streaks have spatial–temporal correlations and mainly focused on the removal of rain streaks in video sequences by exploiting low-rank representation [9, 31, 46] or by analyzing in frequency space [1]. With sparse representation and dictionary learning, some works [7, 39] also formulated single-image rain removal as a matrix decomposition problem where rain streaks can be removed using the sparse prior. Actually, rain streaks in a heavy rain scene do not always have uniform directions or particular patterns, which leads to the failure of the above methods using spatial or temporal correlations. Kim et al. [31] proposed a video deraining and desnowing method based on optical flow and low-rank matrix completion. Their method can handle videos taken from a moving camera well, but it cannot remove all snowflakes in a scene. Sometimes it also has deformation and artifacts on motion objects with large displacement because they do not explicitly detect the moving objects. Many snow/rain removal methods often fail for heavy snow/rain and dynamic scene. The reason is that the existing methods often cannot differentiate moving objects and snowflakes/rain streaks. In Sect. 7.1, we introduce a snowflake removal algorithm based on the global low-rank characteristics of the scene background and the local low-rank characteristics of the matching moving objects. In Sect. 7.2, we introduce a rain and snow removal algorithm based on matrix decomposition. There are four layers, namely the background layer, the sparse rain and snow layer, the dense rain and snow layer and the moving object layer, making the rain and snow removal model adaptable to different rain and snow scenes. Experimental results show that our proposed algorithms perform better than the state-of-the-art methods for snow and rain removal.

Researchers also try to remove rain/snow from single images. A single-image decomposition-based rain removal method was proposed by [29] and [26]. Rain streaks are detected using a method based on morphological component analysis (MCA) and dictionary learning. It separates a rain image into low- and high-frequency parts using a bilateral filter to get the initial set of candidate rain streaks. Recent methods mainly adopted deep learning technique for single-image rain removal. [56] first improved the traditional linear rain model and then developed a recurrent deep learning net to simultaneously learn the binary rain streak map, the appearance of rain streaks, and the clean background. The recurrent mechanism works well for heavy rain but may cause over deraining. [16] found that rain mainly appears in the high frequency of a rainy image. Thus, they first decomposed the image into base layer and detail layer by using a low-pass filter and then applied a CNN to remove rain streaks in the detail layer. To remove rain with different densities, [58] presented a multi-stream densely connected convolutional neural network to estimate rain-density information and then to remove the corresponding rain streaks by incorporating the rain-density label. Though this method can remove rain streaks with a certain density, it may fail for heavy rain scenes containing various of densities at different scene depths. Since heavy rain scene has different types of rain streaks, and it is not easy to remove all the rain streaks in one stage. Therefore, [33] divided the rain removal into multiple stages and proposed a recurrent squeeze-and-excitation net [25] for single-image rain removal. The recurrent mechanism can clearly remove all the rain, but it also causes the degradation of image edges. Inspired by the non-local means filter for image restoration [4, 12, 22, 30], introduced an enhanced encoder–decoder framework using non-local neural networks [53] for single-image rain removal. Depending on deep learning technique and synthetic rain dataset, recent single-image rain removal methods either use a recurrent mechanism to gradually learn the mapping between clear images and rainy images or focus on designing various loss functions to supervise the learning process. One limitation of these deep learning-based methods is that they are trained on particular rain density while often need to derain the image in heavy rain scenes containing different types of rain streaks. Though the recurrent mechanism can clearly remove the rain streaks, it inclines to cause artifacts. In Sect. 7.4, we propose a dually connected deraining net using pixel-wise attention, for single-image rain removal in one stage. In contrast to previous works, our deraining net is dually connected from shallow layers to deep ones for accurately localizing different types of rain streaks. Besides, we add pixel-wise attention to the encoded rain features for preserving image details.

7.2 Snowflake Removal for Videos via Global and Local Low-Rank Decomposition

The research works on eliminating dynamic weather conditions in the literature have been mainly focused on rain removal, and few papers discuss the snowflake removal. A method for rain removal usually is not directly suitable for snow removal. For example, the photometric model in [18] is established by regarding the raindrops as lines and assuming that all raindrops have the same size and fall at almost the same velocity relative to the camera. However, snowflakes have dynamic characteristics that their directions, intensities, and shape are unfixed. Although raindrops and snowflakes have some common characteristics, falling snow has its own features that should be carefully considered. Compared with raindrops, snowflake is not transparent and the motion velocity is slow. Thus, we first develop a specific and effective snowflake removal approach.

In this section, we propose a snowflake removal method based on low-rank decomposition. As shown in Fig. 7.1, our model takes both global and local low-rank properties into consideration for stationary background and moving foreground extraction, respectively. We also consider the features of snowflakes, such as their white color, small size, and the increase limitation in pixel values from background to snowflake (please see the Appendix for details), for better separating snowflakes from moving foreground.

7.2.1 Problem Model

Let $F = \{f_k\}_{k=1}^K$ be the video sequences with K frames, which are corrupted by falling snow. As demonstrated by Fig. 7.1, we model each frame f_k as a combination of stationary background, moving foreground, and the falling snow,

$$f_k = st_k + mv_k + sn_k. \tag{7.1}$$

Generally, $ST = \{st_k\}_{k=1}^K$ describes stationary background free from the occlusions by the moving foreground $MV = \{mv_k\}_{k=1}^K$ and the snowflakes $SN = \{sn_k\}_{k=1}^K$. For snow removal, MV should be also extracted from the corruption by SN. By making each frame of the video into one column vector and aligning them in one matrix, we have $I = [vec(f_1)|\cdots|vec(f_K)]$, $S = [vec(st_1)|\cdots|vec(st_K)]$, $M = [vec(mv_1)|\cdots|vec(mv_K)]$ and $E = [vec(sn_1)|\cdots|vec(sn_K)]$. Following the descriptions mentioned above, all column vectors in S have similar latent image structures, so that it should be a low-rank matrix. The matrix M will be also low rank after a proper transformation. Since the falling snow in one video generally occludes only a small fraction of each frame of the video, we can regard E as a sparse matrix. As a result, the snow removal problem can be viewed as a robust PCA problem:

$$\begin{aligned} &\min_{S,M,E} rank(S) + \lambda_1 \text{rank}(\eta(M)) + \lambda_2 \|E\|_0 \\ &s.t. \quad I = S + M + E. \end{aligned} \tag{7.2}$$

Here, the operator $\eta()$ denotes some transformations on local components of videos, such as translations and rotations, to align every same moving component appeared in each frame of the video. It makes the possibility to extract the latent similar structures of dynamic foreground.

Different from previous work [31] that models the snowflake removal as a completion problem (which means they think that the snowflake occluded pixels lose all their information), we treat the snowflake occluded pixels as a combination of snowflake and its background. We model it as a decomposition problem based on its physical property [18] (which shows that the snowflake polluted pixel values are the weighted summation of background and following snowflake. Please refer to the first paragraph in the Appendix of this paper for details).

Since the rank and the ℓ_0-norm are nonconvex and discontinuous, the above problem (7.2) is NP hard and cannot be directly tractable. A large number of literatures

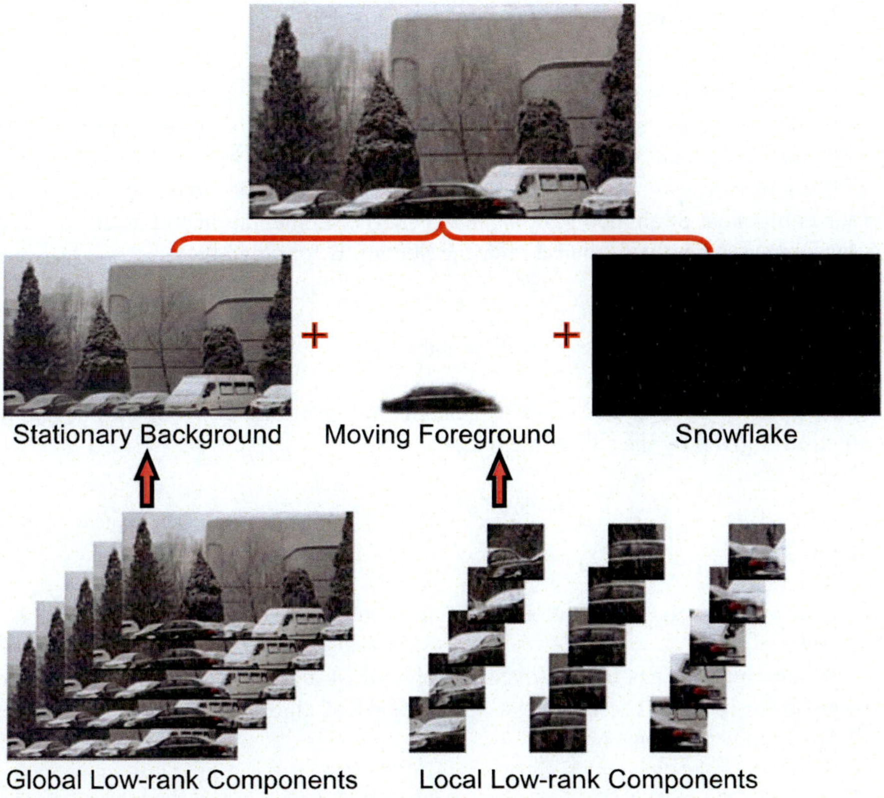

Fig. 7.1 Illustration of the general pipeline of our snowflake removal method via low-rank decomposition. The main idea is to decompose an input video into stationary background, moving foreground, and snowflakes by the two steps of low-rank decomposition: (1) extracting stationary background with global low-rank components in whole frames; (2) extracting moving foreground with local low-rank decomposition on patch matched components. In this step, snowflake features such as their white color, small size, and the increase limitation in pixel values from background to snowflake are considered to separate the falling snow and the moving objects. Finally, the snow removed video is generated by combining the stationary background and the moving foreground. © [2018] IEEE. Reprinted, with permission, from [Jiandong Tian, Zhi Han, Weihong Ren, Xiai Chen, and Yandong Tang, "Snowflake Removal for Videos via Global and Local Low-rank Decomposition," IEEE Transactions on Multimedia, Vol.20, Issue: 10, pp. 2659–2669, 2018]

have proved that the nuclear norm is a relatively tight convex relaxation of the rank and the recover results are fairly nice [5, 6, 24]. Here we use the nuclear norm and the ℓ_1-norm to replace the original rank and ℓ_0-norm, then we obtain the following optimization problem:

$$\begin{aligned} &\min_{S,M,E} \|S\|_* + \lambda_1 \|\eta(M)\|_* + \lambda_2 \|E\|_1 \\ &s.t. \quad I = S + M + E. \end{aligned} \tag{7.3}$$

7.2.2 *Algorithm*

It is difficult to get the solution of (7.3) in one step. In this paper, the solving procedure is divided into two steps: 1) the stationary background S is firstly extracted as the low-rank component of I by the augmented Lagrangian method (ALM) algorithm [35]; 2) from $I - S$, the moving foreground M is secondly extracted as the low-rank component of aligned regions, in which, the segmentation and patch-matching algorithms are involved (please refer the Section B for details).

7.2.2.1 Stationary Background Extraction

The stationary background S is a low-rank component of the given video I, and the solution is given by the subproblem as:

$$\begin{aligned} &\min_{S,L} \|S\|_* + \lambda \|L\|_1 \\ &s.t. \quad I = S + L. \end{aligned} \tag{7.4}$$

Here, L contains the information of moving foreground and falling snow (i.e., $L = M + E$).

In literatures, it has been proved that the ALM algorithm can solve the convex problem fairly well. Therefore, we adopt the ALM algorithm to solve the problem (7.4). It can be written as:

$$\begin{aligned} \mathcal{L}(S, L, Y, \mu) = &\|S\|_* + \lambda\|L\|_1 \\ &+ \langle Y, I - S - L\rangle + \tfrac{\mu}{2} \|I - S - L\|_F^2 , \end{aligned} \tag{7.5}$$

where Y is the Lagrange multiplier, μ is a positive scalar.

The ALM algorithm iteratively estimates both the Lagrange multiplier and the optimal solutions by iteratively minimizing the Lagrange function. Here we minimize the function against only one of the two unknowns S and L at a time because of the difficulty in solving it directly:

$$\begin{aligned} S_{i+1} &= \arg\min_S(S, L_i, Y_i), \\ L_{i+1} &= \arg\min_L(S_{i+1}, L, Y_i), \\ Y_{i+1} &= Y_i + \mu_i \mathcal{L}(S_{i+1}, L_{i+1}). \end{aligned} \tag{7.6}$$

For spelling out the solutions, we apply the singular value threshold (SVT) algorithm, which is proposed in [5] for solving the nuclear norm penalized problems. For $\tau \geq 0$, the singular value shrinkage operator D_τ is defined by

$$[D_\tau(A)]_{\mathrm{ij}} = \mathrm{sgn}(A_{ij})(\left|A_{ij}\right| - \tau)_+ .$$

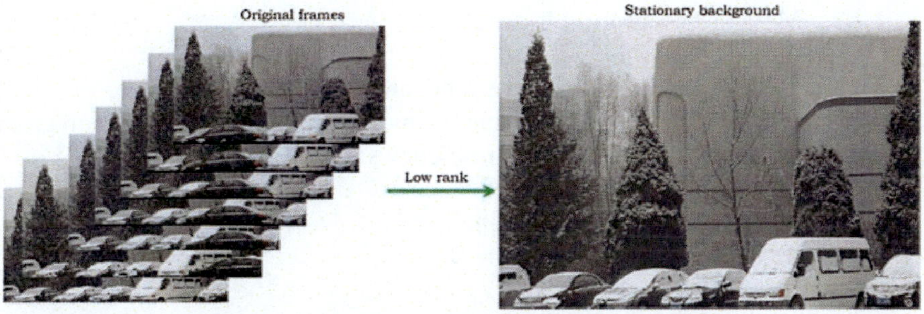

Fig. 7.2 Stationary background extraction result. © [2018] IEEE. Reprinted, with permission, from [Jiandong Tian, Zhi Han, Weihong Ren, Xiai Chen, and Yandong Tang, "Snowflake Removal for Videos via Global and Local Low-rank Decomposition," IEEE Transactions on Multimedia, Vol.20, Issue: 10, pp. 2659–2669, 2018]

By using the SVT, the solutions to each step of (7.6) can be written as follows:

$$\begin{aligned} &(U, \Sigma, V) = \text{svd}(I - L_i + \mu_i^{-1} Y_i), \\ &S_{i+1} = U D_{\mu_i^{-1}}[\Sigma] V^T, \\ &L_{i+1} = D_{\lambda\mu_i^{-1}}[I - S_{i+1} + \mu_i^{-1} Y_i]. \end{aligned} \tag{7.7}$$

where svd$(\cdot)$ denotes the singular value decomposition operator.

The above derivation is based on the assumption that the videos are captured by static cameras. But it is easy to extend to motion background by employing pre-processing alignment methods, e.g., [61] to compensate for the background motion caused by camera moving. The entire algorithm for stationary background extraction is summarized as Algorithm 1, and an example of extracted stationary background is shown in Fig. 7.2.

Algorithm 7.1 Stationary Background Extraction

Input:$I = [vec(f_1)| \cdots |vec(f_K)] \in R^{N\times K}$, Y_0, S_0, L_0;
$\lambda > 0, \mu_0 > 0, \rho > 1$
while not converged do
$(U, \Sigma, V) = svd(I - L_i + \mu_i^{-1} Y_i)$;
$S_{i+1} = U D_{\mu_i^{-1}}[\Sigma] V^T$;
$L_{i+1} = D_{\lambda\mu_i^{-1}}[I - S_{i+1} + \mu_i^{-1} Y_i]$;
$Y_{i+1} = Y_i + \mu_i(I - S_{i+1} - L_{i+1})$; $\mu_{i+1} = \rho\mu_i$;
end while
Output: the stationary background matrix S and the mixture matrix L that contains moving foreground and falling snowflake

7.2.2.2 Moving Foreground Extraction

After the background extraction, the remaining part L is the mixture of moving foreground and falling snow. Following the model (7.3), the solution of foreground extraction is given by the subproblem as follows:

$$\begin{aligned} &\min_{M,E} \|\eta(M)\|_* + \lambda \|E\|_1 \\ &s.t. \quad L = M + E. \end{aligned} \tag{7.8}$$

The difficulty of dealing with moving foreground is aligning moving components together to grab their low-rank structure. As the object motions are various among frames of the video, it is impractical to estimate $\eta()$ globally. As a result, patch-based method is used to jointly remove the falling snow from the video. L is firstly reshaped from the matrix form to the frame sequence form, i.e., $MS = \{ms_k\}_{k=1}^K$. $MS = MV + SN$ is the mix of moving objects and falling snow. Here, we apply the method in [61] to segment these dynamic components. Then, we exclude the falling snow area from the moving objects. In this step, we consider the properties of snowflakes, such as the small size and the increase limitation in pixel values from background to snowflake (please see Appendix A for details). Then the moving foreground area is $MV = \{mv_k\}_{k=1}^K$.

For each frame mv_k, we generate blocks as the minimum bounding rectangle of moving objects. For one block, say, $p_{i,j,k}$, where the subscript indexes refer to the ith patch block which is centered at pixel j in the kth frame. We seek the block of pixels in each frame that is similar to the block in the current frame. There exist extensive patch-matching algorithms for motion estimation with efficient results [14, 27, 44]. We employ the simple and effective MAD (mean absolute difference) criterion in this paper,

$$\begin{aligned} \underset{x_0,y_0}{l\,MAD}(x, y) = \frac{1}{n_x \times n_y} \sum_{i=0}^{W-1} \sum_{j=0}^{H-1} &|p_k(x_0 + i, y_0 + j) \\ &- p_{k+1}(x_0 + x + i, y_0 + y + j)| \,. \end{aligned} \tag{7.9}$$

Eq. (7.9) refers to a $n_x \times n_y$ pixels block. (x_0, y_0) and $(x_0 + x, y_0 + y)$ denote the current block and the frame to be compared with, respectively.

Here, we suppose that m patches are found to be similar in the temporal domain. Then we put all the pixels of matched blocks as each column of a $(n_x \times n_y) \times m$ matrix $P_{j,k}$. In this way, P is defined as

$$P_{j,k} = (p_{1,j,k}, p_{2,j,k}, \ldots, p_{m,j,k}), \quad p_{i,j,k} \in R^2 \tag{7.10}$$

Then, the following formula in matrix form holds:

$$P_{j,k} = M_{j,k} + E_{j,k} \tag{7.11}$$

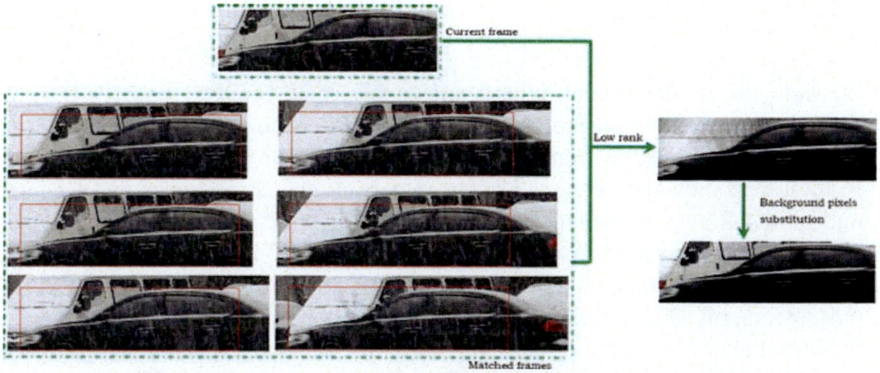

Fig. 7.3 Illustration on obtaining the snowflakes removed moving foreground by applying the local low-rank decomposition on matched blocks. © [2018] IEEE. Reprinted, with permission, from [Jiandong Tian, Zhi Han, Weihong Ren, Xiai Chen, and Yandong Tang, "Snowflake Removal for Videos via Global and Local Low-rank Decomposition," IEEE Transactions on Multimedia, Vol.20, Issue: 10, pp. 2659–2669, 2018]

where $M_{j,k}$ denotes the latent clear patch matrix, i.e., the moving objects, and $E_{j,k}$ denotes the falling snow.

As all column vectors in $M_{j,k}$ have similar latent image structures, $M_{j,k}$ should be a low-rank matrix. Now the subproblem of this step can be also viewed as a robust PCA problem:

$$\begin{aligned} &\min_{M,E} \|M\|_* + \lambda \|E\|_1 \\ &s.t. \quad P = M + E. \end{aligned} \tag{7.12}$$

By applying the proposed Algorithm 1 to matched patches, we can effectively remove the snowflakes in front of moving foreground. Figure 7.3 demonstrates the procedure of obtaining the snowflake-removed moving foreground by applying the local low-rank decomposition on matched blocks. Because pixels inside patches but outside the moving object M do not have the low-rank property (different in each block), they will be substituted by the background pixels to avoid artificial effects. Final snowflakes removed videos are generated by pasting moving foreground to stationary background. To sum up, we provide the entire algorithm of snowflake removal in Algorithm 2.

Algorithm 7.2 Snow Removal

Input:$I = [vec(f_1)|\cdots|vec(f_K)] \in R^{N\times K}$.
Solve subproblem (7.4) by Algorithm 1 and get the result of S and L;
Extract moving foreground M from L;
Get aligned blocks P by the patch matching algorithm;
Solve subproblem (7.12) by Algorithm 1 and get the result of M and E.
Output: Generate snow removed results by pasting the moving foreground M to the stationary background S.

7.2.3 Experiments

We show the results of our and comparison snowflake removal methods in several scenarios with different degrees of snow falling. After snowflake removal, the scenes become clear and the visibility is much improved by our method.

Figure 7.4 shows some snowflake removal results on a static video, which contains mid snow falling. The snowflake can be removed by the proposed method. We also compare our result with it by [48]. The method in [48] is simple and effective for static camera and stationary scenes. It extracts pixels in video sequences with the most saturation. When a test video is about a low saturation scene or the snowflake is heavy, the method cannot remove all the snowflakes or get blur results. The scene depth for the scenario varies largely. Therefore, the size of snowflakes varies from small to large. There is a glare due to the snowflake being close to lens and camera

Fig. 7.4 Zoom-in figures with moderate snow falling and their snowflake removal results. first row: Original images; second row: Our snowflake removal results; third row: Snowflake removal results by [48]. © [2018] IEEE. Reprinted, with permission, from [Jiandong Tian, Zhi Han, Weihong Ren, Xiai Chen, and Yandong Tang, "Snowflake Removal for Videos via Global and Local Low-rank Decomposition," IEEE Transactions on Multimedia, Vol.20, Issue: 10, pp. 2659–2669, 2018]

exposure integration effect (see the lower left red rectangle image in the first row in Fig. 7.4). It is removed by our method but retained by the method of [48].

Figure 7.5 shows results on some challenging videos includes more foreground objects, motion background, and heavy snowflakes. Generally, our results have little decrease in performance for these challenging videos, but are still better than

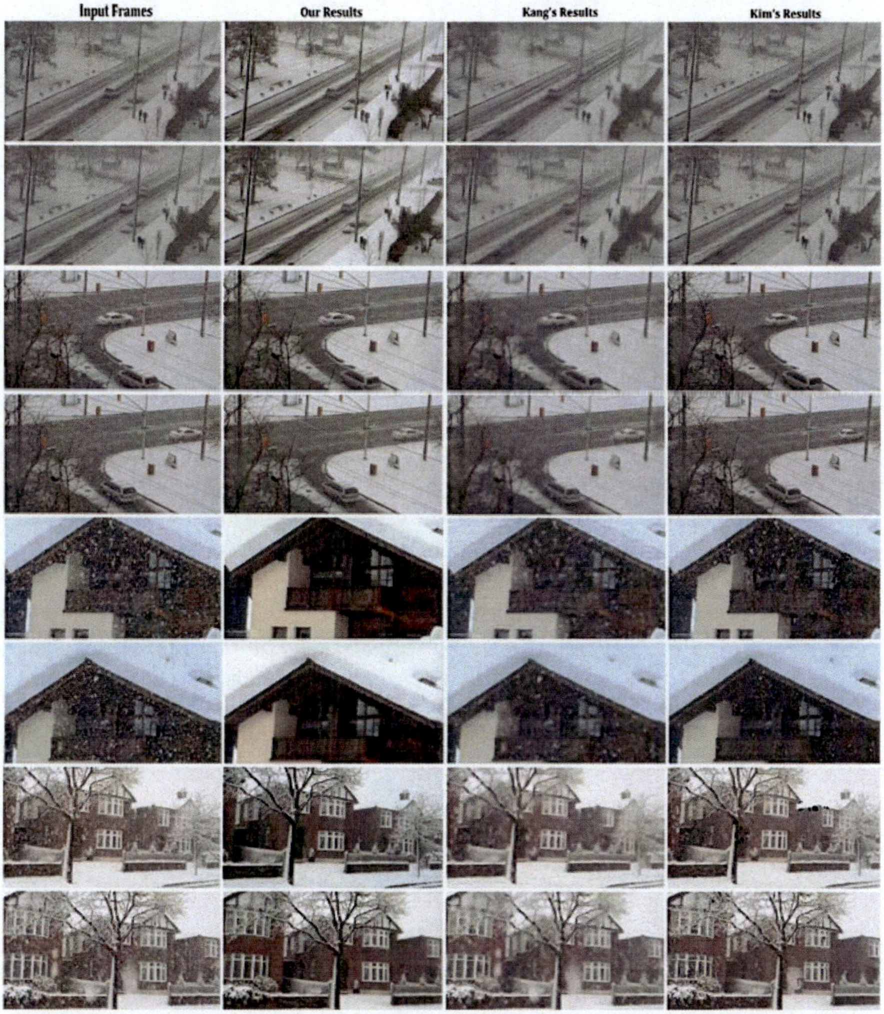

Fig. 7.5 More comparison results on some challenging videos. For each video, we show two frames. The first video contains more moving objects that include cars and pedestrians; the second video contains a fast-moving car and a slow car; the third and the fourth videos are captured by moving cameras, and the snow falling is heavy. © [2018] IEEE. Reprinted, with permission, from [Jiandong Tian, Zhi Han, Weihong Ren, Xiai Chen, and Yandong Tang, "Snowflake Removal for Videos via Global and Local Low-rank Decomposition," IEEE Transactions on Multimedia, Vol.20, Issue: 10, pp. 2659–2669, 2018]

Fig. 7.6 Comparison of two synthetic snowflake videos which are used for quantitative analysis. © [2018] IEEE. Reprinted, with permission, from [Jiandong Tian, Zhi Han, Weihong Ren, Xiai Chen, and Yandong Tang, "Snowflake Removal for Videos via Global and Local Low-rank Decomposition," IEEE Transactions on Multimedia, Vol.20, Issue: 10, pp. 2659–2669, 2018]

Table 7.1 SSIM and PSNR comparisons for the two synthetic snowflake videos

	Sakaino [48]	Kang [29]	Kim [31]	Ours
Video1 (SSIM)	0.8192	0.8963	0.8555	0.9723
Video1 (PSNR)	23.8006	27.5255	26.5437	29.2197
Video2 (SSIM)	0.8020	0.8014	0.7929	0.9540
Video2 (PSNR)	22.8949	24.6913	24.7875	31.0505

comparison results. The results by [29] are blur and have lots of snowflake residue. The results by [31] are less clear than our results for the first video and have distortion at the rear side of the fast-moving car in the second video. For the rest of two videos, the snowflakes in their results are not completely removed.

Figure 7.6 shows snowflake removal results of two synthetic snowing scenes: Video1 (upper) and Video2 (lower). Our results are close to the ground truth while other methods either fail to removal all snowflakes or get blur results. Table 7.1 shows the quantitative comparisons for these two videos using average SSIM and PSNR values. From the table, we can see that our method performs the best.

Unlike rain is semi-transparent, snowflake is white and not transparent, which may cause more serious occlusions. In this section, we propose a low-rank-based snowflake removal framework. Both global structure (static background) and local appearance (moving objects) are effectively unified in our method. The proposed framework contains two steps of low-rank decomposition. The first step is to extract stationary background with the whole frames. The second step focuses on local blocks obtained by patch matching for removing snowflakes in front of moving foreground. The experimental results for various falling snow scenes demonstrate the effectiveness of our method. The limitation of this method is that the current block-wise alignment algorithm cannot process more challenging situations. If a foreground object motion has large occlusion, fast scale variation, or is not rigid motion, block-wise alignment becomes less effective over frames and the low-rank decomposition cannot differentiate moving objects and snowflakes well. These challenging problems will be addressed in Sect. 7.2 by the proposed multi-label matrix decomposition algorithm.

7.3 Video Desnowing and Deraining Based on Matrix Decomposition

The existing snow/rain removal methods often fail for heavy snow/rain and dynamic scene. One reason for the failure is due to the assumption that all the snowflakes/rain streaks are sparse in snow/rain scene. The other is that the existing methods often cannot differentiate moving objects and snowflakes/rain streaks well. In this section, we propose a model based on matrix decomposition for video desnowing and deraining to solve the problems mentioned above. We divide snowflakes/rain streaks into two categories: sparse ones and dense ones. With background fluctuations and optical flow information, the detection of moving objects and sparse snowflakes/rain streaks is formulated as a multi label Markov random Fields (MRFs). As for dense snowflakes/rain streaks, they are considered to obey Gaussian distribution. The snowflakes/rain streaks, including sparse ones and dense ones, in scene backgrounds are removed by low-rank representation of the backgrounds. Meanwhile, a group sparsity term in our model is designed to filter snow/rain pixels within the moving objects. Experimental results show that our proposed model performs better than the state-of-the-art methods for snow and rain removal.

Figure 7.7 shows the composition of a general rain scene: low-rank backgrounds, foregrounds, sparse rain streaks, and dense rain streaks. Sparse rain streaks usually have large intensity fluctuations, consistent directions and particular shapes, and they can be detected by the existing methods. However, dense rain streaks, with no obvious properties, lead to wrong detection. This may be the reason why the existing methods do not perform well for heavy rain. In a rain scene, it is also difficult to separate moving objects and sparse rain streaks. Thus, most of the existing methods often cause deformations and artifacts on moving objects.

For this issue, we propose a model based on matrix decomposition for video deraining or desnowing under heavy rain and dynamic scene. We divide rain streaks into two categories: sparse ones and dense ones. With background fluctuations and optical flow information, the detection of sparse rain streaks and moving objects is formulated as multi-label MRFs. As for dense rain streaks, they are assumed to obey Gaussian distribution. All the rain streaks, including sparse ones and dense ones, in scene backgrounds are removed through low-rank representation of the backgrounds. After the detection of moving objects, a group sparsity term is designed to filter rain pixels within the moving objects.

7.3.1 General Model

Let $I = [I_1, I_2, ..., I_n] \in R^{m\times n}$ be a video sequence with n frames, which is corrupted by rain or snow, and $I_j \in R^m$ denotes the jth frame of the video sequence. The ith pixel in the jth frame is denoted as I_{ij}. The input video I can be regarded as a combination of B, F, S_s and S_d, i.e.,

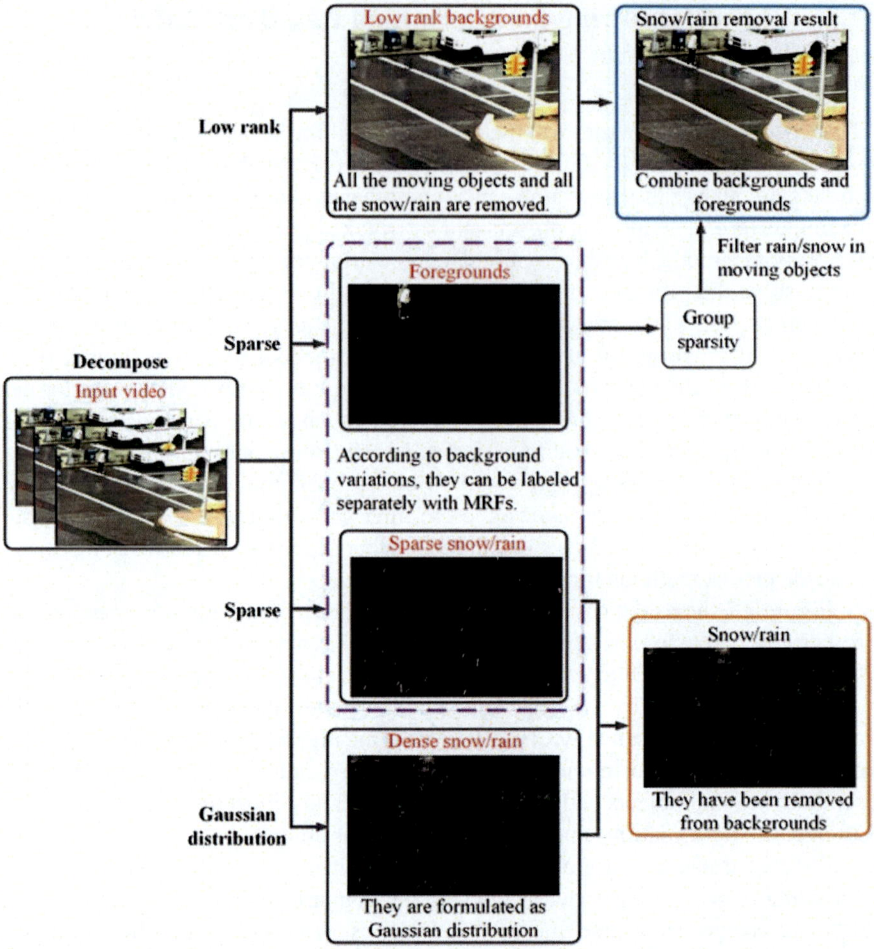

Fig. 7.7 Composition of a general snow/rain scene: Low-rank backgrounds, foregrounds, sparse snow/rain, and dense snow/rain. Sparse snow/rain usually has large intensity fluctuations, consistent directions, and particular shapes, and they can be detected by the existing methods. However, dense snow/rain mislead the detection. © [2017] IEEE. Reprinted, with permission, from [Weihong Ren, Jiandong Tian, Zhi Han, Antoni Chan, and Yandong Tang, "Video Desnowing and Deraining Based on Matrix Decomposition," IEEE Conference on Computer Vision and Pattern Recognition, pp.4210–4219, 2017]

$$I = B + F + S_s + S_d, \tag{7.13}$$

where $B \in R^{m\times n}$ is a matrix with the same size of I, which denotes the clear backgrounds; $F \in R^{m\times n}$ represents the intensity fluctuations caused by foregrounds; $S_s \in R^{m\times n}$ and $S_d \in R^{m\times n}$ denote the intensity fluctuations caused by sparse rain streaks and dense rain streaks, respectively.

Backgrounds intensity should be unchanged, and thus, they are linearly correlated with each other, which forms a low-rank matrix B. We impose the following constraint on B:

$$rank\,(B) \leq \kappa,$$

where κ is a constant, which constraints the complexity of the background model.

The intensity fluctuations over video sequences are mainly caused by sparse outliers, i.e., moving objects F and sparse rain streaks S_s, and they are regarded as sparse matrices.

For S_d, we regard it as the fluctuations caused by dense rain streaks. Actually, it may contain fluctuations caused by background noise and illumination variations. In this paper, we simply assume it as Gaussian distribution.

$$P\,(S_d) = \frac{1}{\left(\sqrt{2\pi}\sigma_d\right)^N} \exp\left(-\frac{\|S_d\|_F^2}{2\sigma_d^2}\right), \tag{7.14}$$

where σ_d is the standard deviation of the Gaussian distribution, and N is the number of image pixels.

The foregrounds, namely moving objects, have similar structures in a video and can be formulated via group sparsity. Referring to [5, 61], we detect and remove rain streaks through solving a matrix decomposition problem:

$$\begin{aligned} &\min_{B,F,S_s,S_d} \frac{1}{2{\sigma_d}^2} \|S_d\|_F^2 + \eta \cdot rank\,(B) + \lambda_1 \|S_s + F\|_0 \\ &+ \|\mathcal{P}\,(F)\|_G, \\ &s.t. \quad I = B + F + S_s + S_d, \end{aligned} \tag{7.15}$$

where σ_d, η and λ_1 are regulation parameters. The proper choice of these parameters will be discussed in the following part. After obtaining intensity fluctuations F, the operator $\mathcal{P}$ first extracts foregrounds from I, and then it applies block matching on the foregrounds. The pseudomatrix norm $\|\mathcal{P}\,(F)\|_G$ makes foregrounds group sparse after the operation of $\mathcal{P}$.

In order to deal with dynamic scenes taken by a moving camera, we first align neighboring frames to a target frame before the target frame is derained or desnowed. Let $I_j \circ \tau_j$ denote the frame after the transformation parameterized by vector τ_j. Then, the proposed decomposition (7.13) becomes $I \circ \tau = B + F + S_d + S_s$, and $I \circ \tau = [I_1 \circ \tau_1, I_2 \circ \tau_2, \ldots, I_n \circ \tau_n]$. We also use nuclear norm to replace the rank operator in (7.15) [45], and the final model can be written as:

$$\begin{aligned} &\min_{B,F,S_s,\tau} \frac{1}{2\sigma_d^2} \|I \circ \tau - B - F - S_s\|_F^2 + \eta \cdot \|B\|_* \\ &+ \lambda_1 \|S_s + F\|_0 + \|\mathcal{P}\,(F)\|_G. \end{aligned} \tag{7.16}$$

7.3.2 Moving Objects and Sparse Rain Streaks Modeling with MRFs

Moving objects in a rain scene are difficult to handle. If a frame is directly filtered without knowing the exact locations of moving objects in it, deformations and artifacts are usually caused on the moving objects (Fig. 7.10). To avoid this problem, not like other methods [18, 31, 50, 59], we explicitly detect the moving objects with MRFs for group sparsity filtering.

Equation(7.16) is intractable to optimize due to the existence of sparse outliers S_s and F. Let $M \in \{0, 1\}^{m\times n}$ be a two-valued matrix denoting the sparse outliers support (S_s and F):

$$M_{ij} = \begin{cases} 0, \text{if} I_{ij} \text{is background,} \\ 1, \text{if} I_{ij} \text{is sparse outlier.} \end{cases} \tag{7.17}$$

We set $E = I \circ \tau - B$ to represent background fluctuations. As long as $E_{ij} \neq 0$, i.e., $M_{ij} \neq 0$, we must have $E_{ij} = F_{ij}$ or $E_{ij} = (S_s)_{ij}$ to minimize (7.16). Thus, (7.16) can be written as:

$$\min_{B,M} \frac{1}{2\sigma_d^2} \|(1 - M) \odot E\|_F^2 + \lambda_1 \|M\|_1 + \eta \cdot \|B\|_* + \|\mathcal{P}(F)\|_G, \tag{7.18}$$

where $\odot$ denotes the pixel-wise multiplication. When B and τ are fixed, the optimization for M is determined by λ_1 and the background variations E. Equation(7.18) can be further written as:

$$\min_{M} \sum_{ij} \left(\lambda_1 - \frac{1}{2\sigma_d^2}(E_{ij})^2\right) \cdot M_{ij}. \tag{7.19}$$

When dependency between neighbors is considered, (7.19) can be rewritten as a first-order MRFs:

$$\min_{M} \sum_{ij} \left(\lambda_1 - \frac{1}{2\sigma_d^2}(E_{ij})^2\right) \cdot M_{ij} + \sum_{ij} \sum_{kl \in \mathcal{N}_{ij}} \gamma \cdot \psi(M_{ij} - M_{kl}), \tag{7.20}$$

where γ is a regulation parameter, $\mathcal{N}_{ij}$ is the set of 4-connected neighbors of M_{ij}, and

$$\psi(x) = \begin{cases} 0, \text{if} x = 0, \\ 1, \text{if} x \neq 0. \end{cases} \tag{7.21}$$

Through (7.20), backgrounds and sparse outliers are separated. However, it is still difficult to differentiate F and S_s.

Through analysis of abundant rain and snow video sequences, we find that intensity fluctuations caused by moving object are more drastic than those caused by rain

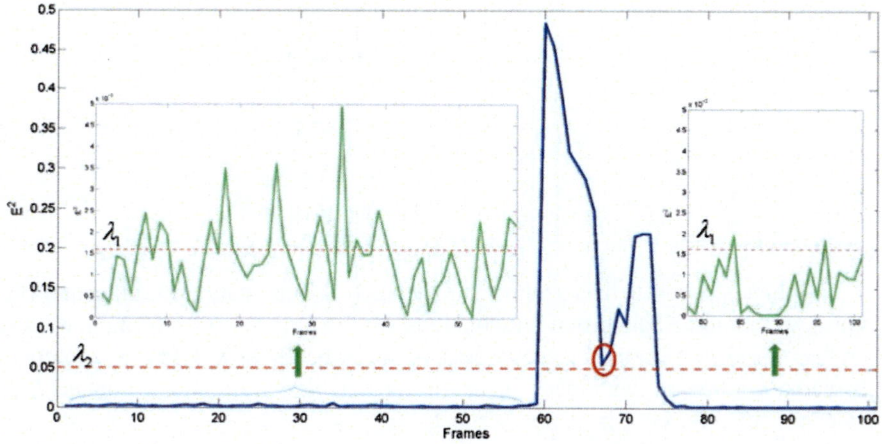

Fig. 7.8 In both ends of the plot, intensity fluctuations in backgrounds vary irregularly due to rain, and there is a rough threshold λ_1 to distinguish rain pixels and backgrounds. Another threshold denoted as λ_2, caused by a moving object, is available to differentiate moving object and backgrounds. © [2017] IEEE. Reprinted, with permission, from [Weihong Ren, Jiandong Tian, Zhi Han, Antoni Chan, and Yandong Tang, "Video Desnowing and Deraining Based on Matrix Decomposition," IEEE Conference on Computer Vision and Pattern Recognition, pp. 4210–4219, 2017]

streaks or snowflakes. Figure 7.8 shows the plot of background fluctuations E in 100 consecutive frames for a pixel at particular position. In both ends of the plot, the intensity fluctuations in the backgrounds vary irregularly due to rain, and there is a threshold λ_1 to distinguish sparse outliers and the backgrounds. This analysis is reflected in (7.20). From frame 60 to frame 75, a moving object passes through backgrounds and causes great intensity shifts.

Another threshold denoted as λ_2 is available to differentiate the moving object and rain streaks, even though rain streaks, marked in a red circle in Fig. 7.8, sometimes appear in the moving object and decrease the intensity shifts. In order to differentiate F and S_s, we rewrite M in (7.17) as:

$$\overline{M}_{ij} = \begin{cases} 0, \text{if} I_{ij} \text{is background}, \\ 1, \text{if} I_{ij} \text{is sparse rain/snow } S_s, \\ 2, \text{if} I_{ij} \text{is movingobject } F. \end{cases} \tag{7.22}$$

Meantime, we also rewrite (7.20) as a multi-label MRFs:

$$\min_{\overline{M}} E_d + \sum_{ij} \sum_{kl \in \mathcal{N}_{ij}} \gamma \cdot \psi(\overline{M}_{ij} - \overline{M}_{kl}), \tag{7.23}$$

where

$$E_d = \begin{cases} \frac{1}{2\sigma_d^2}\left(E_{ij}\right)^2 \quad if\overline{M}_{ij} = 0, \\ \left|\lambda_1 - \frac{1}{2\sigma_d^2}\left(E_{ij}\right)^2\right|, \text{if}\overline{M}_{ij} = 1, \\ \left|\lambda_2 - \frac{1}{2\sigma_d^2}\left(E_{ij}\right)^2\right|, \text{if}\overline{M}_{ij} = 2. \end{cases} \tag{7.24}$$

To make λ_1 robust to background changes such as illumination variation and video noise, we combine background fluctuations E with optical flow information. Based on the optical flow method [36], we first estimate the 2D flow field $\mathbf{u}(x, y)$ between two adjacent frames in a video. Then, we warp the next frame I_{k+1} to obtain $\bar{I}_k$ and calculate the difference between I_k and $\bar{I}_k$.

$$\begin{aligned} \bar{I}_k &= I_{k+1}\left(\mathbf{x} + \mathbf{u}(x, y)\right), \\ I_r &= \left|I_k - \bar{I}_k\right|. \end{aligned} \tag{7.25}$$

Since rain streaks are too small and move too fast to affect the optical flow estimation, we use I_r to guide MRFs for optimizing $\overline{M}$.

$$E_d = \begin{cases} \frac{1}{2\sigma_d^2}\left(E_{ij}\right)^2, \quad \text{if}\overline{M}_{ij} = 0, \\ \left|\lambda_1 + \lambda_r - \frac{1}{2\sigma_d^2}\left(E_{ij}\right)^2 - \omega_1\left(I_r\right)_{ij}^2\right|, \text{if}\overline{M}_{ij} = 1, \\ \left|\lambda_2 - \frac{1}{2\sigma_d^2}\left(E_{ij}\right)^2\right|, \text{if}\overline{M}_{ij} = 2, \end{cases} \tag{7.26}$$

where ω_1 controls the weight between background variations and optical flow information. The thresholds λ_1 and λ_2 are related to background variations E, and λ_r is related to optical information. The proper choice of these parameters will be discussed in the next section.

In Fig. 7.9, rain streaks and moving objects can be accurately detected with Markov random fields. With the locations of moving objects, a group sparsity term is particularly designed to filter rain pixels within them. This process can effectively avoid deformations and artifacts of the moving objects.

7.3.3 Group Sparsity for Moving Objects

With local or non-local filtering algorithms, the existing methods usually remove rain within moving objects in a pixel-wise manner, and this process often causes blur and artifacts on the moving objects. Since image patches have non-local similarity in

Fig. 7.9 Detected sparse outliers with Markov random field. With local spatial information I_r and **u**, moving objects and sparse rain streaks or snowflakes can be detected. © [2017] IEEE. Reprinted, with permission, from [Weihong Ren, Jiandong Tian, Zhi Han, Antoni Chan, and Yandong Tang,"Video Desnowing and Deraining Based on Matrix Decomposition," IEEE Conference on Computer Vision and Pattern Recognition, pp. 4210–4219, 2017]

spatial and temporal space [13, 27], we design a group sparsity term to filter moving objects in a patch-wise manner.

Based on $\overline{M}$ obtained from (7.23), the operator $\mathcal{P}(F)$ first extracts foregrounds from I. Then, it divides foregrounds into overlapped patches and groups similar patches together within specific search windows centered on reference patches. For a target frame, similar patches are only searched in its two neighboring frames, i.e., three successive frames. For more details of patch-matching, please refer to [11]. The pseudomatrix norm $\|\cdot\|_G$ seeks for group sparsity of a patch set. After the patch set is processed, we put image patches back through a weighted average approach. $(\mathcal{P}(F))_{i,j}$ is a patch set in $\mathcal{P}(F)$, and its reference patch is the jth patch in the ith frame. Let $Y = (\mathcal{P}(F))_{i,j}$ and $Y = [y_1, y_2, ..., y_l] \in R^{h \times l}$, where $y_1, y_2, ..., y_l$ are the similar pathes in foregrounds extracted from three successive frames. According to the group/structured sparsity [40, 62], the group sparsity term $\|Y\|_G$ can be solved by

$$(U, \Sigma, V) = \underset{U,\Sigma,V}{\arg\min} \left\|Y - U\Sigma V^T\right\|_F^2 + \mu\sum_{i=1}^{r}\sigma_i, \tag{7.27}$$

where $\Sigma = \text{diag}\{\sigma_1, \sigma_2...\sigma_r\}$ $(r = \text{rank}(Y))$. Equation(7.27) can be interpreted by singular value decomposition (SVD) from a bilateral variance estimation perspective [13]. Since rain streaks are randomly distributed in spatial and temporal field, the rain pixels within the moving objects can be removed by (7.27).

7.3.4 Algorithm

The objective function (7.16) is nonconvex and is hard to obtain the solution in one step. Hence, we adopt an alternating algorithm that separates the energy minimization over B, τ, F, and S_s. The step for B and τ is a convex problem, and the SOFT-INPUT algorithm [41] [45] has been proved effective for nuclear norm optimization problem. The step for S_s and F is solved by (7.23) to obtain $\overline{M}$. Based on $\overline{M}$, the operation $\mathcal{P}$ on F then filters rain within foregrounds by (7.27). The final rain removal result is obtained by combining clear backgrounds and filtered foregrounds. The overall algorithm is presented in Algorithm 7.3.

7.3.4.1 Reconstruct Background *B*

When sparse outliers support M (i.e., $\overline{M}$) and transformation τ are fixed, according to (7.18), the minimization of objective function (7.16) can be transformed into:

$$\min_{B,\tau} \frac{1}{2\sigma_d^2} \|(1 - M) \odot (I \circ \tau - B)\|_F^2 + \eta \cdot \|B\|_*. \tag{7.28}$$

Referring to [41], the solution to B in (7.28) can be given by SVD:

$$B = \text{svd}\Big((1 - M) \odot (I \circ \tau) + M \odot \tilde{B}\Big)_\alpha = U\Sigma_\alpha V^T, \tag{7.29}$$

where $\Sigma_\alpha = diag\left[(d_1 - \alpha)_+, ..., (d_r - \alpha)_+\right]$, $t_+ = \max(t, 0)$, and $\Sigma = diag$ $[d_1, ..., d_r]$. In the computation process, $\tilde{B}$ can be arbitrary initialized, and B is computed by iteratively using (7.29). The parameters η and σ_d^2 in (7.28) control the complexity of the scene background. A larger η gives a smoother scene background. In our algorithm, a rough estimation of rank $\kappa = \sqrt{n}$ is firstly given, and n denotes the number of frames that we select to reconstruct the scene background. In this paper, we set $n = 15$ for light rain scene and $n = 31$ for heavy rain. Then we start from a large η, and experimentally η is initialized to be the second largest singular value of I. The parameter σ_d^2 is set to 0.5 to estimate the dense rain streaks. After each iteration of (7.29), if rank $(B) \leq \kappa$, we reduce η by a factor $\theta_1 = 1/\sqrt{2}$ and repeat (7.29) until $rank\ (B) > \kappa$.

In (7.28), $\tau = [\tau_1, \tau_2, ..., \tau_n] \in R^{p\times n}$ are the parameterizations of all the transformations. When the change in τ is small, this constraint can be approximated by linearizing the current estimate of τ [45]. For $\Delta\tau \in R^{p\times n}$, let $I \circ (\tau + \Delta\tau) \approx I \circ \tau + J_\tau \Delta\tau$, where J_τ is the Jacobian matrix of the images. This linearization leads to a convex optimization problem over $\Delta\tau$, when B and M are fixed:

$$\begin{aligned} &\Delta\tau^* \leftarrow \arg\min_{\Delta\tau} \|(1 - M) \odot (I \circ \tau - B + J_\tau \Delta\tau)\|_F^2 \\ &\tau \leftarrow \tau + \Delta\tau^*. \end{aligned} \tag{7.30}$$

7.3.4.2 Detect Continuous Outliers M (i.e., $\overline{M}$)

The parameter ω_1 in (7.26) controls the weight between background fluctuations and optical flow information, and λ_1, λ_2, λ_r are related to sparse outliers. Typically, we set $\omega_1 = 0.5$, because I_r provides much information for S_s detection. Background variation $\lambda_1 = 4.5\sigma^2$, and $\lambda_2 = 12.5\sigma^2$, where σ^2 is estimated online by the variance of $(I \circ \tau - B)_{ij}$. Similarly, $\lambda_r = 12.5{\sigma_r}^2$, where ${\sigma_r}^2$ is estimated by I_r. In Sect. 7.3.2, Eq.(7.16) is transformed into (7.23) for differentiating F and S_s. When B and τ are given, Eq.(7.23) is a standard form of MRFs with multiple labels, which can be exactly solved by [3] [32]. When $\overline{M}$ is obtained,

$$M = \begin{cases} 0, \text{if} \overline{M} = 0, \\ 1, \text{if} \overline{M} \neq 0. \end{cases} \tag{7.31}$$

Algorithm 7.3 Overall Algorithm for Rain or Snow Removal

Input: Rain or Snow video sequence $I = [I_1, I_2, ..., I_n] \in R^{m \times n}$
Output: Rain or Snow removal sequence $I^* \leftarrow [I_1^*, I_2^* \ I_n^*]$
Initialize: Transformation τ, backgrounds B, Support $\overline{M}$
1. **Repeat**
2. Update τ through (7.30)
3. Reconstruct B through (7.29)
4. Estimate λ_1, λ_2, and λ_r in (7.26)
5. Optimize $\overline{M}$ through (7.23) with MRFs
6. **Until** convergence
7. Filter rain/snow within Foregrounds through (7.27)
8. Combine Foregrounds and backgrounds to obtain I^*

7.3.4.3 Filter Rain Pixels Within Moving Objects

In Sect. 7.3.4.1, rain streaks in scene backgrounds are removed by low-rank representation of the backgrounds. The key problem is to filter the rain pixels within moving objects. Based on sparse outliers $\overline{M}$ obtained in Sect. 7.3.4.2, Eq.(7.27) employs three successive frames to seek for group sparsity over foreground patches, and it is regarded as a standard matrix factorization problem. Since rain streaks are randomly distributed in spatial and temporal field, the rain pixels within the moving objects can be removed by (7.27). The technique proposed by [13] is adopted to solve (7.27), and the parameter μ is generally set to 5.

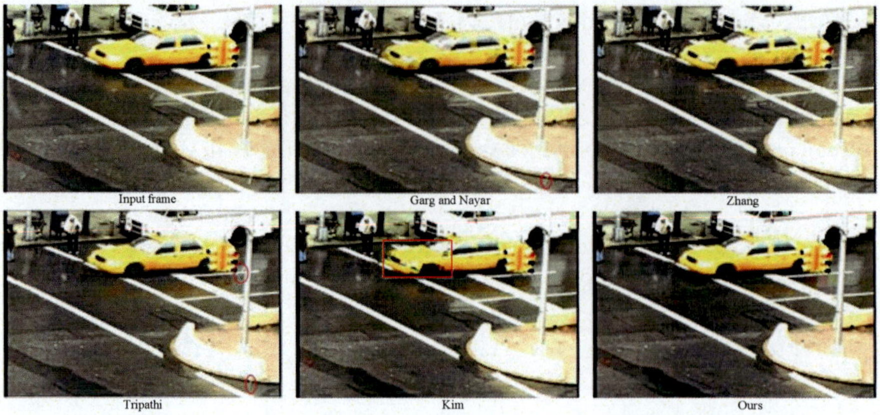

Fig. 7.10 Rain removal results of a dynamic scene. In our model, the moving object is filtered using group sparsity, and it avoids deformation and artifacts in restored frames. © [2017] IEEE. Reprinted, with permission, from [Weihong Ren, Jiandong Tian, Zhi Han, Antoni Chan, and Yandong Tang, "Video Desnowing and Deraining Based on Matrix Decomposition," IEEE Conference on Computer Vision and Pattern Recognition, pp. 4210–4219, 2017]

Fig. 7.11 Rain removal results of a heavy rain scene. Our method can remove almost all the rain streaks. Since the rain is too heavy, the method causes blur on the background. © [2017] IEEE. Reprinted, with permission, from [Weihong Ren, Jiandong Tian, Zhi Han, Antoni Chan, and Yandong Tang, "Video Desnowing and Deraining Based on Matrix Decomposition," IEEE Conference on Computer Vision and Pattern Recognition, pp. 4210–4219, 2017]

7.3.5 Experiments

In order to evaluate the performance of our proposed model, various rain and snow video sequences are employed. They contain illumination variations, camera motions, moving objects, etc.

In Fig. 7.10, we show a dynamic rain scene. There are several moving objects in the rain scene with different velocities and directions. For the fast-moving object,

Table 7.2 PSNR comparisons for five synthetic rain scenes

Scene	Rain Videos	Garg, 2004	Zhang, 2006	Tripathi, 2012	Kim, 2015	Ours
Street1	29.37	28.61	24.25	26.49	30.96	**31.40**
Street2	28.80	29.00	24.27	25.71	29.03	**29.62**
Campus	29.13	29.18	24.01	27.47	29.82	**30.45**
Human4	30.17	28.62	24.33	23.00	30.79	**31.35**
Bike	29.95	33.05	28.93	30.40	33.21	**33.54**

namely the car, the method of [50] and ours can handle well, but other methods degrade the edges of the moving object severely. The disadvantage of the method of [50] is that snowflakes cannot be removed clearly in the background. With low-rank representation of the background, our model can remove almost all the snowflakes. Meanwhile, a group sparsity model is designed to filter rain streaks within moving objects, and it avoids deformation and artifacts of the moving objects in restored frame.

Figure 7.11 is a heavy rain scene taken with a smart phone across a window, and it is challenging to remove all the rain streaks. Since the rain is too heavy, all the methods cause blur on the background. However, our method can remove almost all the rain streaks with 31 frames to reconstruct scene backgrounds and detect moving objects. Similar to our method, Kim et al. [31] employ 5 frames to remove rain with low-rank matrix completion, but the heavy rain limits their method's performance. The comparison seems unfair for Kim's method, but this is not the case. Kim's method does not use a explicit way to detect moving objects, and it will cause blur and artifacts on moving objects if more than 5 frames are employed. Figure 7.12 shows rain removal results of two synthetic rain scenes: *Street1* and *Street2*. The rain videos are synthesized using the technique in [20], and we just show the results of Kim [31] and ours since the PSNR values of other methods are relatively lower (Tab. 7.2). Though Kim's method can remove most of the rain streaks, it fails to handle wide and bright rain streaks. Our method divide rain streaks into sparse ones and dense ones, and all the snowflakes can be detected and removed through our model.

Tab. 7.2 shows the quantitative comparisons for five synthetic rain videos using average PSNR value of a video. *Street1*, *Street2* and *Campus* are taken with a moving smart phone, and other two videos are chosen from [55]. Our method performs the best from the perspective of PSNR values, the reason is that our method can remove all the rain streaks. Due to improper filtering algorithms, the methods in [18, 50, 59] perform not good for the synthetic heavy rain scenes. Kim's method in [31] achieves high PSNR values, but it fails to remove all the rain streaks (Fig. 7.12).

In this section, we divide rain streaks into sparse ones and dense ones, and model them separately in a matrix decomposition framework. The division makes our model more effective to handle heavy rain than other methods. Mislead by rain streaks, it is difficult to detect and filter moving objects. What is worse, improper filtering

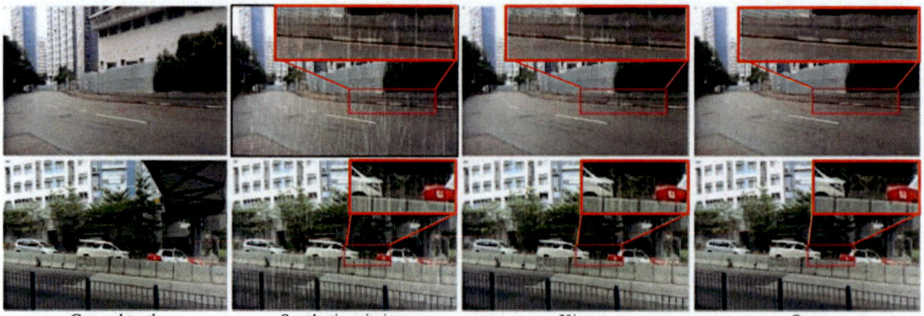

Fig. 7.12 Comparison of two synthetic rain videos. Kim's method [31] can remove most of the rain streaks, but it fails to remove wide and bright rain streaks. However, our method can handle all the rain streaks well due to its different models for rain streaks. © [2017] IEEE. Reprinted, with permission, from [Weihong Ren, Jiandong Tian, Zhi Han, Antoni Chan, and Yandong Tang, "Video Desnowing and Deraining Based on Matrix Decomposition," IEEE Conference on Computer Vision and Pattern Recognition, pp.4210–4219, 2017.].

algorithm will cause deformations and artifacts on the moving objects. To avoid this problem, we first explicitly detect moving objects with MRFs and then design a group sparsity term to filter rain pixels within them. The experiments on real and synthetic video sequences demonstrate that our method performs better than other state-of-the-art methods. Currently, our method can remove all the rain streaks in the scene, but it may impair dynamic textures such as waves, swaying trees, and crowed streets. In other words, our method may fail to separate dynamic textures and moving objects. To decompose rain video into more layers, such as moving object, sparse rain, dense rain, and dynamic texture, may be a good choice, but it needs to extract more features for segmentation. The processing of dynamic textures and complex scenes will be solved by using a deep learning technique in Sect. 7.4.

7.4 Dually Connected Single-Image Deraining Net Using Pixel-Wise Attention

With the development of deep learning technology, some single-image rain and snow removal algorithms have appeared in recent years, but the removal effect is still difficult to match the video removal algorithm. This is because the heavy rain or snow weather scenes are not only blocked by rain and snow, but also blurred by rain and snow. Therefore, it is difficult to detect and deal with the rain and snow that are deep in the scene. To address the above issues, we propose a dually connected deraining net using pixel-wise attention with only simple MSE loss, for single-image rain removal in one stage. Overall, the deraining net has an encoder–decoder architecture by jointly using skip sum connection and skip concatenation connection. The dual connections enable the deraining net to contain more information flow and

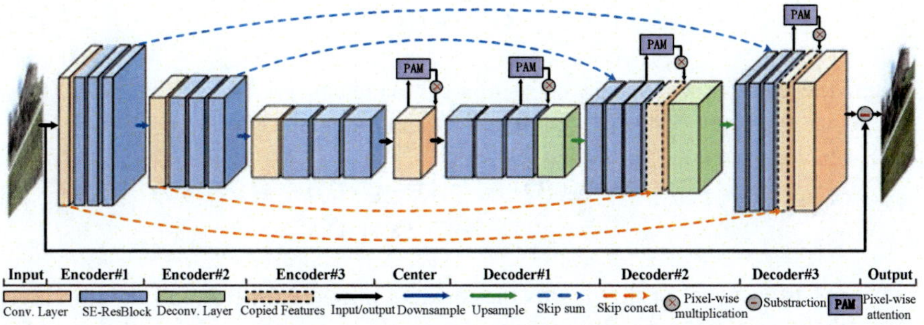

Fig. 7.13 Overall architecture of the deraining net. The backbone of the deraining net is an encoder–decoder net, which follows the residual learning architecture. It takes a rainy image as input and produces the corresponding residual rain map. © [2020] IEEE. Reprinted, with permission, from [Weihong Ren, Jiandong Tian, Qiang Wang, and Yandong Tang, "Dually Connected Deraining Net Using Pixel-Wise Attention," IEEE Signal Processing Letters, Vol.27, pp.316–320, 2020.].

thus can allow it to accurately discriminate and localize the rain streaks. Also, we apply pixel-wise attention to decode rain features, which can help to learn the final residual rain map.

7.4.1 Dually Connected Deraining Net

Formally, the observed rainy image $\mathbf{O}$ can be seen as a linear combination of the clean background layer $\mathbf{B}$ and the rain streak layer $\mathbf{R}$,

$$\mathbf{O} = \mathbf{B} + \mathbf{R}. \tag{7.32}$$

In heavy rain scenes, rain streaks have different appearances as the distance to camera changes. In addition, the accumulation of the overlapped rain streaks will cause attenuation and scattering. Inspired by the haze model [15, 43], the rain model can be further extended as:

$$\mathbf{O} = \sum_{i=1}^{N} \boldsymbol{\alpha}_i \otimes \mathbf{B} + (1 - \boldsymbol{\alpha}_i) \otimes \mathbf{R}_i, \tag{7.33}$$

where $\boldsymbol{\alpha}_i$ represents the scene transmission map in layer i, and $\mathbf{R}_i$ is its corresponding brightness of the rain streaks. The operation $\otimes$ is the pixel-wise multiplication, and N is the number of rain layers. Based on model (7.33), the recurrent mechanism for rain removal may be a good choice, but it will cause artifacts (see Fig. 7.17c).

In this section, we aim to remove rains streaks in one stage and preserve image details in the meanwhile. Both model (7.32) and (7.33) can be reformulated as

$$\mathbf{O} = \mathcal{F}(\mathbf{B}), \tag{7.34}$$

where the operator $\mathcal{F}$ maps the clear image $\mathbf{B}$ to the observed rainy image $\mathbf{O}$. Suppose $\mathcal{F}$ has its inverse operator $\mathcal{G}$, and then the clean image can be obtained by

$$\begin{aligned}\mathbf{B} = \mathcal{G}(\mathbf{O}) &= -(\mathbf{O} - \mathcal{G}(\mathbf{O})) + \mathbf{O} \\ &= -\mathcal{R}(\mathbf{O}) + \mathbf{O},\end{aligned} \tag{7.35}$$

where $\mathcal{R}$ is an operator that learns the residual rain streaks from all the rain layers. Note that rain streaks usually distribute sparsely in a scene which is different from haze/fog causing global attenuation. Hence, learning the residual mapping $\mathcal{R}$ is more efficient than directly learning the mapping $\mathcal{G}$, and this is the reason why our deraining net follows the residual learning architecture (see Fig. 7.13).

The overall architecture of our deraining net follows the residual learning, which is shown in Fig. 7.13. It takes the rainy image as input and produces the corresponding clear image after the mapping of three encoded, one centered, and three decoded blocks. Each encoded/decoded block has 1 convolutional/deconvolutional layer and 3 SE-ResBlocks. The centered block contains the encoded features which are further enhanced by a convolutional layer. The symmetric encoded and decoded blocks are dually connected with the skip sum connection and the skip concatenation connection. Besides, all the decoded features as well as the centered features are weighted by the learnable pixel-wise attention module (denoted as "PAM").

SE-ResBlock SE-ResBlock built on ResNet [23] was proposed by [25], which can adaptively recalibrate channel-wise feature responses by explicitly modeling independencies between channels and thus can enhance the representational power of a basic module. It was used as the encoded block for classification and has achieved significant improvements for the existing deep architectures. Figure 7.14a shows the SE-ResBlock where the SE block generates channel-wise attention for the learned features from the two preceding convolutional layers. The "Scale" operator then recalibrates feature responses using the attentive vector learned from the SE block.

Dually Skip Connections Our deraining net follows the encoder–decoder neural network [37, 47], and it can transfer a rainy image to a residual map containing rain streaks and negative image details. The convolutional layer in each encoded block is connected to its symmetric decoded block with a skip concatenation which provides spatial information for rain detection. Besides, output features of each encoded block are added to the input of its corresponding decoded block with a skip sum connection, which can combine the appearance information (e.g., rain brightness and image details) from shallow layer with semantic information (e.g., rain density) for the final rain removal. In summary, the dually skip connections promote the information flow from the shallow layers to deep layers, which helps the deraining net to accurately discriminate and localize the rain streaks at different scene depths.

Pixel-wise Attention Attention mechanism has been widely used for high-level computer vision tasks, e.g., image captioning [8, 57], pose estimation [10], image classification [51], and salience detection [60], but few works applied it to image

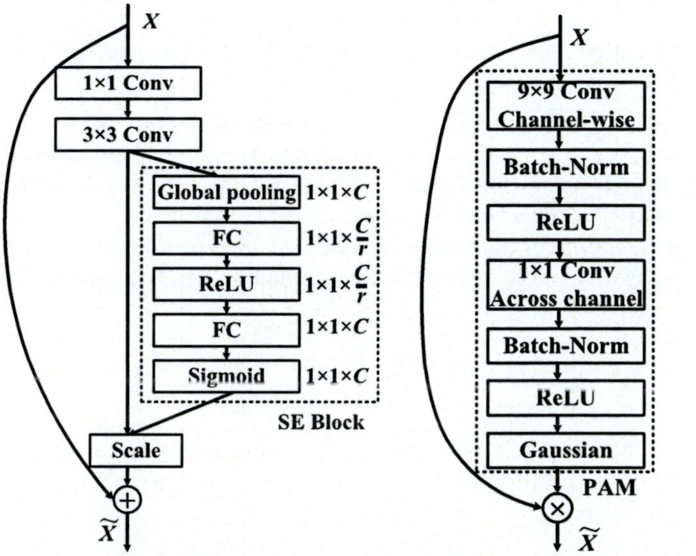

Fig. 7.14 Illustration of SE-ResBlock and Pixel-wise attention module. The operators ⊕ and ⊗ represent pixel-wise sum and pixel-wise multiplication, respectively. © [2020] IEEE. Reprinted, with permission, from [Weihong Ren, Jiandong Tian, Qiang Wang, and Yandong Tang, "Dually Connected Deraining Net Using Pixel-Wise Attention," IEEE Signal Processing Letters, Vol.27, pp. 316–320, 2020]

restoration task. Generally, rain streaks are sparse in the scene, and we thus can highlight the rain regions and avoid degradations of image edges using attention mechanism. Different from the high-level tasks focusing on global and semantic information, image restoration should pay attention to the pixel-level changes, and this is the reason why we design a pixel-wise attention mechanism. Figure 7.14 shows how to apply pixel-wise attention mechanism to input features.

We denote the features at current layer as $\mathbf{X} \in \mathbb{R}^{W \times H \times C}$. Then, spatial attention is generated by a channel-wise convolutional operator followed by an activation function f

$$\mathbf{m}_s^c = \mathbf{W}_s^c * \mathbf{X}^c + \mathbf{b}_s^c, \quad \mathbf{a}_s^c = f\left(\mathbf{m}_s^c\right), \tag{7.36}$$

where $*$ denotes the convolution operation, and $\mathbf{W}_s^c$ and $\mathbf{b}_s^c$ represent the convolutional filters and bias parameters for channel c in $\mathbf{X}$, respectively. The response $\mathbf{a}_s^c \in \mathbb{R}^{W \times H}$ characterizes the spatial correlations for each feature map in $\mathbf{X}$. To further integrate information from all the channels, a 1×1 convolutional operator together with an activation function f is then applied to the spatial response $\mathbf{a}_s$

$$\mathbf{m}_p = \mathbf{W}_p * \mathbf{a}_s + \mathbf{b}_p, \quad \mathbf{a}_p = f\left(\mathbf{m}_p\right), \tag{7.37}$$

where $\mathbf{W}_p$ and $\mathbf{b}_p$ are the convolutional filters and bias parameters, and $\mathbf{a}_p$ aggregates all the information along and across channels. The final weighted features $\widetilde{\mathbf{X}}$ can be obtained by

$$\widetilde{\mathbf{X}} = \mathbf{X} \otimes \mathbf{g}_p, \quad \mathbf{g}_p = \text{Gaussian}\left(\mathbf{a}_p\right) = e^{-\mathbf{a}_p^2}, \tag{7.38}$$

where $\mathbf{g}_p$ is final the attention map in range [0, 1], which is obtained by gating the response map $\mathbf{a}_p$ using a Gaussian operator.

The loss function for training is the mean square error (MSE) between the output derained image $\hat{\mathbf{B}}_i$ and its corresponding groundtruth $\mathbf{B}_i$, which can be formulated as

$$\mathcal{L} = \frac{1}{N} \sum_{i=1}^{N} \|\hat{\mathbf{B}}_i - \mathbf{B}_i\|_2^2, \tag{7.39}$$

where N is the input batch size. Here, we do not use the popular adversarial loss [21] or perceptual loss [28] for our task, since we find that they contribute little to our deraining network.

7.4.2 Experiments

In this section, we compare our deraining net with the state-of-the-art methods on both synthetic and real-world datasets. Also, an ablation study of network architecture and extended experiments on snow removal are presented. We directly take four public available datasets (i.e., *DDN-data* [17], *DID-MDN-data* [58], *Rain100L* and *Rain100H* [56]) to train our model. *DDN-data* [17] has 1000 clean images, and each image is used to generate 14 rainy images with different rain directions and densities. *DID-MDN-data* provided by [58] consists of 12,000 training images and 4000 test images where each image is assigned a label (i.e., light, medium, and heavy) based on its corresponding rain-density level. Both *Rain100L* and *Rain100H* are synthesized by [56]. *Rain100L* has only one type of rain streaks consisting of 200 image pairs for training and the other 100 pairs for test, while *Rain100H* is more challenging and it contains five different rain directions containing 1800 images for training and 100 images for test. For the real-world dataset, some images are released by [17, 56, 58], some are downloaded from the Internet and also some are captured by ourselves (see the removal results in n).

For parameter settings of the deraining net, we set kernel sizes to 3 × 3 for all the convolutional layers in encoded and decoded blocks. For model training, we randomly generate 10 patch pairs with a size of 128 × 128 for each training image pair. The entire network is then trained on an Nvidia TITAN Xp GPU using Pytorch framework with a batch size of 20. The RMSprop optimizer [49] is used for training, where the initial learning rate is set to 0.001 and linearly decreases as the number of iterations increases. The network converges after approximately 100 epochs of training.

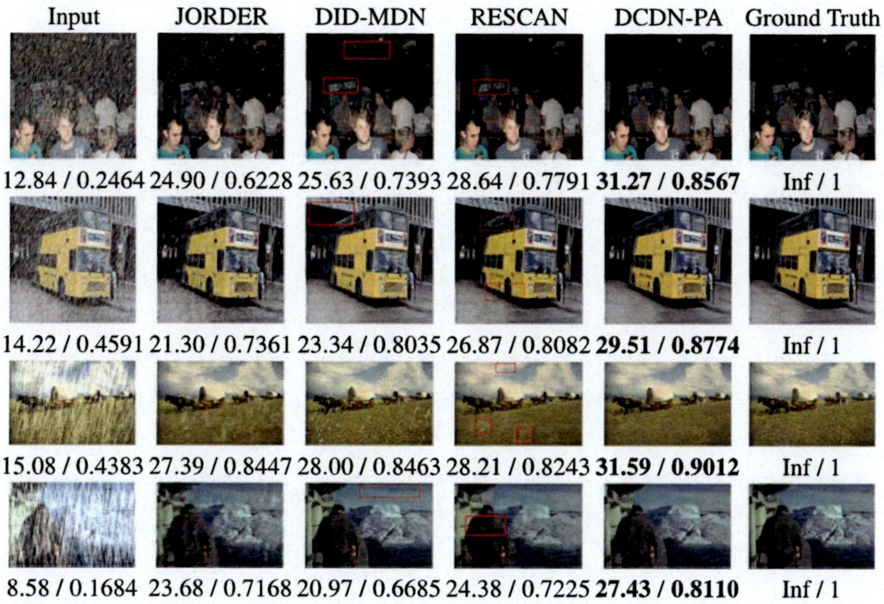

Fig. 7.15 Rain removal results on sample images from *DID-MDN-data* and *Rain100H*, respectively. JORDER [56] and RESCAN [33] use recurrent mechanism for rain removal but they tend to cause blur and artifacts, while DID-MDN [58] leaves too much rain streaks unremoved. © [2020] IEEE. Reprinted, with permission, from [Weihong Ren, Jiandong Tian, Qiang Wang, and Yandong Tang, "Dually Connected Deraining Net Using Pixel-Wise Attention," IEEE Signal Processing Letters, Vol.27, pp.316–320, 2020]

7.4.2.1 Comparisons on Synthetic Datasets

We denote our method as DCDN-PA (dually connected deraining net with pixel-wise attention), and compare it with six recent deraining methods LP [34], DDN [17], JORDER [56], DID-MDN [58], RESCAN [33], and SPANet [52] on four benchmark datasets. The peak signal-to-noise ratio (PSNR) and structure similarity index (SSIM) [54] are used as our metrics to evaluate fthe removal results (Fig. 7.15).

Table 7.3 presents the quantitative results of different methods on the four benchmarks. Overall, our method performs better than the other methods on the four datasets. For example, our DCDN-PA model can improve the PSNR metric achieved by the current best-performing model RESCAN by an average of 1.19db, 1.58db, 0.44db, and 2.75db, respectively, on the four datasets. SPANet [52] used SSIM as the loss function, so it has relatively high SSIM values. Both JORDER [56] and RESCAN [33] use the recurrent mechanism to remove rain in multiple stages and thus can achieve relatively better results. Though estimating rain density as additional information for rain removal, DID-MDN [58] performs slightly worse than other methods since a single label cannot accurately represent the rain density of a whole image. LP [34] adopts decomposition method for rain removal. It performs well for some light rain scenes, but badly for heavy rain scenes.

Table 7.3 Quantitative comparisons on four benchmark datasets in terms of PSNR and SSIM (PSNR / SSIM)

Dataset	Input	LP	DDN	JORDER	DID-MDN	RESCAN	SPANet	DCDN-PA
DDN-data	23.65/0.7534	25.64/0.8360	28.81/0.8874	28.92/0.8795	27.76/0.8771	29.94/0.8936	28.66/**0.9275**	**31.13**/0.9142
DID-MDN-data	22.14/0.6863	21.37/0.7923	27.47/0.8093	29.27/0.8542	28.22/0.8785	31.53/0.8898	29.78/0.9186	**33.11/0.9201**
Rain100L	25.52/0.8140	28.25/0.8763	31.05/0.9156	35.31/0.9672	29.14/0.9244	35.83/0.9675	34.55/0.9688	**36.27/0.9725**
Rain100H	12.13/0.3437	15.64/0.4239	23.34/0.6854	25.37/0.7988	23.17/0.7620	26.36/0.8109	26.34/**0.9043**	**29.11**/0.8691

7.4.2.2 Ablation Study of Network Architecture

In this subsection, we compare different variants of the DCDN-PA architecture. The first baseline is a plain network which is built on the ResBlocks. Besides, it also removes the dually skip connections and all the PAMs in DCDN-PA. To test the effect of SE-ResBlocks, the second baseline then is built by replacing ResBlocks in the plain net with SE-ResBlocks (denoted as "EncDec"). We also test the effect of using different skip connections by adding either the skip sum connection ("EncDec+**Sum**") or the skip concatenation connection ("EncDec+**Concat**") to the "EncDec." To evaluate the effectiveness of the pixel-wise attention module (PAM), we test a DCDN deraining net, by removing all the PAMs in DCDN-PA. Actually, PAM has two attention modules, i.e., spatial attention module (Eq. 7.36) and across-channel attention module (Eq. 7.37). We thus create another two new baselines by removing the channel-wise convolutional layer (denoted as "DCDN+**A**") or by removing the across-channel convolutional layer (denoted as "DCDN+**S**"). As described in [25], SE block also can be regarded as a channel-wise attention which can adaptively recalibrate feature responses. We thus create another new variant by replacing the PAMs in DCDN-PA with SE blocks (denoted as "DCDN+**SE**"). Our DCDN-PA model only applies the PAM to centered and decoded blocks. To evaluate the effectiveness of PAM on the encoded blocks, we also add PAM to the last layer of each encoded block in DCDN-PA (denoted as "DCDN+**All**"). All the variants are tested on the above four datasets.

Table 7.4 reports the average rain removal results. Using any of the attention mechanisms seems to perform better than not using the attention mechanism. For example, for *DID-MDN-data*, DCDN+**S** can improve DCDN by PSNR of 0.49db. In addition, pixel-wise attention model DCDN-PA achieves better results, compare with other attention models DCDN+**SE**, DCDN+**A** and DCDN+**S**. This indicates that the channel-wise attention and the across-channel attention are complementary and either provide useful information. Besides, either skip sum connection or skip concatenation connection contributes to the improvement of rain removal. DCDN+**All** performs worse than DCDN-PA, which indicates that the attention module on encoded blocks contributes little to the rain removal performance. As observed, the plain net can perform better compared with other state-of-the-art methods (see results on *Rain100H*), which demonstrates that the encoder–decoder net is good at handling rain removal problem.

7.4.2.3 Evaluation on Real-World Rain Dataset

We also evaluate our proposed deraining net on the collected real-world rainy images, and the derained results are shown in Figs. 7.16 and 7.17. As observed, JORDER [56] and DID-MDN [58] often fail to remove all the rain streaks in the images (see the second and sixth rows), while RESCAN [33] can remove all the rain streaks but it causes blur and artifacts to the derained images, e.g., the fence marked in red box

Table 7.4 Ablation study of the deraining net architecture

Dataset	Plain	EncDec	EncDec+**Concat**	EncDec+**Sum**	DCDN	DCDN+**SE**	DCDN+**A**	DCDN+**S**	DCDN+**All**	DCDN-PA
DDN-data	29.77/0.8986	30.58/0.9011	30.84 / 0.9062	30.75/0.9100	30.93/0.9090	31.03/0.9113	30.95/0.9141	31.05 /0.9120	31.08/0.9124	**31.13/0.9142**
DID-MDN-data	31.78/0.8876	32.21/0.8953	32.31 / 0.9028	32.37/0.9060	32.46/0.9057	32.44/0.9049	32.51/0.9079	32.95/0.9187	33.01/0.9196	**33.11/0.9201**
Rain100L	34.03/0.9512	34.86/0.9572	34.66 / 0.9605	34.94/0.9636	35.16/0.9670	34.74/0.9617	35.49/0.9682	35.25/0.9670	35.64/0.9691	**36.27/0.9725**
Rain100H	28.01/0.8281	27.94/0.8415	28.71/0.8641	28.90/0.8653	28.98/0.8658	28.96/0.8646	29.03/**0.8705**	29.10/0.8684	28.62/0.8664	**29.11**/0.8691

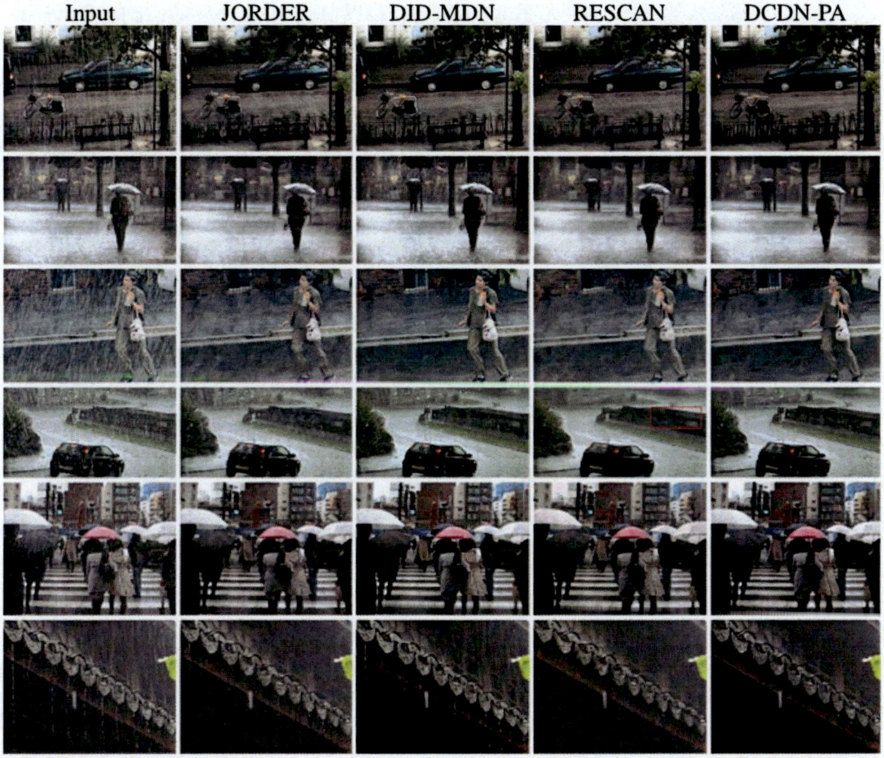

Fig. 7.16 Rain removal results on real-world rainy images. © [2020] IEEE. Reprinted, with permission, from [Weihong Ren, Jiandong Tian, Qiang Wang, and Yandong Tang, "Dually Connected Deraining Net Using Pixel-Wise Attention," IEEE Signal Processing Letters, Vol. 27, pp. 316–320, 2020]

in the fourth row. In contrast, our deraining net can effectively deal with heavy rain scenes containing different rain conditions, such as long rain streaks in the first and sixth rows and cluttered rain streaks in the second row.

7.4.2.4 Extension to Snow Removal

Different from rain streaks, snowflakes can drift in the air and also can fall quickly like rain streaks, which leads to complex shapes and scales of snowflakes. However, snowflakes and rain streaks often cause similar effects on the images, such as white streaks and haze accumulation. In this subsection, we explore whether the deraining model can be directly used for snow removal. Figure 7.18 shows the snow removal results on three different scenes. JORDER [56] can remove some white streaks in the image (see the second row), but it fails to remove the cluttered snowflakes (see

Fig. 7.17 Deraining results of a heavy rain image. **a** Input; **b** DID-MDN [58]; **c** RESCAN [33]; **d** Ours. The heavy rain scene has various rain densities, and thus, the density-aware method DID-MDN [58] tends to under derain the image. The recurrent method RESCAN [33] can remove all the rain streaks, but it degrades the image edges.

the first row). DID-MDN [58] adopted a refinement module to eliminate the haze effect, and thus achieves clear results. However, it leaves too many snowflakes in the restored results. As before, RESCAN [33] causes degradation of image edges for snow removal (see the red boxes in the fourth column). Our proposed network is able to remove snowflakes and also preserve image details. Note that all the models used for snow removal are trained on synthetic rain datasets. Overall, our proposed net has a good generalization ability for both rain and snow removal.

7.4.2.5 Running Times

Running times and parameters are reported in Table 7.5. Our method has relatively more parameters, but only takes $0.27s$ to derain an image of size 512×512. Also, our running time is comparable to other methods (e.g., 0.19s of DDN [17] and 0.2s of DID-MDN [58]). JORDER [56] and RESCAN [33] adopt the recurrent mechanism for rain removal and takes more time.

In this section, we propose an encoder–decoder deraining net for single-image rain removal in heavy scenes. It is only supervised by MSE loss and adopts dually skip connections for rain removal in one stage. The dual connections enable the deraining

Fig. 7.18 Rain removal results on real-world snowy images. © [2020] IEEE. Reprinted, with permission, from [Weihong Ren, Jiandong Tian, Qiang Wang, and Yandong Tang, "Dually Connected Deraining Net Using Pixel-Wise Attention," IEEE Signal Processing Letters, Vol.27, pp. 316–320, 2020]

Table 7.5 Running times of different methods

Size	LP (CPU)	DDN	JORDER	DID-MDN	RESCAN	SPANet	Ours
512×512	932.72 s	0.19 s	1.46 s	0.20 s	0.65 s	1.14 s	0.27 s
Parameters	–	0.0582M	0.024M	0.373M	0.055M	0.284M	0.460M

net to promote information flow between layers and thus can help to localize rain streaks. In addition, the learnable pixel-wise attention can adaptively recalibrate the decoded rain features to preserve more image details. Experimental results prove the superior performance of our method over the existing methods.

Appendix

Separating Moving Objects From Falling Snow

We explore some snowflake features to coarsely separate moving objects from falling snow. It is known that snow is white color and bright. We suppose a snowflake with RGB value vector S =[255 255 255] is passing through a background pixel, whose pixel value vector is denoted by E. If we denote the duration time for a snowflake remains within the background pixel as τ and the camera exposure time as T, the pixel values polluted by the snowflake can be given by $f = \alpha S + (1 - \alpha)E$ [18], where $\alpha = \frac{\tau}{T}$. Therefore, the intensity increased by snowflake is $\nabla I = \alpha(S - E)$. The maximum value of ∇I is 255α when background is totally black (Fig. 7.19).

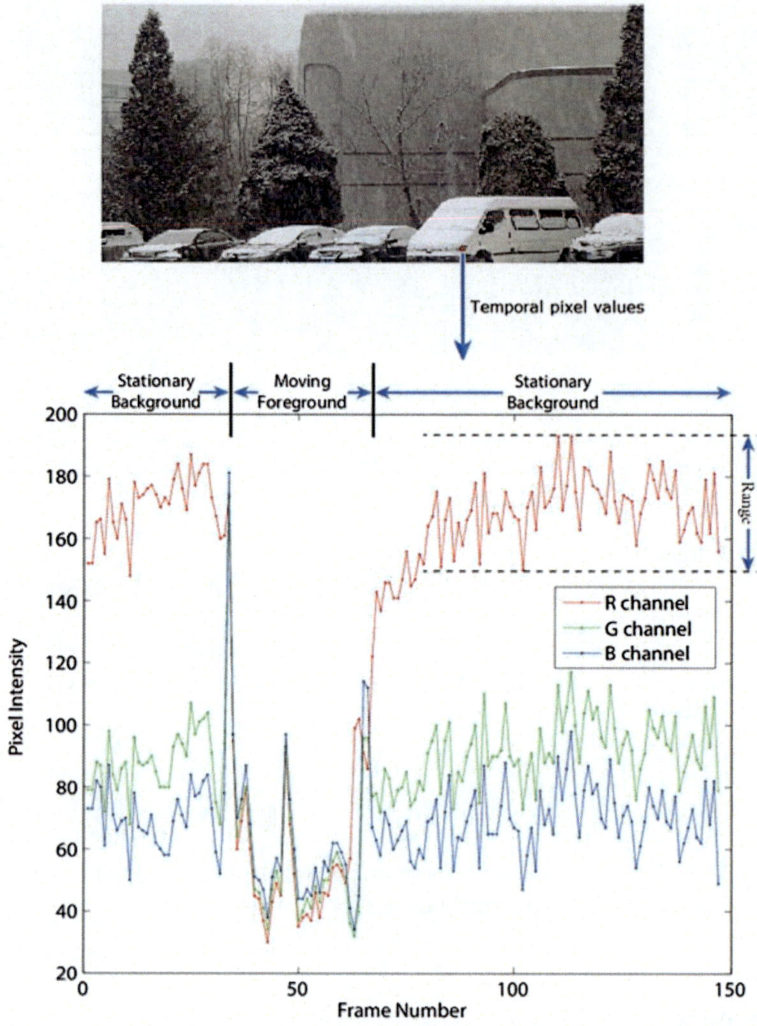

Fig. 7.19 Variations of pixel values caused by falling snow and moving foreground are different. We can find that variations caused by snowflake is limited compared with those caused by moving foreground. © [2018] IEEE. Reprinted, with permission, from [Jiandong Tian, Zhi Han, Weihong Ren, Xiai Chen, and Yandong Tang, "Snowflake Removal for Videos via Global and Local Low-rank Decomposition," IEEE Transactions on Multimedia, Vol. 20, Issue: 10, pp. 2659–2669, 2018]

The literature [18] shows that for a raindrop, the maximum value of τ is approximately 1.18 ms, which is much less than the typical exposure time $T \approx 30$ms of a video camera. Although snowflake sometimes is falling slower than raindrop, τ is still much less than T. If we assume snowflake falling five times slower than raindrop, in this situation, maximum of ∇I is 50. This shows that the intensity variation caused by snowflake has a bound. In other words, compared with the maximum variation values between moving foreground and stationary background, 255, maximum value of ∇I is much smaller. Figure 11 shows the temporal values of a fixed pixel location. From Fig. 11, we can find that variations caused by snowflake are limited compared with those caused by moving foreground. We also noticed that the size of snowflake is much smaller than moving foreground. Based on these two properties, we can separate dynamic regions of falling snowflakes and moving objects.

References

1. P.C. Barnum, S. Narasimhan, T. Kanade, Analysis of rain and snow in frequency space. Int. J. Comput. Vis. **86**(2–3), 256–274 (2010)
2. J. Black, T. Ellis, P. Rosin, A novel method for video tracking performance evaluation, in *Proceedings of the IEEE International Workshop on Visual Surveillance and Performance Evaluation of Tracking and Surveillance (VS-PETS 03)*, pp. 125–132 (2003)
3. A. Blake, P. Kohli, and C. Rother. *Markov Random Fields for Vision and Image Processing*. (MIT Press, 2011)
4. A. Buades, B. Coll, J. Morel, A non-local algorithm for image denoising, in *Proceedings of IEEE Computer Society Conference on Computer Vision and Pattern Recognition*, vol. 2 (2005), pp. 60–65
5. E.J. Candès, X. Li, Y. Ma, J. Wright, Robust principal component analysis? J. ACM **58**(3), 11 (2011)
6. V. Chandrasekaran, S. Sanghavi, P. Parrilo, A. Willsky, Rank-sparsity incoherence for matrix decomposition. SIAM J. Optim. **21**(2), 572–596 (2011)
7. D.-Y. Chen, C.-C. Chen, L.-W. Kang, Visual depth guided color image rain streaks removal using sparse coding. IEEE TCSVT **24**(8), 1430–1455 (2014)
8. L. Chen, H. Zhang, J. Xiao, L. Nie, J. Shao, W. Liu, T. S. Chua. Sca-cnn: Spatial and channel-wise attention in convolutional networks for image captioning, in *Proceedings of IEEE CVPR* (2017), pp. 6298–6306
9. Y. L. Chen, C. T. Hsu, A generalized low-rank appearance model for spatio-temporally correlated rain streaks, in *2013 IEEE International Conference on Computer Vision (ICCV)* (2013), pp. 1968–1975
10. X. Chu, W. Yang, W. Ouyang, C. Ma, A. L. Yuille, X. Wang, Multi-context attention for human pose estimation, in *Proceedings ofp IEEE CVPR* (2017), pp. 5669–5678
11. K. Dabov, A. Foi, V. Katkovnik, K. Egiazarian, Image denoising by sparse 3-d transform-domain collaborative filtering. IEEE Trans. Image Process. **16**(8), 2080–2095 (2007)
12. W. Dong, L. Zhang, G. Shi, X. Wu, Image deblurring and super-resolution by adaptive sparse domain selection and adaptive regularization. IEEE TIP **20**(7), 1838–1857 (2011)
13. W. Dong, G. Shi, X. Li, Nonlocal image restoration with bilateral variance estimation: a low-rank approach. IEEE Trans. Image Process. **22**(2), 700–711 (2013)
14. M. Elad, M. Aharon, Image denoising via sparse and redundant representations over learned dictionaries. IEEE Trans. Image Process. **15**(12), 3736–3745 (2006)
15. R. Fattal, Single image dehazing. ACM TOG **27**(3), 1–9 (2008)

16. X. Fu, J. Huang, X. Ding, Y. Liao, J. Paisley, Clearing the skies: a deep network architecture for single-image rain removal. IEEE TIP **26**(6), 2944–2956 (2017)
17. X. Fu, J. Huang, D. Zeng, Y. Huang, X. Ding, J. Paisley, Removing rain from single images via a deep detail network, in *Proceedings of IEEE CVPR, 2017*, pp. 1715–1723
18. K. Garg, S. K. Nayar, Detection and removal of rain from videos, in *Proceedings of IEEE CVPR, 2004*, pp. 528–535
19. K. Garg, S.K. Nayar, When does a camera see rain? in *IEEE International Conference on Computer Vision* (IEEE, 2005), vol. 2, pp. 1067–1074
20. K. Garg, S.K. Nayar. Photorealistic rendering of rain streaks, in *ACM Transactions on Graphics (TOG), ACM* (2006), vol. 25, pp. 996–1002
21. I.J. Goodfellow, J. Pougetabadie, M. Mirza, B. Xu, D. Wardefarley, S. Ozair, A.C. Courville, Y. Bengio, Generative adversarial nets, in *Proeedings of NeurIPS* (2014), pp. 2672–2680
22. L. Guanbin, H. Xiang, Z. Wei, C. Huiyou, D. Le, L. Liang, Non-locally enhanced encoder-decoder network for single image de-raining, in *Proceedings of ACMMM* (2018)
23. K. He, X. Zhang, S. Ren, J. Sun, Deep residual learning for image recognition, in *Proceedings of IEEE CVPR*, (2016) pp. 770–778
24. W. Hengyou, C. Yigang, H. Zhihai, Z. Ruizhen. Robust generalized low-rank decomposition of multi-matrices for image recovery. IEEE Trans. Multimedia (99), 1–14 (2016)
25. J. Hu, L. Shen, G. Sun, Squeeze-and-excitation networks, in *Proceedings of IEEE CVPR* (2018), pp. 7132–7141
26. D.A. Huang, L.W. Kang, Y.C.F. Wang, C.W. Lin, Self-learning based image decomposition with applications to single image denoising. IEEE Trans. Multimedia **16**(1), 83–93 (2014)
27. H. Ji, C. Liu, Z. Shen, Y. Xu, Robust video denoising using low rank matrix completion, in *IEEE Conference on Computer Vision and Pattern Recognition (CVPR)* (IEEE, 2010), pp. 1791–1798
28. J. Johnson, A. Alahi, L. Feifei, Perceptual losses for real-time style transfer and super-resolution, in *Proceedings of ECCV* (2016), pp. 694–711
29. L.W. Kang, C.W. Lin, Y.H. Fu, Automatic single-image-based rain streaks removal via image decomposition. IEEE Trans. Image Process. **21**(4), 1742–1755 (2012)
30. J.H. Kim, C. Lee, J.Y. Sim, C.S. Kim, Single-image deraining using an adaptive nonlocal means filter, in *2013 20th IEEE International Conference on Image Processing (ICIP)* (2013), pp. 914–917
31. J.H. Kim, J.Y. Sim, C.S. Kim, Video deraining and desnowing using temporal correlation and low-rank matrix completion. IEEE TIP **24**, 2658–2670 (2015)
32. V. Kolmogorov, R. Zabin, What energy functions can be minimized via graph cuts? IEEE Trans. Pattern Anal. Mach. Intell. **26**(2), 147–159 (2004)
33. X. Li, J. Wu, Z. Lin, H. Liu, H. Zha, Recurrent squeeze-and-excitation context aggregation net for single image deraining, in *Proceedings of ECCV* (September 2018)
34. Y. Li, R.T. Tan, X. Guo, J. Lu, M.S. Brown, Rain streak removal using layer priors, in *Proceedings of IEEE CVPR* (2016), pp. 2736–2744
35. Z. Lin, M. Chen, Y. Ma, The augmented lagrange multiplier method for exact recovery of corrupted low-rank matrices, in *Technical Report UILU-ENG-09-2215, UIUC* (2009)
36. C. Liu, J. Yuen, A. Torralba, SIFT flow: dense correspondence across scenes and its applications. IEEE Trans. Pattern Anal. Mach. Intell. **d33**(5), 978–994 (2011)
37. J. Long, E. Shelhamer, T. Darrell, Fully convolutional networks for semantic segmentation, in *Proceedings of IEEE CVPR* (2015), pp. 3431–3440
38. D.G. Lowe, Object recognition from local scale-invariant features, in *The Proceedings of the Seventh IEEE International Conference on Computer Vision*, vol. 2 (IEEE 1999), pp. 1150–1157
39. Y. Luo, Y. Xu, H. Ji, Removing rain from a single image via discriminative sparse coding, in *Proceedings of IEEE ICCV* (2015, December)
40. J. Mairal, F. Bach, J. Ponce, G. Sapiro, A. Zisserman, Non-local sparse models for image restoration, in *2009 IEEE 12th International Conference on Computer Vision* (IEEE, 2009), pp. 2272–2279

41. R. Mazumder, T. Hastie, R. Tibshirani, Spectral regularization algorithms for learning large incomplete matrices. J. Mach. Learn. Res. **11**, 2287–2322 (2010)
42. D.W. Murray, B.F. Buxton, Scene segmentation from visual motion using global optimization. IEEE Trans. Pattern Anal. Mach. Intell. **2**, 220–228 (1987)
43. S.G. Narasimhan, S.K. Nayar, Vision and the atmosphere. IJCV **48**(3), 233–254 (2002)
44. M.T. Orchard, G.J. Sullivan, Overlapped block motion compensation: an estimation-theoretic approach. IEEE Trans. Image Process. **3**(5), 693–699 (1994)
45. Y. Peng, A. Ganesh, J. Wright, W. Xu, Y. Ma, RASL: robust alignment by sparse and low-rank decomposition for linearly correlated images. IEEE Trans. Pattern Anal. Mach. Intell. **34**(11), 2233–2246 (2012)
46. W. Ren, J. Tian, Z. Han, A. Chan, Y. Tang, Video desnowing and deraining based on matrix decomposition, in *Proceedings of IEEE CVPR* (2017), pp. 4210–4219
47. O. Ronneberger, P. Fischer, T. Brox, U-net: convolutional networks for biomedical image segmentation, in *Proceedings of ICMICCAI* (2015), pp. 234–241
48. H. Sakaino, A semitransparency-based optical-flow method with a point trajectory model for particle-like video. IEEE Trans. Image Process. **21**(2), 441–450 (2012)
49. T. Tieleman, G. Hinton, Lecture 6.5-rmsprop: divide the gradient by a running average of its recent magnitude. COURSERA Neural Netw. Mach. Learn. **4**(2), 26–31 (2012)
50. A.K. Tripathi, S. Mukhopadhyay, Video post processing: low-latency spatiotemporal approach for detection and removal of rain. IET Image Process. **6**(2), 181–196 (2012)
51. F. Wang, M. Jiang, C. Qian, S. Yang, C. Li, H. Zhang, X. Wang, X. Tang, Residual attention network for image classification, in *Proceedings of IEEE CVPR* (2017), pp. 3156–3164
52. T. Wang, X. Yang, K. Xu, S. Chen, Q. Zhang, R.W. Lau, Spatial attentive single image deraining with a high quality real rain dataset, in *Proceedings of IEEE CVPR* (2019), pp. 12270–12279
53. X. Wang, R.B. Girshick, A. Gupta, K. He, Non-local neural networks, in *Proceedings of IEEE CVPR* (2018), pp. 7794–7803
54. Z. Wang, A.C. Bovik, H.R. Sheikh, E.P. Simoncelli, Image quality assessment: from error visibility to structural similarity. IEEE TIP **13**(4), 600–612 (2004)
55. Y. Wu, J. Lim, M.-H. Yang, Online object tracking: a benchmark, in *Proceedings of the IEEE Conference on Computer Vision and Pattern Recognition* (2013), pp. 2411–2418
56. W. Yang, R. T. Tan, J. Feng, J. Liu, Z. Guo, S. Yan, Deep joint rain detection and removal from a single image, in *Proceedings of IEEE CVPR* (2017), pp. 1357–1366
57. Q. You, H. Jin, Z. Wang, C. Fang, J. Luo, Image captioning with semantic attention, in *Proceedings of IEEE CVPR* (2016), pp. 4651–4659
58. H. Zhang, V.M. Patel, Density-aware single image de-raining using a multi-stream dense network, in *Proceedings of IEEE CVPR* (2018), pp. 695–704
59. X. Zhang, H. Li, Y. Qi, W. K. Leow, T.K. Ng, Rain removal in video by combining temporal and chromatic properties, in *IEEE International Conference on Multimedia and Expo* (IEEE, 2006), pp. 461–464
60. X. Zhang, T. Wang, J. Qi, H. Lu, G. Wang, Progressive attention guided recurrent network for salient object detection, in *Proceedings of IEEE CVPR* (2018), pp. 714–722
61. X. Zhou, C. Yang, W. Yu, Moving object detection by detecting contiguous outliers in the low-rank representation. IEEE Trans. Pattern Anal. Mach. Intell. **35**(3), 597–610 (2013)
62. D. Zoran, Y. Weiss, From learning models of natural image patches to whole image restoration, in *2011 IEEE International Conference on Computer Vision (ICCV)* (IEEE, 2011), pp. 479–486

Chapter 8
Single-Image Dehazing

Abstract Image dehazing is a useful technique which can eliminate the bad effect of haze on images and enhance the performances of image/video processing algorithms in hazy weather. In this chapter, we first propose a single-image dehazing method. We estimate the initial transmission properly based on latent region segmentation and refine the estimated initial transmission by an objective function with a novel weighted L1-norm regularization term. The half-quadratic splitting minimization method is employed to solve this optimization problem. We also define an evaluation function to estimate the reliable global atmospheric light. With the refined transmission map and atmospheric light, we recover the haze-free image by the haze imaging model. We then propose a retinex-based decomposition model for a hazy image and a novel end-to-end image dehazing network. In the model, the illumination of the hazy image is decomposed into natural illumination for the haze-free image and residual illumination caused by haze. Based on this model, we design a deep retinex dehazing network (RDN) to jointly estimate the residual illumination map and the haze-free image. Our RDN consists of a multiscale residual dense network and a U-Net with channel and spatial attention mechanisms. The multiscale residual dense network can simultaneously capture global contextual information from small-scale receptive fields and local detailed information from large-scale receptive fields to precisely estimate the residual illumination map caused by haze. And in the dehazing U-Net, we apply the channel and spatial attention mechanisms in the skip connection to achieve a trade-off between overdehazing and underdehazing by automatically adjusting the channel-wise and pixel-wise attention weights so as to obtain a refined haze-free image.

8.1 Introduction

In nature, visible light is often scattered due to the non-uniformity of the light propagation medium, and the main visible light scattering environment includes fog, haze, and underwater. Scattering often destroys image quality, reduces image clarity and contrast, blurs target features, produces color deviation, and makes scene information unable to be accurately expressed. Because most of the computer vision algorithms

J. Tian, *All Weather Robot Vision*, Research on Intelligent Manufacturing,
https://doi.org/10.1007/978-981-16-6429-8_8

are developed based on clear input images, their performance can be greatly reduced or even invalid in fog, haze, and underwater environment. As an important means to eliminate the influence of visible light scattering, image restoration algorithm in visible light scattering environment can significantly improve the clarity, contrast and texture details of the image, recover the true color of the image, and effectively improve the accuracy and stability of computer vision algorithm in fog, haze, and underwater environment. In Chap. 8 and Chap. 9, we will introduce image defogging and underwater image descattering, respectively.

At present, most of the scattering environment image restoration algorithms that conform to the physical mechanism are based on the fog environment imaging model proposed by McCartney in 1976 [22], and later improved by Narasimhan and Nayar [25], which can be expressed as:

$$I(x) = t(x)J(x) + (1 - t(x))A, \tag{8.1}$$

where I represents the foggy image taken by the camera; J represents the clear image without fog to be recovered; A is a global constant parameter, which represents the ambient light in the atmosphere, also known as skylight; t is the transmissivity of the propagation medium, which reflects the components of the reflected light of the scene that is not affected by the scattering and reaches the camera. The model assumes that the skylight is homogeneous, then $t(x)$ is inversely proportional to the depth of the scene, which can be expressed as:

$$t(x) = e^{-\beta d(x)}, \tag{8.2}$$

where β is the skylight scattering coefficient, and $d(x)$ is the depth from the surface point of each scene to the camera.

We can see that there are three unknowns in the model Eq. 8.1, so image restoration in scattering environment is a sick problem. The core problem is how to solve the transmissivity (or scene depth) efficiently and accurately, which is a very challenging task.

He et al. [10] proposed a famous dark channel priori to estimate the scene depth of foggy images. By using the dark channel priori, the algorithm first generates an initial transmittance map and then uses the soft matting algorithm to optimize the initial transmittance. The defogging effect of this method is very significant, but the disadvantage of this method is that the calculation speed of soft matting algorithm is slow. In order to improve the computing speed, He et al. [11] further proposed an edge keeping smoothing factor: guided filtering. The filter can avoid the gradient reversal near the strong edge. By using guided filtering to optimize the initial transmittance, the computational speed of the algorithm is significantly improved, and the quality of the recovery results is similar to that of the soft matting method. Berman et al. proposed an improved color-line fog removal algorithm, which is called haze line. It is assumed that a fog-free image is approximately generated by hundreds of intrinsic colors. Through statistical observation, we can get that the radiation intensity values in the fog image under RGB color space coordinate system are all along the straight line between the color intrinsic radiation intensity value and the atmospheric light intensity value, and it shows the distance to the atmosphere light intensity value with the increase and decrease of the concentration of the mist and target distance. Therefore, the algorithm can use the ratio of the radiation intensity of each pixel to the radiation intensity of the original color certificate to determine the initial transmittance, and then use the regularization method to optimize the transmittance. In most cases, the algorithm can get ideal and natural color defogging results, but the algorithm often fails when the atmospheric brightness is too much higher than the scene target brightness.

8.2 Single-Image Dehazing by Latent Region Segmentation

In this section, we propose a transmission estimation method based on latent region segmentation to estimate the proper initial transmission from single image. To optimize the initial transmission, we define a novel objective function. In this objective function, the weighted L_1-norm regularization term is used to smooth contextual regions of transmission map while preserving the salient edges. Considering of both intensity similarity and spatial similarity within a local patch, our weight function in the regularization term can inhibit the halo artifacts efficiently. We adopt the half-quadratic splitting minimization method [35, 39] to solve the optimization problem iteratively and achieve our refined transmission map. To estimate the global atmospheric light, we define a novel evaluation function. With the refined transmission and reliable atmospheric light, we can restore haze-free results from single image. Our method is compared with three state-of-the-art methods and is validated by two image quality assessment methods. We adopt the dynamic range-independent IQA [2] and blind contrast restoration assessment method based on visible edges ratio [9] to evaluate the performances of our method in color space and gray level, respectively. The comparative experimental results and evaluations demonstrate that our method can recover the better results with clear details, low contrast loss, and high contrast.

For image dehazing, the purpose is to recover scene radiance J.According to Eq. 8.1, the scene surface radiance J at each pixel is described as

$$J(x) = \frac{I(x) - A}{t(x)} + A. \tag{8.3}$$

From Eq. 8.3, t and A are also unknowns, so haze removal is a severe ill-posed problem. To estimate the medium transmission, He et al. [10] propose the dark channel prior. This method is based on the observation that at least one of the RGB color channels has very low intensity at some pixels in most of the non-sky regions. They further assume that the transmission in a local patch Ω is constant and define the dark channel J_d of J:

$$J_d(x) = \min_c(\min_{x\in\Omega(y)}(J^c(x)) = 0, \tag{8.4}$$

where c denotes one channel of RGB color space, Ω(y) is a local patch centered at y. Then they take the min operation in the local patch Ω on the haze imaging model Eq. 8.1 among three color channels:

$$\min_c(\min_{x\in\Omega(y)}(\frac{I^c(x)}{A^c})) = t(x)\min_c(\min_{x\in\Omega(y)}(\frac{J^c(x)}{A^c}) + (1 - t(x)). \tag{8.5}$$

After substituting Eq. 8.4 into Eq. 8.5, the estimated transmission t_d by dark channel prior at each pixel is

$$t_d(x) = 1 - \min_c(\min_{x\in\Omega(y)}(\frac{I^c(x)}{A^c})). \tag{8.6}$$

For neglecting the sky regions, the dark channel prior is not available in the large bright areas. The reason is mainly that the intensities of pixels in the sky and other bright regions are quite high, and they do not approach to zero as the assumption in a dark channel prior. According to Eqs. 8.4 and 8.5, we obtain

$$t(x) = \frac{1 - \min_c(\min_{x\in\Omega(y)}((I^c(x))/A^c))}{1 - ((J_d(x))/A)}. \tag{8.7}$$

In bright regions, $J_d(x)$ does not approximate to zero while has high intensity. The denominator in Eq. 8.7 is not equal to 1 as in Eq. 8.6, but has low value. Therefore, in bright regions, the transmission $t(x)$ should be higher than $t_d(x)$. If we adopt the loose bound of dark channel prior and simply set $J_d(x)$ to be zero as Eq. 8.6, in bright regions the transmission will be underestimated. From Eq. 8.3, the underestimated transmission will result in over-saturated pixels.

Though in non-bright regions $J_d(x)$ conforms to dark channel prior, the patch-wise min-operator on $I^c(x)/A^c$ in Eq. 8.6 will lead to overestimated transmission. For example, in one patch, if the pixel (which has the minimum value) and the central point belong to different objects, the term $I^c(x)/A^c$ in Eq. 8.6 will be lower than its true value and the transmission of current pixel will be higher than its accurate value. From Eq. 8.3, the overestimated transmission will lead the intensity of pixels in dehazing results to be lower than the true scene surface radiance.

8.2.1 Estimating Transmission

Transmission estimation is a significant procedure in dehazing. The effect of dehazing mainly depends on the accuracy of transmission estimation. Both underestimating and overestimating transmissions will result in undesirable dehazing results. In this section, to solve the problems mentioned above, according to the dark channel prior and another intrinsic prior, we propose a transmission estimation method based on latent region segmentation to obtain proper initial transmission. We know that the transmission ranges from zero to one, as $0 \leq t(x) \leq 1$. According to Eq. 8.1, we derive

$$t^c(x) = \mid \frac{A - I^c(x)}{A - J^c(x)} \mid, \tag{8.8}$$

In practice, most pixels in bright regions (e.g. sky, bright water or glass surfaces, the headlights of vehicles and some white objects) can be a little brighter than the global atmospheric light. Therefore, in bright regions, $I^c(x)$ are higher than A, and we have

$$t^c(x) = \frac{I^c(x) - A}{\mid A - J^c(x) \mid}, \tag{8.9}$$

To avoid the transmission being underestimated in bright regions, we do not assume the haze-free radiance J to be zero as in [10]. According to the priors that in bright regions $J(x)$ has high value and $J(x) \leq 255$, we have the inequality

$$t^c(x) = \frac{I^c(x) - A}{\mid A - J^c(x) \mid} \geq \frac{I^c(x) - A}{\mid A \quad 255 \mid}. \tag{8.10}$$

Supposing that A is given, we take the maximum value of $t^c(x)$ as the lower bound of transmission as follows:

$$t_b(x) = \max_c(t^c(x)) = \frac{max_c(I^c(x)) - A}{\mid A - 255 \mid} \tag{8.11}$$

To avoid the overestimated transmission, we do not use the patch-wise min-operator as in [10]. For in non-bright regions $I^c(x)$ are lower than the value of A, at each pixel we can get another inequality

$$t^c(x) = \frac{A - I^c(x)}{\mid A - J^c(x) \mid} \geq \frac{A - I^c(x)}{\mid A - min_c(J^c(x)) \mid}. \tag{8.12}$$

Since in non-bright regions the dark channel prior is available, $\min_c(J_c(x))$ should equal to zero. The inequality Eq. 8.12 can be rewritten as follows:

$$t^c(x) \geq \frac{A - I^c(x)}{A} = 1 - \frac{I^c(x)}{A}. \tag{8.13}$$

Still supposing that A is given and taking the maximum value in inequality Eq. 8.13 as the lower bound of $t(x)$ in non-bright regions, we have

$$t_n(x) = \max_c(t^c(x)) = 1 - \frac{\min_c(I^c(x))}{A}. \tag{8.14}$$

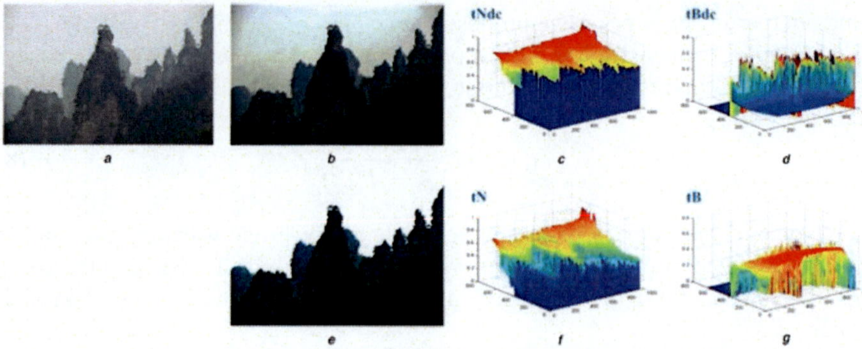

Fig. 8.1 Comparing the tough transmission of our latent region segmentation-based transmission estimation method with that of dark channel method **a** input hazy image, **b** haze-free result by dark channel prior, **c** 3D scatter plot of transmission in non-bright region by dark channel prior, **d** 3D scatter plot of transmission in bright region by dark channel prior, **e** haze-free result by our method, **f** 3D scatter plot of transmission in non-bright region by our method, **g** 3D scatter plot of transmission in bright region by our method. Reproduced from [Tong Cui, Jiandong Tian, Ende Wang, and Yandong Tang, "Single Image Dehazing by Latent Region Segmentation Based Transmission Estimation and Weighted L1-norm Regularization," IET Image Processing, Vol.11, Issue 2, 2017, pp. 145–154.] by permission of John Wiley & Sons Ltd.

To sum up, the initial transmission of the whole image can be described as follows:

$$t_o(x) = min\{\max\{\frac{\max_c(I^c(x) - A)}{\mid A - 255 \mid}, 1 - \frac{\min_c(I^c(x))}{A}\}, 1\}. \tag{8.15}$$

Equation 8.15 can separate the scene into bright and non-bright regions latently with smooth edges and correct the transmissions in the two sorts of regions, respectively. It also keeps the transmission values within the reasonable range. If the transmission of any pixel is higher than 1, this pixel will be over-saturated.

Figure 8.1*b* and *e* illustrates the dehazing results by the dark channel prior and our latent region segmentation-based transmission estimation method, respectively. We can observe that in non-bright regions the dehazing effect of Fig. 8.1*e* is superior to Fig. 8.1*b*, and in bright regions, our method also produces fewer halo artefacts than the dark channel prior. The scatter plots in Fig. 8.1 aim at pixel wisely displaying the differences between the rough transmission obtained by the dark channel prior and our method in both non-bright and bright regions. Figure 8.1*c* and *f* illustrate the scatters plots of transmission in non-bright regions by the two methods. Figure 8.1*d* and *g* illustrates the scatter plots of transmission in bright regions by the two methods. We can see that our pixel-wise transmission values are much lower in the non-bright regions and higher in the bright regions than the values of patch-wise transmission estimated by dark channel prior at most pixels. The scatters plots demonstrate that our latent region segmentation-based transmission estimation method can efficiently overcome the problems of overestimation and underestimation by dark channel prior in the whole image.

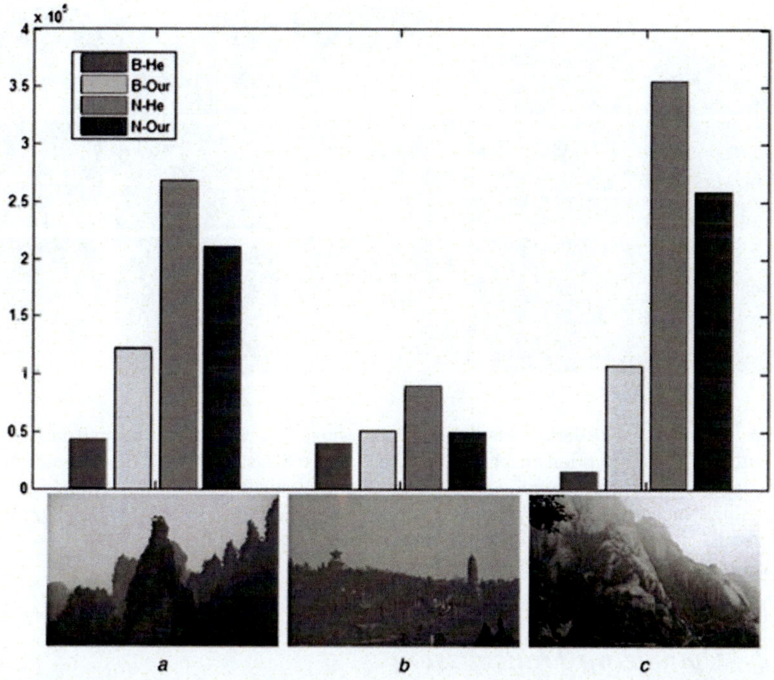

Fig. 8.2 Transmission statistical histograms. In groups a, b, and c, the first and second bars denote the sums of transmission values in bright regions by dark channel prior and our method, respectively, the third and fourth bars denote the sums of transmission values in non-bright regions by dark channel prior and our method, respectively. Reproduced from [Tong Cui, Jiandong Tian, Ende Wang, and Yandong Tang, "Single Image Dehazing by Latent Region Segmentation Based Transmission Estimation and Weighted L1-norm Regularization," IET Image Processing, Vol.11, Issue 2, 2017, pp. 145–154.] by permission of John Wiley & Sons Ltd.

The histograms in Fig. 8.2 are transmission statistical histograms. In each group, the first denotes the sum of transmission values in bright regions by dark channel prior; the first bar denotes the sum of transmission values in bright regions by our method; the third and fourth bars denote the sums of transmission values in non-bright regions by dark channel prior and our method, respectively. The statistical histograms further validate that our transmission estimation method can increase the transmission values when they are underestimated in bright regions and reduce the transmission values when they are overestimated in non-bright regions.

Figure 8.3 displays the comparative results by the dark channel prior and our method. From top to bottom, we compare the performances of the two algorithms in the sky region, bright water surface and dense fog, respectively. In the sky region, our dehazing result has fewer halo artefacts. On the water surface and in the dense hazy regions, our results are clearer. In non-bright regions, our results have more prominent dehazing effects than that of the dark channel.

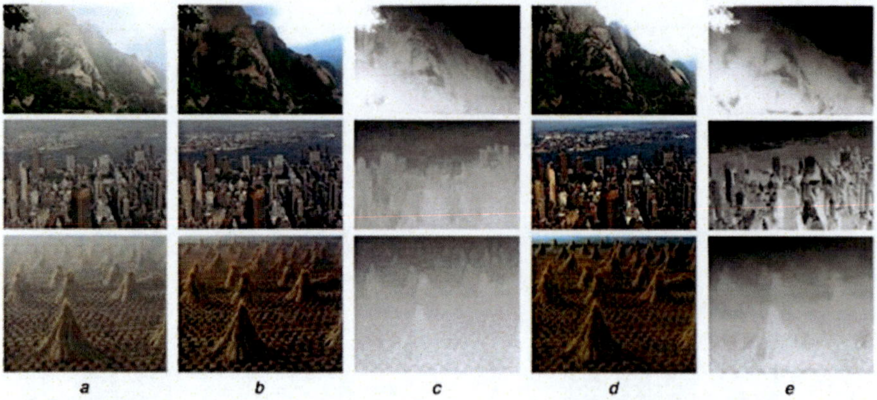

Fig. 8.3 Comparing with dark channel **a** input hazy image, **b**, **c** Haze-free result and the transmission map with dark channel prior, respectively, **d**, **e** Haze-free result and the transmission map with our method, respectively. Reproduced from [Tong Cui, Jiandong Tian, Ende Wang, and Yandong Tang, "Single Image Dehazing by Latent Region Segmentation Based Transmission Estimation and Weighted L1-norm Regularization," IET Image Processing, Vol.11, Issue 2, 2017, pp. 145–154.] by permission of John Wiley & Sons Ltd.

8.2.2 *Optimizing Transmission*

In Sect. 8.2.1, we obtain proper initial transmission by our latent region segmentation-based transmission estimation method. However, the pixel-wise transmission will lead the result to be distorted in color and make haze-free result look like a canvas.

We need to refine the rough pixel-wise transmission map for dehazing. Based on the observation that the pixels in a local patch often share similar depth value in common texture regions and the depth on the edges usually has drastic variation, we propose an objective function with a weighted L_1-norm regularization term to refine the rough transmission map. The half-quadratic splitting minimization method is used to solve the optimization problem iteratively.

8.2.2.1 Cost Function with Weighted L_1-norm Regularization Term

Though the pixels in the same object commonly share similar depth, on the edges between objects the depth will abruptly leap and lead to significant halo artefacts in the dehazing results. To restrain the halo artefacts, we constrain a weight function W on the gradient of rough transmission. $W(x)$ plays a switch role of the constraint on the gradient of rough transmission at pixel x. When the gradient of depth (rough transmission) at x is large, $W(x)$ should be small, and vice versa. When $W(x)$=0, the corresponding contextual constraint of $\nabla t(x)$ will be canceled. We describe this process as follows:

$$W(x) \circ \nabla t(x) \simeq 0, \tag{8.16}$$

where $\nabla t(x) = (\nabla_x t(x), \nabla_y t(x))$ denotes the horizontal and vertical gradient of rough transmission, and $\circ$ denotes the dot product operator.

The weight function $W(x)$ is defined as follows:

$$W(x) = e^{-\sum_{y\in\omega(x)}(\sum_{c\in\{r,g,b\}}|I_y^c - I_x^c|/(1-d_c(y)))/\sigma^2} \tag{8.17}$$

where $\omega(x)$ is a local patch centered at x, I is an input image with haze, $d_c(y)$ denotes the distance of any point y in $\omega(x)$ from the central point x, σ is a scale parameter. Our weight function considers both intensity similarity (depth difference) and spatial similarity (distance). Like most researchers did, we use color differences $| I_y^c - I_x^c |$ to replace depth differences which cannot be achieved directly. From Eq. 8.16 and Eq. 8.17, we can find that, if the color difference $\sum_{y\in\Omega(x)}(\sum_{c\in\Omega\{r,g,b\}} | I_y^c - I_x^c) |)$ is large, the $W(x)$ will be small and can guarantee the gradient of transmission $\nabla t(x)$ to be large enough. In this condition, W can restrain the gradient reversal artefacts near the salient edges. If $\sum_{y\in\Omega(x)}(\sum_{c\in\Omega\{r,g,b\}} | I_y^c - I_x^c) |)$ is small, the $W(x)$ will be large and can guarantee the gradient of transmission $\nabla t(x)$ to be small. W can smooth the contextual regions of transmission map. The distance function $d_c(y)$ is defined as follows:

$$d_c(y) = 1 - e^{-(n_i^2+n_j^2)/(2\cdot\delta^2)}, \tag{8.18}$$

where $d_c(y) \in [0, 1)$, generally $\delta = 3$, the size of $\omega(x)$ in our experiments usually is 11×11, n_i and n_j denote the local vertical coordinate and horizontal coordinate of pixel y in $\omega(x)$, respectively, $n_i = \{-5, -4, -3, -2, -1, 0, 1, 2, 3, 4, 5\}$ (from bottom to top), $n_j = \{-5, -4, -3, -2, -1, 0, 1, 2, 3, 4, 5\}$ (from left to right), the local coordinate of central point x in each $\omega(x)$ is (0,0). In Eq. 8.17, the distance $d_c(y)$ is inversely proportional to $W(x)$, when the distances of some pixels are quite large from x, our distance function can ensure these pixels have little effects on $W(x)$, and vice versa.

In the whole image, we compactly describe the discrete regularization on $\nabla t(x)$ as $|| W \circ \nabla t(x) ||_1$. To obtain the optimal transmission, we give an optimization function with the L_1-norm regularization as follows:

$$t = \underset{t}{\operatorname{argmin}}\{\frac{\lambda}{2} || t - t_o ||_2^2 + || W \circ \nabla t ||_1\}, \tag{8.19}$$

where the first part is the data term that measures the fidelity of $t(x)$ to the original transmission $t_o(x)$ derived from Eq. 8.15, the second part is the weighted L_1-norm regularization term, λ is a parameter to seek a balance between the two terms.

8.2.2.2 Transmission Calculation

We optimize transmission $t(x)$ through minimizing objective function Eq. 8.19 with half-quadratic splitting minimization method. Xu et al. [39] and Wang et al. [35] used this splitting scheme to solve two other different convex problems. We introduce the auxiliary variable u and transform Eq. 8.19 to a new optimization function as follows:

$$(t, u) = \underset{t,u}{\operatorname{argmin}}\{\frac{\lambda}{2} \parallel t - t_o \parallel_2^2 + \parallel W \circ \nabla t \parallel_1 + \frac{\beta}{2} \parallel u - \nabla t \parallel_2^2\}, \tag{8.20}$$

where β is a weight, $u = (u_h, u_v)$, u_h and u_v are the horizontal and vertical gradients, respectively.

We assume that t is fixed in Eq. 8.20 and minimize u at each pixel by following function:

$$u(x) = \underset{u(x)}{\operatorname{argmin}}\{\parallel W(x) \circ \nabla t(x) \parallel_1 + \frac{\beta}{2} \parallel u(x) - \nabla t(x) \parallel_2^2\}, \tag{8.21}$$

for $W(x)$, β and $\nabla t(x)$ are given, we can solve Eq. 8.21 as

$$u(x) = max(| \nabla t(x) | - \frac{W(x)}{\beta}, 0) \cdot sign(\nabla t(x)), \tag{8.22}$$

where sign($\cdot$) is a sign function. Then we optimize t with fixing u derived from Eq. 8.22 by minimizing the function below:

$$t = \underset{t}{\operatorname{argmin}}\{\frac{\lambda}{2} \parallel t - t_o \parallel_2^2 + \frac{\beta}{2} \parallel u - \nabla t \parallel_2^2\}. \tag{8.23}$$

The optimized t can be calculated by the equation below:

$$\frac{\lambda}{\beta}(t - t_o) + \nabla_x^T \otimes (\nabla_x t - u_h) + \nabla_y^T \otimes (\nabla_y t - u_v) = 0, \tag{8.24}$$

where ∇_x^T and ∇_y^T denote the transposed editions of directional gradient operators and $\otimes$ expresses the convolution operator. The closed-form solution for this least-squares minimization problem can be speeded up by fast Fourier transform. We can compute the optimized transmission t_f directly as follows:

$$t_f = F^{-1}(\frac{(\lambda/\beta)F(t_o) + \overline{F(\nabla_x)} \circ F(u_h) + \overline{F(\nabla_y)} \circ F(u_v)}{(\lambda/\beta) + \overline{F(\nabla_x)} \circ F(\nabla_x) + \overline{F(\nabla_y)} \circ F(\nabla_y)}), \tag{8.25}$$

where $F(\cdot)$ is the Fourier transform and $F^{-1}(\cdot)$ is its inverse version, $\overline{F(\cdot)}$ is its complex conjugate version.

Algorithm 1 describes the solution of Eq. 8.20. In the iteration process, β is increasing from $\beta_0 = 1$ to $\beta_{max} = 2^6$, multiplied by $\kappa = 2$ for each iteration. The parameter setting can ensure a good balance between efficiency and performance.

Algorithm 1 Solving Eq. 8.20

Input: Initial transmission $t_o(x)$, weight $W(x)$
Output: Optimized transmission $t_f(x)$
Initialisation: λ,β_o, β_{max} and κ.
1: while $\beta < \beta_{max}$ do
2: Solve Eq. 8.22 for u,
3: Solve Eq. 8.25 for t_f,
4: $\beta = \kappa\beta$,
5: end while
6: return t_f .

a b c d e

Fig. 8.4 Comparing our transmission refinement algorithm with guided image filter **a** input hazy image, **b** refined transmission with guided image filter, **c** refined transmission with our weighted L1-norm regularization method, **d** haze-free result with map (**b**), **e** haze-free result with map (**c**). Reproduced from [Tong Cui, Jiandong Tian, Ende Wang, and Yandong Tang, "Single Image Dehazing by Latent Region Segmentation Based Transmission Estimation and Weighted L1-norm Regularization," IET Image Processing, Vol.11, Issue 2, 2017, pp. 145–154.] by permission of John Wiley & Sons Ltd.

We compared our refined transmission maps with that using guided image filter. Figure 8.4 shows that our refined transmission maps are smoother and our haze-free results are also clearer. We mark the regions where our advantage is obvious with rectangles.

8.2.3 Recovering the Scene Radiance

Removing haze from single image by Eq. 8.3 and estimating transmission by our method both require an accurate global atmospheric light A. In this section, we introduce a new strategy to estimate the reliable global atmospheric light and recover our haze-free results with haze imaging model.

8.2.3.1 Estimating the Atmospheric Light

From Eq. 8.15, we know that the estimated transmission $t_o(x)$ and atmospheric light A have a close relationship. Inaccurate estimation of A will undermine the accuracy of transmission estimation and the quality of recovered scene radiance J. The atmospheric light A in Eq. 8.1 is often estimated as the brightest color in an image, such as [7, 34]. In practice, using the bright objects may lead to undesirable selection of the atmospheric light. In most cases, the optimal global atmospheric light is a little darker than the brightest pixels in an image. He et al. [10] first picked the top 0.1% brightest pixels in the dark channel, then selected the pixels with the highest intensity in input image I as the atmospheric light. Kim et al. [14] proposed a hierarchical searching method based on the quad-tree subdivision. They divide the region, which has the highest score (the average pixel value subtracted by the standard deviation of the pixel values within each region), into four smaller rectangular regions continually until the size of the selected region is smaller than a prescribed threshold.

We define a scoring formula based on a prior to estimate the atmospheric light. The prior is that the atmospheric light of an outdoor haze image usually appears in the dense haze areas or sky areas. In these areas, the contrasts of pixels are very low and the textures of images are very sparse. So we define the scoring formula as follows:

$$E_A(x) = c(x) \cdot ds(x)(x \in p_b), \tag{8.26}$$

where p_b indicates the set of the top 0.1% brightest pixels in the dark channel, $c(x)$ indicates the contrast value at each x; $ds(x)$ indicates the texture density in a square patch $\omega(x)$ centered at the pixel x. Generally, the size of $\omega(x)$ is 11×11, and $ds(x)$ can be expressed as follows:

$$ds(x) = \frac{1}{\mid \omega(x) \mid} \sum_{y \in \omega(x)} I_e(y) \tag{8.27}$$

where $\mid \omega(x) \mid$ expresses the number of the pixels in $\omega(x)$, I_e is a binary map which denotes the Sobel edge detection result of an input image. In I_e, 1 denotes the edge pixels, 0 denotes the non-edge pixels. We can employ the L_0-norm to calculate the sum of $I_e(x)$ in each $\omega(x)$.

In our strategy, the pixel whose score $E_A(x)$ is minimal in p_b is selected as the center point of the atmospheric light area $\omega_A(x)$ [its size is same with $\omega(x)$]. Then we get the maximum channel value $\max_c\ (I^c(y))$ for each pixel $y \in \omega_A(x)$ from the input image I. Finally, we compute the average value of $\max_c\ (I^c(y))$ as the atmospheric light intensity A. The red rectangles in Fig. 8.5 illustrate the atmospheric light regions.

Fig. 8.5 Rectangles represent the alternative region of atmospheric light. From top to bottom, the input hazy images, the dehazing results obtained by our method and the corresponding optimal transmission. Reproduced from [Tong Cui, Jiandong Tian, Ende Wang, and Yandong Tang, "Single Image Dehazing by Latent Region Segmentation Based Transmission Estimation and Weighted L1-norm Regularization," IET Image Processing, Vol.11, Issue 2, 2017, pp. 145–154.] by permission of John Wiley & Sons Ltd.

8.2.3.2 Recovering the Scene Radiance

With the refined transmission map and the estimated atmospheric light, the ill-posed single-image dehazing problem can be solved. We rewrite Eq. 8.1 as follows:

$$I(x) = t_f(x)J(x) + (1 - t_f(x))A, \tag{8.28}$$

then we recover the scene radiance as follows:

$$J(x) = \frac{I(x) - A}{\max(t_f(x), \varepsilon)} + A. \tag{8.29}$$

We illustrate the framework of our single-image dehazing algorithm in Fig. 8.6. Some of our recovered haze-free results are shown in Figs. 8.7 and 8.5.

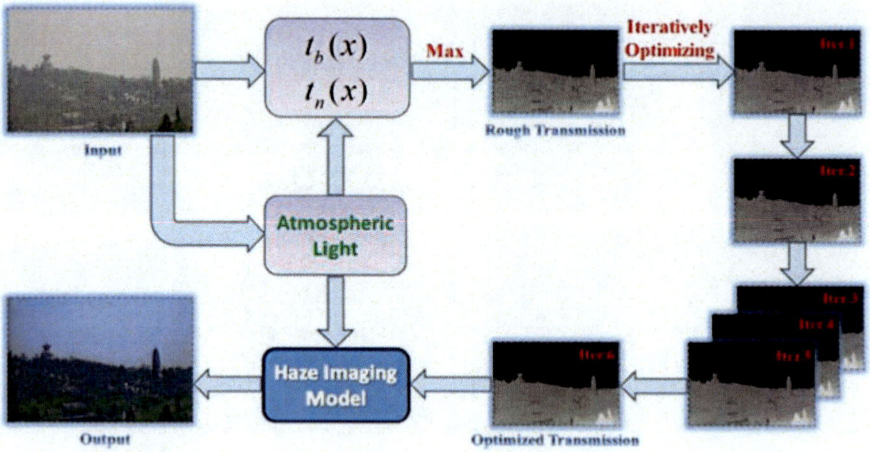

Fig. 8.6 Flowchart of our single-image dehazing algorithm

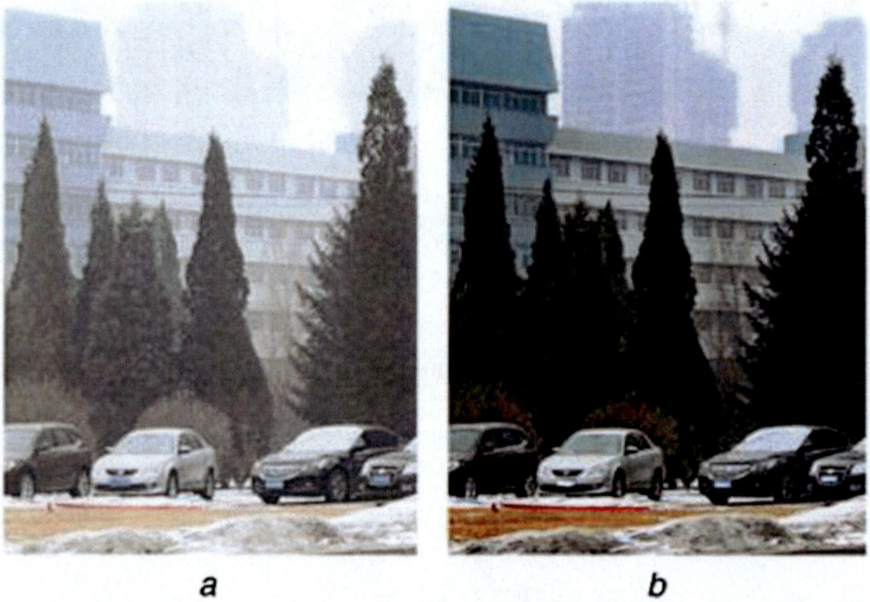

Fig. 8.7 Haze removal with our method **a** hazy input image, **b** dehazed image. Reproduced from [Tong Cui, Jiandong Tian, Ende Wang, and Yandong Tang, "Single Image Dehazing by Latent Region Segmentation Based Transmission Estimation and Weighted L1-norm Regularization," IET Image Processing, Vol.11, Issue 2, 2017, pp. 145–154.] by permission of John Wiley & Sons Ltd.

8.2.3.3 Implementing Details

To make our algorithm easier to be understood, here we introduce some of parameters and implement details. We fix σ in Eq. 8.19 to 10 for all results illustrated in this paper. The sizes of local patches ω in Eq. 8.17 and Eq. 8.27 are both 11×11; λ in Eq. 8.19 is generally set as 2. We solve Eq. 8.20 by 6 iterations, β is increasing from $\beta_0 = 1$ to $\beta_{\max} = 26$, multiplied by $\kappa = 2$ for each time. In Eq. 8.27, to decide which patch is sparse enough to express the atmospheric light region, we hope to achieve edge points as much as possible, so we select a low value 0.001 as the threshold for edge detection. ε in Eq. 8.29 is set as 0.1 for avoiding division by zero. Note that, the parameters mentioned above are all constant for the input hazy images.

8.2.4 Experimental Results

In this section, we display some of our dehazing results and compared the performances of our method with that of other three state-of-the-art methods [10, 14, 24] in computational cost and quality of results. Two image assessment methods are employed to evaluate the performances of the dehazing methods in color space and gray level, respectively.

8.2.4.1 Example Results

Figures 8.7 and 8.5 display some examples of our dehazing results and the corresponding optimal transmission maps. In these examples, we can see that our method can efficiently restore haze-free images with high contrast and vivid color information. Moreover, in the sky region, the halo artefacts of our results are also quite few. From bottom row of Fig. 8.5, we can see that our transmission maps are quite consistent with perception and approximate to the depth map.

8.2.4.2 Qualitative Comparisons

In Fig. 8.8, we compared our results with those of [10, 14, 24] in quality. We obtain some results of other algorithms from the authors' personal homepages and others through running the programs posted by the authors. Comparing our method with that in [10], in some details, especially in bright regions, our results are better than theirs. For instance, compared Figs. 8.8(1.b) and (2.b) with Figs. 8.8(1.e) and (2.e), the red carriage in dense haze, the sky and the distant tree regions in our results are more evident and have higher contrast. The tower crane disappears in Fig. 8.8(3.b), but it is restored in Fig. 8.8(3.e) clearly. In Fig. 8.8(4.b), the veil on the wall is still obvious, while in Fig. 8.8(4.e), we can see the clear red wall. From Figs. 8.8(1.c)–(4.c), we learn that the method in [24] can restore a haze-free image with high quality and fine

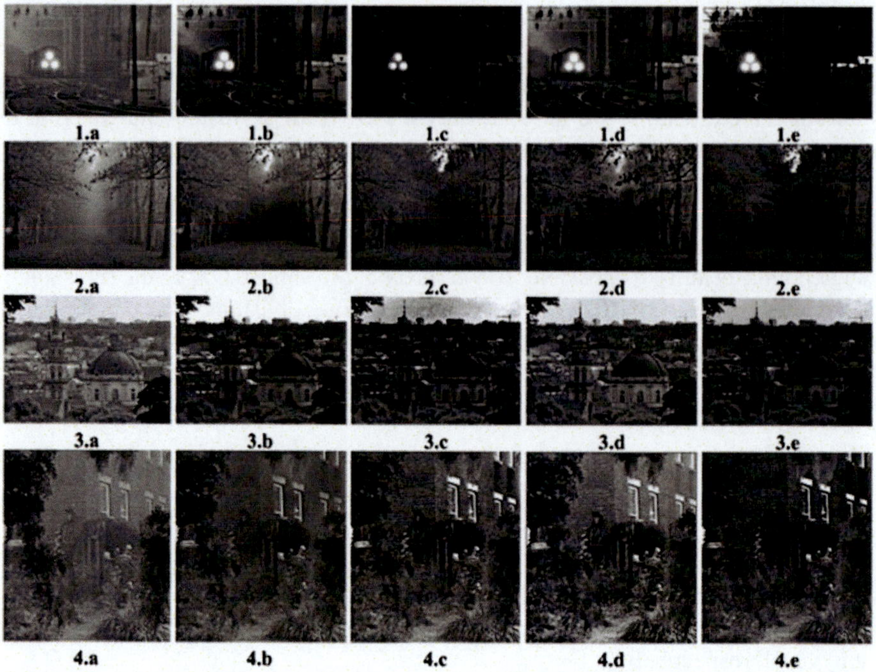

Fig. 8.8 From 1.a to 4.a, the original images with haze; from 1.b to 4.b, the restored outcomes by He's algorithm [10]; from 1.c to 4.c, the outputs of Meng's algorithm [14]; from 1.d to 4.d, the haze-free results of Kim's algorithm [24]; from 1.e to 4.e, our dehazing effects. Reproduced from [Tong Cui, Jiandong Tian, Ende Wang, and Yandong Tang, "Single Image Dehazing by Latent Region Segmentation Based Transmission Estimation and Weighted L1-norm Regularization," IET Image Processing, Vol.11, Issue 2, 2017, pp. 145–154.] by permission of John Wiley & Sons Ltd.

edge details. However, because they employ the weak constraint on the transmission of the whole image and choose different values of the atmospheric airlight intensity in the three channels, their results usually suffer from supersaturation and hue errors. Comparing the results of our method with that produced by the algorithm of [14] in the last two rows, our dehazing results are superior to theirs in most cases. However, we find that the objects in our results (especially the white and gray ones) near to the lens usually suffer from the supersaturation. This is due to that our method uses color differences to replace depth differences, and these objects have similar color with the dense hazes while the haze on them are quite light. Our method will process such objects as the dense haze. Consequently, the contrast of these pixels is excessively and inaccurately enhanced, such as the small white houses in Fig. 8.8(1.e).

8.2.4.3 Quantitative Evaluation

We processed the experimental images on a PC with 3.1 GHz Intel Core i5-2004 processor and 4G memory We compare our computational cost with the three algorithms mentioned above. The data of other algorithms are obtained through running the programs posted by the authors on the web. The programs of [10, 14, 24] and ours are implemented with MATLAB language, only the program of [24] is implemented with $C++$ language. Processing a 600×400 pixel image, our method needs about 0.8-1 s. Processing the same size images, the algorithm in [10] needs 10–20 s. The dark channel algorithm with guided image filter will take 1-1.5 s. Moreover, the algorithm in [24] needs 2.5-4 s. The algorithm in [14] is a real-time algorithm, its processing speed can reach about 35 fps.

We employ the dynamic range-independent IQA [2] to evaluate the contrast restoration capacity of our method in color space like in [1]. The dynamic range-independent IQA compares images with radically different dynamic ranges and evaluates the quality of contrast restoration pixel-wise. The three evaluation indicators of this metric include: loss of visible contrast (green)-contrast that was visible in the input image becomes invisible in the output image, amplification of invisible contrast (blue) contrast that was invisible in input image becomes visible in the output image, and reversal of visible contrast (red)- contrast that is visible in both images, but has different contrast polarity. The loss of visible contrast P_{loss}, amplification of invisible contrast P_{ampl}, and reversal of visible contrast P_{rev} are computed as follows:

$$\begin{aligned}P_{\text{loss}} &= 1 - \prod_{k=1}^{N}\prod_{i=1}^{M}(1 - \hat{P}_{\text{loss}}^{k,l})\\ P_{\text{ampl}} &= 1 - \prod_{k=1}^{N}\prod_{i=1}^{M}(1 - \hat{P}_{\text{ampl}}^{k,l})\\ P_{\text{rev}} &= 1 - \prod_{k=1}^{N}\prod_{i=1}^{M}(1 - \hat{P}_{\text{rev}}^{k,l})\end{aligned} \tag{8.30}$$

where k and l are the spatial band and orientation indices, respectively, $P_{\text{loss}}^{k,l}$, $P_{\text{ampl}}^{k,l}$ and $P_{\text{rev}}^{k,l}$ mean the conditional probabilities of loss of visible contrast, amplification of invisible contrast and reversal of visible contrast, respectively, $\hat{P}$ denotes conditional probability which is filtered by the corresponding cortex filter (for details about IQA please refer to [2]).

Figure 8.9 shows the comparative haze-free and IQA results of the four algorithms. We use the online implementation of IQA metric at the authors' home page [2] to produce the evaluation results. The bottom-left table of Fig. 8.9 shows the comparative ratios of the pixels yielded by the dynamic range-independent IQA. We only record the pixels whose probability scale is higher than 70% to reduce the possibility of misclassification. From the left bottom table of Fig. 8.9, we can find that our

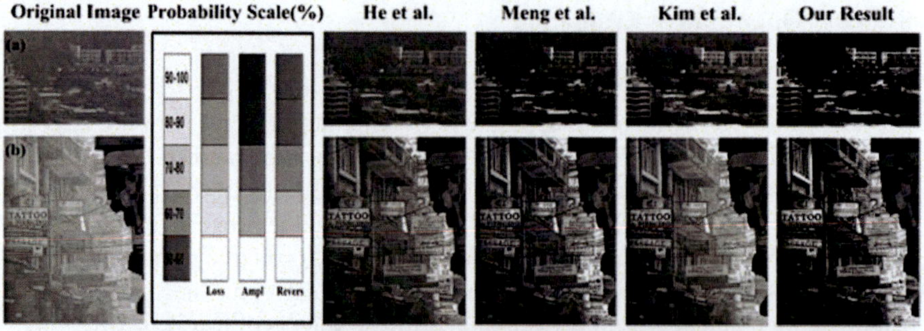

		Revers(%)	Ampl(%)	Loss(%)
He et al.	a	0	1.61	0.019
	b	6.29	28.30	0.84
Meng et al.	a	0. 75	73.70	0.002
	b	13.26	53.67	0.54
Kim et al.	a	0. 60	74.47	0.006
	b	10.54	30.95	1.27
Our Method	a	13.85	74.89	0.001
	b	37.35	41.87	0.27

Fig. 8.9 Evaluation of the results using dynamic range independent IQA. Top two lines present the comparative results obtained by dark channel prior [10], Meng et al. [24], Kim et al. [14] and our method. Bottom two lines show the results after applying the dynamic range-independent IQA between input and output images. Left bottom table illustrates the evaluation indicator values on the two group images obtained by each method. We arbitrarily chose green for loss of visible contrast (loss), blue for amplification of invisible contrast (ampl), and red for reversal of visible contrast (revers). Reproduced from [Tong Cui, Jiandong Tian, Ende Wang, and Yandong Tang, "Single Image Dehazing by Latent Region Segmentation Based Transmission Estimation and Weighted L1-norm Regularization," IET Image Processing, Vol.11, Issue 2, 2017, pp. 145–154.] by permission of John Wiley & Sons Ltd.

method can get almost all the best evaluation results on Fig. 8.9a, b except the "contrast amplification" on Fig. 8.9b (Our method gets the second-best evaluation result). According to the data, we see that our dehazing results have lower contrast loss ratio, higher contrast amplification, and higher reversal ratios than other three algorithms. As a general interpretation, contrast loss (green) is related to image blurring, contrast amplification (blue), and reversal (red) are connected to image sharpening [2]. The evaluation results objectively reflect that our restored results have better sharpening and more texture information in most cases.

We use the visible edges ratio image assessment method proposed in [9] to evaluate the performances of the four compared methods in gray level. The two indicators of this assessment method are the ratio of new visible edges (e) and ratio of average gradient ($\overline{r}$).

Table 8.1 Ratio of new visible edges (e) and ratio of average gradient ($\bar{r}$) obtained by the four compared methods on the images in Fig. 8.9

		[10]	[12]	[14]	Our method
Fig. 8.9a	e	0.94	2.24	2.01	2.29
	$\bar{r}$	1.69	2.85	3.34	3.51
Fig. 8.9b	e	0.25	0.17	0.19	0.21
	$\bar{r}$	1.87	2.67	2.25	2.71

The e denotes the increased rate of visible edges after haze removal and is described as follows:

$$e = \frac{n_J - n_I}{n_I}, \tag{8.31}$$

where n_I and n_J denote the cardinal number of the visible edges in the input hazy image I and the restored haze-free image J, respectively. The larger e is, the ability of method to restore edges is stronger.

The $\overline{r}$ uses the gradients of visible edges in the restored image to represent the restoration degree of the image edge and texture information.$\overline{r}$ is described as follows:

$$\overline{r} = e^{[(1/n_J)\sum_{i\in\psi_J} logr_i]}, \tag{8.32}$$

where $r_i = \Delta J / \Delta I$,$\Delta I$ and ΔJ are the gradients of I and J, respectively, r_i denotes the set of visible edges of J. A larger $\overline{r}$ means that the corresponding dehazing method has better capacity of edge preservation than others.

In Table 8.1, we show the ratio of new visible edges e and ratio of average gradient $\overline{r}$ obtained by the four compared methods on the images in Fig. 8.9. From Table 8.1, our method gets the best results of the ratio of average gradient on both Fig. 8.9a and b and gets the best and second-best results of the ratio of new visible edges on Fig. 8.9a and b. We can draw the conclusion that the performance of our method is better than that of other methods in most cases.

In this section, we propose an efficient single-image dehazing method. From some intrinsic priors, we derive a lower bound of transmission, which is used to estimate the initial transmission. This method can efficiently solve the problems of transmission overestimation in non-bright regions and transmission underestimation in bright regions by the dark channel prior. A novel objective function including with weighted L_1-norm regularization is proposed to refine this initial transmission map. For both intensity similarity and spatial similarity of transmission are considered, our weight function can efficiently restrain the halo artefacts. The half-quadratic splitting minimization method is employed to solve the optimization problem iteratively. We also define an evaluation function to estimate the global atmospheric light. With the refined transmission and accurate atmospheric light, we can restore haze-free results from single image. Our method is compared with three state-of-the-art algorithms and is validated by two image assessment methods both in color space and gray level.

The comparative experimental results and evaluations demonstrate that our method can recover the better results with clear details, low contrast loss and high contrast in most cases.

Nevertheless, sometimes our method makes pixels supersaturated. It is due to that some objects (especially the white and gray ones) near to the lens have similar color with the dense haze, but the haze on these objects are quite light. In these conditions, for we use color differences to replace depth differences which cannot be achieved directly, our method cannot reliably distinguish which pixel is heavily haze-contaminated and which is not. Then the method may process these objects as the dense haze and give these pixels very low transmission. Consequently, the contrast of pixels in these regions is excessively and inaccurately enhanced, such as shown in Fig. 8.7, the ground in Fig. 8.8(1.d) and some pixels in our result of Fig. 8.9b. For solving this problem, maybe more intrinsic priors from the multi-spectral images should be required. The processing speed of our algorithm also should be accelerated for real-time dehazing.

8.3 Deep Retinex Network for Single-Image Dehazing

In recent years, due to the powerful feature representation and end-to-end nonlinear mapping capacities for deep convolutional neural networks, many deep learning-based haze removal methods have been proposed. These methods include fully data-driven deep networks (e.g., encoder-decoder networks [23], GAN [18], GridNet [20], EPDN [28], and DADN [33]), fusion-based deep networks (GFN [29], DM2F-Net [6], and LAP-Net [19]), and scattering model-based deep networks such as DehazeNet [3], AOD-Net [16], and DCDPN [40]. Fully data-driven deep networks usually have limitations in interpretability and generalization. Fusion-based deep networks such as GFN, LAP-Net, and DM2F-Net need to learn three, four, and five kinds of confidence maps (weights), respectively. The imbalance among these maps is likely to cause image over-saturation or residual haze issues. Moreover, the number of the confidence maps and the pattern of each model usually requires artificial selection which usually cannot contains all models. For example, although DM2F-Net includes five models, it does not contain a pattern for the model $J = I/R$. The scattering model-based deep networks need to simultaneously predict the transmission map and atmospheric light. If the estimation is not accurate, the non-joint estimation of two physical parameters may further amplify the error when applied together, and the final recovered images usually suffer from color distortion or residual haze issues. In addition, the scattering model is a simplified model. Because the absorption, radiation, and multiple scattering of atmospheric particles are ignored, it usually cannot accurately describe the real-world hazy imaging. In view of the above analysis, image dehazing is still a challenging problem, especially the removal of heavy haze.

Retinex theory [15] is a very generic and simple theory. It separates an image into a reflectance image R multiplied by an illumination image L. It is widely used in the field of image lighting processing. For image dehazing, the essence of this problem is to get the transmittance which is closely related to illumination. If we can restore the light transmittance to an approximately normal state, the problem of image dehazing can be effectively solved. Recently, Galdran et al. [8] also provided mathematical proof of the dual relationship between image dehazing and retinex theory. In this section, we deduce a retinex-based dehazing decomposition model that does not rely on assumptions and priors and design a deep retinex network for image haze removal. Compared with the scattering model-based network, RDN can avoid errors caused by simplified scattering model. Additionally, our RDN has better generalization ability than the fully data-driven network, and it has no dependence on prior information. In addition, the powerful nonlinear fitting ability of our RDN avoids the drawback of traditional retinex-based methods, which require the artificial selection of the surrounding function and its parameters. For the detailed structure of our RDN, generally, the small-scale receptive field easily retains global semantic information, and the large-scale receptive field tends to preserve detailed local information. Therefore, we first design a multi-scale residual dense network to learn the residual illumination map that contains both local (e.g., shadow and highlight) and global information (e.g., color distribution and average brightness). Then, we obtain a coarse haze-free image with our retinex-based dehazing decomposition model. Finally, the coarse haze free image is fed into a dehazing refinement subnetwork to obtain a fine haze-free image. The dehazing refinement subnetwork is a typical U-Net framework consisting of residual dense blocks. Since the influence of haze on each pixel of a hazy image is different, we design a novel channel and spatial attention mechanism to achieve a trade-off between over-dehazing and under-dehazing by automatically adjusting the channel-wise and pixel-wise attention weights.

8.3.1 Retinex-Based Dehazing Model

In this section, we present our retinex-based image dehazing decomposition model and a deep retinex network for image dehazing.

In 1971, Edwin H. Land suggested that the human color perception system perceives object color to remain relatively constant under varying illumination conditions. He formulated retinex theory [15] to explain this phenomenon. According to retinex theory, an image I is a reflectance image R multiplied by an illumination image L. That is, each pixel value in the image can be calculated by:

$$I(x, y) = R(x, y) * L(x, y), \tag{8.33}$$

where $*$ denotes an element-wise multiplication.

Retinex theory has been widely applied in the field of image illumination processing [5, 21, 30, 38], in which it is essential to obtain the reflectance image R and the illumination image L from image I. For image haze removal, the essence of this problem is to obtain the transmittance, which is closely related to illumination, and it is not necessary to obtain a real reflection image R. We only need to obtain a dehazed image with natural illumination. That is, if we can restore the light transmittance to an approximately normal state, the problem of image dehazing can be effectively solved. Thus, we adopt retinex theory for image dehazing, and we deduce a retinex-based dehazing decomposition model as follows.

From retinex theory, a hazy image can be described as:

$$I_h(x, y) = R(x, y) * L_h(x, y), \tag{8.34}$$

where $L_h(x, y)$ denotes the illumination map of a hazy image. This map is an equivalent illumination map influenced by haze absorption and scattering. Similarly, a dehazed image can be formulated as:

$$I_{dh}(x, y) = R(x, y) * L_{dh}(x, y), \tag{8.35}$$

where $L_{dh}(x, y)$ denotes a natural illumination map. The reflectance R is illumination invariant. Due to atmospheric scattering in a hazy scene, the illumination values of the hazy image are larger than those of the dehazed image. From Eq. 8.34 and Eq. 8.35, we can obtain

$$I_h(x, y) = R(x, y) * L_{dh}(x, y) * \frac{L_h(x, y)}{L_{dh}(x, y)}, \tag{8.36}$$

let $L_r(x, y) = L_h(x, y)/L_{dh}(x, y)$. $L_r(x, y)$ denotes the absorbed and scattered light caused by haze. Further, we rewrite the above equation as:

$$I_h(x, y) = I_{dh}(x, y) * L_r(x, y). \tag{8.37}$$

That is, the hazy image I_h is the dehazed image I_{dh} multiplied by the residual illumination map L_r. We call this model as retinex-based dehazing decomposition model.

8.3.2 Network Architecture

Our network consists of a multi-scale subnetwork for estimating the residual illumination map and a channel and spatial attention U-Net for haze removal. Figure 8.10 shows the overall structure of our network.

Residual illumination map estimation network (RIMEN). Multiresolution image representation and multi-scale feature representation are widely used in both low-level and high-level computer vision tasks. Usually, the small-scale receptive field tends to preserve global contextual information, and the large-scale receptive field tends to retain local detailed information. The residual illumination map contains both local (e.g., shadows and highlights) and global information (e.g., color distribution and average brightness). Thus, to capture both local and global information, we use a multi-scale network to learn the residual illumination map. The RIMEN consists of three branches, and each branch contains a downsampling layer, a residual dense

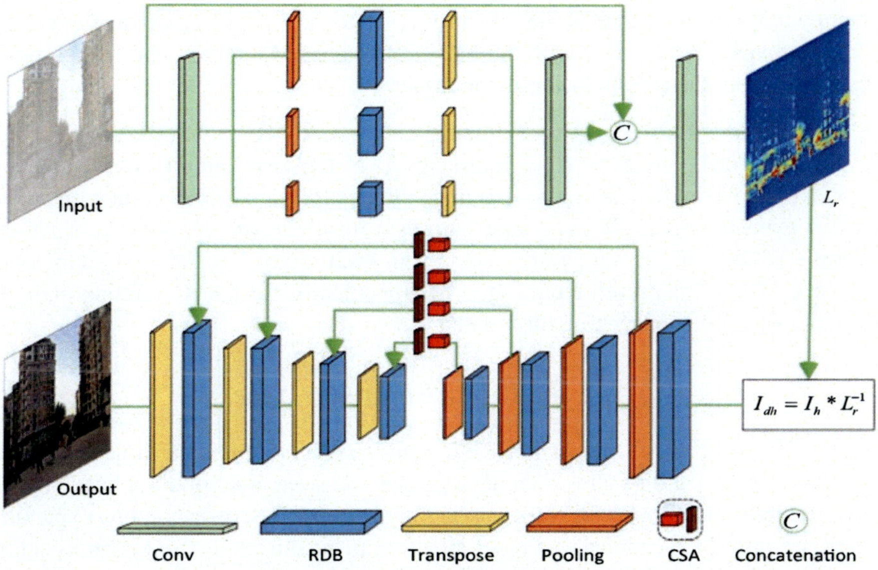

Fig. 8.10 An overview of our network. Our RDN consists of two modules: a residual illumination map estimating submodule using a multi-scale residual dense network and an image dehazing submodule using a channel and spatial attention U-Net. © [2021] IEEE. Reprinted, with permission, from [Pengyue Li, Jiandong Tian, Yandong Tang, Guolin Wang, and Chengdong Wu, "Deep Retinex Network for Single Image Dehazing," IEEE Transactions on Image Processing, Vol.30, pp. 1100–1115, 2021]

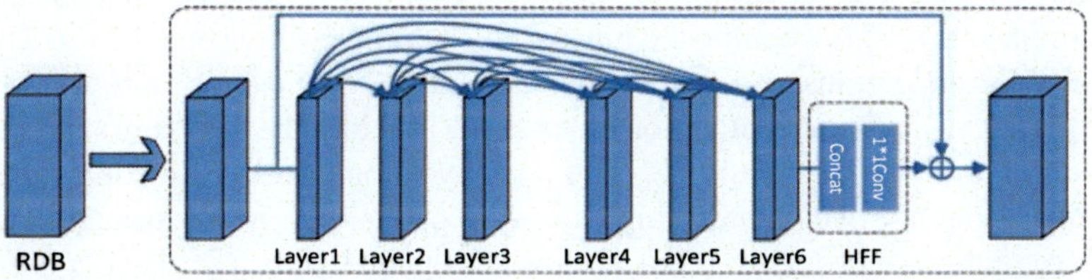

Fig. 8.11 Illustration of the RDB in Fig. 8.10. © [2021] IEEE. Reprinted, with permission, from [Pengyue Li, Jiandong Tian, Yandong Tang, Guolin Wang, and Chengdong Wu, "Deep Retinex Network for Single Image Dehazing," IEEE Transactions on Image Processing, Vol. 30, pp. 1100–1115, 2021]

block, and a transpose convolution layer. Specifically, we first downsample an input hazy image to 1/2 scale, 1/4 scale, and 1/8 scale. Then, the multi-scale images from the three branches are, respectively, fed into the corresponding residual dense block (RDB) [41] to capture multi-scale features. The overview of the RDB is shown in Fig. 8.11. The dense block is first used to extract features from the multi-scale images. Then, hierarchical feature fusion (HFF) and residual learning (RL) are sequentially conducted in the RDB. The RDB makes full use of all the convolution layer features to promote the network representation ability. After multi-scale feature learning, the feature maps from the three scales are upsampled to the same size, and the same-sized feature maps concatenated with the hazy image are fed into a general convolution layer to output the residual illumination map.

Channel and spatial attention dehazing U-Net (CASDUN). For the image dehazing subnetwork, we introduce a four-stage U-Net [31] to simultaneously capture contextual information with a contracting path and location information with a symmetric expanding path. To use all the features from all the convolution layers, we still use the RDB as the feature extraction unit in each stage. The detailed architecture of the CASDUN is presented in Fig. 8.10. Specifically, for the contracting path, the output of the ith stage can be formulated as:

$$U_c^i = f_c(U_c^{i-1}), i = 1, 2, 3, 4, \tag{8.38}$$

where U_c^0 is the input of the first RDB, and $f_c(\cdot)$ is a composite function of two consecutive operations: a residual dense block and a maximal pooling layer (MPL). In the skip connection between the contracting path and the corresponding expanding path, we introduce channel and pixel attention mechanisms to automatically adjust the channel-wise and pixel-wise attention weights. For the expanding path, the output of the jth stage can be described as:

$$U_e^j = f_e([U_e^{j-1}, U_{ca}^{9-j}]), j = 5, 6, 7, 8, \tag{8.39}$$

where U_e^4 is the output of the contracting path, U_{ca}^i is the ith-stage output of the contracting path after attention operation, $[U_e^{j-1}, U_{ca}^{9-j}]$ denotes the concatenation operation, and $f_e(\cdot)$ is a composite function of two consecutive operations: a residual dense block and a transpose operation.

The channel and spatial attention block (CSA) enables the dehazing subnetwork to give different levels of attention to each pixel by automatically adjusting the channel and pixel attention weights. In this approach, pixels with heavy haze are assigned large weights to thoroughly remove haze, and pixels with light haze are assigned small weights to avoid oversaturated results. Motivated by previous studies, such as [12] and [37], we designed a novel channel and spatial attention block. Notably, unlike the squeeze-and-excitation block [12], which only uses average pooling to gather global information, we use both global average pooling and maximal pooling to capture global common and distinctive information, respectively. Additionally, to enable the channel to directly use weights and avoid dimensionality reduction issues, we substitute the gating structure with depth-wise separable convolution to predict the weight of each channel independently. Moreover, unlike the CBAM block in [37], we use depth-wise separable convolution to produce a spatial attention map that preserves location information and avoids information loss caused by the pooling layer. Specifically, the channel and spatial attention block adjusts the pixel-wise weight through channel attention and spatial attention steps. As shown in Fig. 8.12, the channel attention first transforms global information into a 1D channel attention map $\mathbf{M}_c \in R^{C\times1\times1}$ using global maximal/average pooling and depth-wise separable

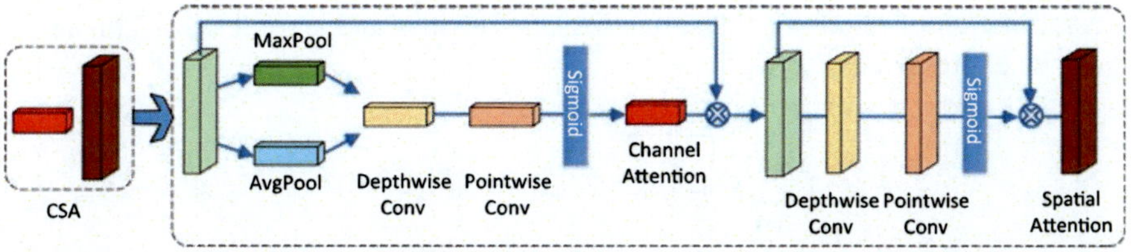

Fig. 8.12 Illustration of the CSA in Fig. 8.10.© [2021] IEEE. Reprinted, with permission, from [Pengyue Li, Jiandong Tian, Yandong Tang, Guolin Wang, and Chengdong Wu, "Deep Retinex Network for Single Image Dehazing," IEEE Transactions on Image Processing, Vol.30, pp. 1100–1115, 2021]

convolution (DSC). This step can be written as:

$$\mathbf{M_c} = f_c(z_a^c, z_m^c, w) = \sigma(W_2\delta(W_1 z_a^c) + W_2\delta(W_1 z_m^c)), \tag{8.40}$$

where δ and σ denote the ReLU and sigmoid functions, respectively; $W_1 \in R^{c\times 1}$ and $W_2 \in R^{1\times c}$ are DSC weights; and z_a^c and z_m^c denote the average-pooled features and maximal-pooled features, respectively. Then, the final output of the channel attention step is obtained by:

$$\mathbf{F}_c = \mathbf{M}_c * \mathbf{F}, \tag{8.41}$$

where $*$ denotes element-wise product between the output $\mathbf{F}$ of the residual dense block and the channel attention map $\mathbf{M}_c$. The spatial attention step aggregates the channel information of the feature map using the depth-wise separable convolution to produce a 2D spatial attention map $\mathbf{M}_s \in R^{H\times W}$. This step can be represented as:

$$\mathbf{M}_s = f_s(\mathbf{F}_c, w) = \sigma(\mathbf{W}_4\delta(\mathbf{W}_3\mathbf{F}_c), \tag{8.42}$$

where δ and σ denote the ReLU and sigmoid functions; $\mathbf{W}_3$ and $\mathbf{W}_4$ are DSC weights. The output of the spatial attention step is calculated by:

$$\mathbf{F}_s = \mathbf{M}_s * \mathbf{F}_c, \tag{8.43}$$

where $*$ denotes element-wise multiplication between the spatial attention map $\mathbf{M}_s$ and the channel attention output $\mathbf{F}_c$.

8.3.3 Network Optimization

Our RDN is optimized in an end-to-end fashion using the illumination error loss function L_M and the dehazed image error loss function L_D. L_M is a weighted combination of pixel-wise absolute loss, SSIM loss and smooth loss. This relation can be expressed as:

$$L_M = \sum_{i=1}^{N} \omega_{\text{am}} L^i_{\text{am}} + \omega_{\text{ssm}}(1 - L^i_{\text{ssm}}) + \omega_{\text{smm}} L^i_{smm}, \tag{8.44}$$

where L^i_{am}, L^i_{ssm}, and L^i_{smm} denote the absolute loss, the SSIM loss, and the smooth loss, respectively. ω_r, ω_{ss}, and ω_{sm} are the corresponding weights. The smooth loss is calculated by:

$$L^i_{smm} = \sum_p (\partial_x L^p_r)^2 + (\partial_y L^p_r)^2, \tag{8.45}$$

where L^p_r indicates the residual illumination value of pixel p and ∂_x and ∂_y are partial derivatives in horizontal and vertical directions in an image, respectively.

For the dehazing subnetwork, we not only use the pixel-wise absolute loss and the SSIM loss to constrain network learning, but also apply the edge-preserving loss function to improve the edge of the dehazed image. The dehazed image error loss function L_D is expressed as:

$$L_D = \sum_{i=1}^{N} \omega_{\text{ad}} L^i_{\text{ad}} + \omega_{\text{ssd}}(1 - L^i_{\text{ssd}}) + \omega_{\text{egd}} L^i_{\text{egd}}, \tag{8.46}$$

where L^i_{ad}, L^i_{ssd}, and L^i_{egd} denote the absolute loss, the SSIM loss, and the edge loss, respectively. ω_r, ω_{ss}, and ω_{sm} are the corresponding weights. The edge loss function is defined as:

$$L^i_{\text{egd}} = \sum_{w,h} ||E_{\text{canny}}(I_{\text{dh}})_{w,h} - E_{\text{canny}}(I_{\text{gt}})_{w,h}||_1, \tag{8.47}$$

where $E_{canny}()_{w,h}$ denotes Canny edge detection.

To prevent overfitting, early stopping is used to end training upon reaching convergence in validation loss. In addition, we adopt a stage-wise learning strategy to train our network. First, we optimize only the residual illumination map estimation subnetwork (RIMEN) without updating the dehazing subnetwork. When the RIMEN converges, we start to optimize the dehazing subnetwork without updating the RIMEN. After performing learning for each subnetwork separately, we adjust the entire network by jointly optimizing the two networks. The proposed RDN was implemented based on the Pytorch framework. The entire network was optimized using the Adam optimizer. We set the initial learning rate to $5 * 10^{-3}$ and the mini-batch size to 32 for training. We choose $\omega_{\text{am}} = 1$, $\omega_{\text{ssm}} = 0.2$, and $\omega_{\text{smm}} = 0.3$ to estimate the illumination map and $\omega_{ad} = 1$, $\omega_{\text{ssd}} = 0.5$, and $\omega_{\text{egd}} = 0.0002$ to optimize the dehazed subnetwork. During training, the learning rate is periodically decreased at a certain epoch, and validation is performed for each epoch.

8.3.4 Experiments

We first compare our method with the representative methods on three synthetic hazy datasets, three real-world hazy datasets, and some real-world hazy images that we collected to show the superiority of our method. Then, we conduct ablation analysis to demonstrate the effectiveness of core module in our network and the effectiveness of each loss function.

8.3.4.1 Datasets and Metrics

From our dehazing decomposition model, we expand the existing datasets with hazy image pairs to new datasets with hazy image triplets, including hazy images, haze-free images, and residual illumination maps. For image dehazing, the haze enables an image to brighten. When the hazy image and the ground truth are normalized, the values of L_r computed by Eq. 8.37 are usually greater than one. It usually cannot be directly saved as an image for learning. To treat with this problem, we use a color map to represent the change of illumination from minimum (blue) to maximum (red). The datasets used for training and testing include three synthetic datasets (HDB, RESIDE-ITS, and RESIDE-OTS) and three real-world datasets (I-HAZE, O-HAZE, and DENSE-HAZE). The HDB dataset provided by Zhang et al. [40] contains 4000 synthetic images for training, and 400 images for testing. RESIDE synthesized by Li et al. [17] contains 13,990 synthetic indoor hazy images (ITS) and 72,500 synthetic outdoor hazy images (OTS) in the training set. The testing set contains 500 indoor scenes and 500 outdoor scenes. The real-world hazy datasets I-HAZE and O-HAZE used in the NTIRE 2018 image dehazing challenge contain 30 indoor real-world hazy images and 45 outdoor real-world hazy images, respectively, for training and testing. DENSE-HAZE used in the NTIRE 2019 dehazing challenge contains 55 real-world hazy images for training and testing. In addition, we collected real-world hazy images from the Internet for testing. As the corresponding ground truth of these images is usually not available, the performance of dehazing methods on these real-world images was assessed by visual evaluation. Like most papers, we adopt PSNR [13] and SSIM [36] metrics to evaluate the quantitative performance of each method for the synthetic and real-world datasets.

8.3.4.2 Quantitative and Qualitative Evaluation

We compare the performance of our method with that of twelve state-of-the-art methods: (a) the DCP method [10] (CVPR'09), (b) DehazeNet [3] (TIP'16), (c) an all-in-one dehazing network (AOD-Net) [16] (ICCV'17), (d) a gated fusion network (GFN) [29] (CVPR'18), (e) a densely connected pyramid dehazing network (DCPDN) [40] (CVPR'18), (f) a patch quality comparator for haze removal (PQC) [32] (TIP'18), (g) a patch map-based dehazing network (PMS-Net) [4] (CVPR'19), (h) an enhanced pix2pix dehazing network (EPDN) [28] (CVPR'19), (i) a deep multi-model fusion

Table 8.2 Quantitative results of different state-of-the-art methods evaluated on synthetic and real-world hazy datasets. © [2021] IEEE. Reprinted, with permission, from [Pengyue Li, Jiandong Tian, Yandong Tang, Guolin Wang, and Chengdong Wu, "Deep Retinex Network for Single Image Dehazing," IEEE Transactions on Image Processing, Vol. 30, pp. 1100–1115, 2021]

Dataset	HDB		RESIDE-ITS		RESIDE-OTS		I-HAZE		O-HAZE		DENSE-HAZE	
Metric	PSNR	SSIM	PSNR	SSIM	PSNR	SSIM	PSNR	SSIM	PSNR	SSIM	PSNR	SSIM
DCP	12.957	0.646	16.167	0.754	13.031	0.662	11.988	0.528	13.265	0.576	11.510	0.413
DehazeNet	19.430	0.857	21.088	0.872	22.262	0.909	14.575	0.688	15.770	0.697	10.173	0.366
AOD-Net	17.809	0.800	17.576	0.810	19.420	0.888	15.344	0.704	16.446	0.684	11.628	0.385
GFN	21.283	0.877	22.085	0.898	20.915	0.856	14.310	0.631	16.916	0.777	14.086	0.505
DCPDN	26.883	0.952	18.987	0.840	19.708	0.827	15.327	0.731	15.159	0.769	10.705	0.350
PQC	18.829	0.863	20.075	0.877	18.729	0.891	14.476	0.696	17.045	0.819	11.148	0.421
EPDN	23.481	0.919	24.431	0.925	19.817	0.882	15.789	0.749	16.574	0.810	12.852	0.477
GridNet	21.812	0.899	32.160	0.983	30.860	0.981	16.531	0.768	16.753	0.766	9.978	0.369
DADN	24.439	0.924	25.295	0.942	26.441	0.9597	17.125	0.804	17.973	0.855	12.559	0.499
FFA-Net	24.730	0.922	33.025	0.987	30.587	0.976	15.256	0.696	21.364	0.869	15.431	0.533
Ours	31.082	0.981	32.414	0.985	31.173	0.983	24.854	0.877	25.236	0.913	20.541	0.673

Table 8.3 Compared with DM2F-NET on RESIDE and O-Haze

Models	DM2F-NET [6]	Ours
RESIDE	34.29/0.984	31.793/0.984
O-Haze	25.188/0.777	25.236/0.913

network (DM2F-NET) [6] (CVPR'19), (j) a grid dehazing network (GridNet) [20] (ICCV'19), (k) a domain adaptation dehazing network (DADN) [33] (CVPR'20), and (l) a feature fusion attention network (FFA-Net) [27]. Table 8.2 displays the quantitative results of the above methods on three synthetic testing datasets and three real-world benchmark datasets. The best and second-best scores are highlighted in red and blue, respectively. As observed, our method considerably surpasses other methods on two synthetic datasets and three real-world datasets, except that it is slightly inferior to FFA-Net on the synthetic dataset ITS. Specifically, the PSNR values of our method are better than those of the second-best method with margins of 4.199 dB, 0.313 dB, 7.729 dB, 3.872 dB, and 5.110 dB on HDB, OTS, I-HAZE, O-HAZE, and DENSE-HAZE, respectively, and the SSIM values of our method also outperform those of the second-best method with gains of 2.9%, 0.2%, 7.3%, 4.4%, and 14.0% on five datasets, respectively. Although the code of DM2F-NET[6] is not published, the PSNR and SSIM values of this method on some synthetic and real databases are given in [6], so we compare the average values of our method on these databases with those of DM2F-NET in Table 8.3. As observed, although our method is slightly inferior to DNET on the synthetic dataset SOTS, our method outperforms DM2F-NET on the real-word dataset O-HAZE, with gains of 0.048 dB on PSNR and 13.6% on SSIM.

Visual comparison of synthetic hazy images is shown in Figs. 8.13 and 8.14. These three hazy images are derived from HDB, ITS, and OTS, respectively. The dehazed results of DehazeNet, AOD-Net, DCPDN, PQC, and GridNet on the indoor/outdoor

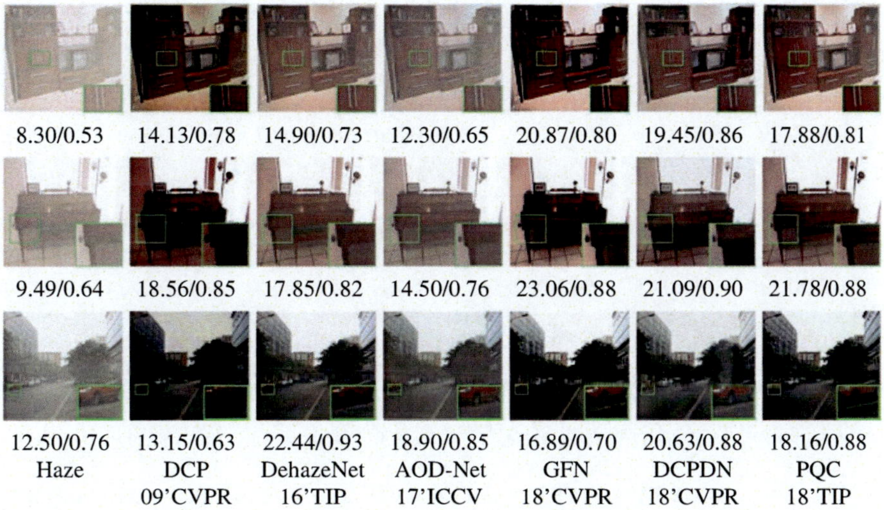

Fig. 8.13 Dehazing results on the synthetic datasets.© [2021] IEEE. Reprinted, with permission, from [Pengyue Li, Jiandong Tian, Yandong Tang, Guolin Wang, and Chengdong Wu, "Deep Retinex Network for Single Image Dehazing," IEEE Transactions on Image Processing, Vol.30, pp. 1100–1115, 2021]

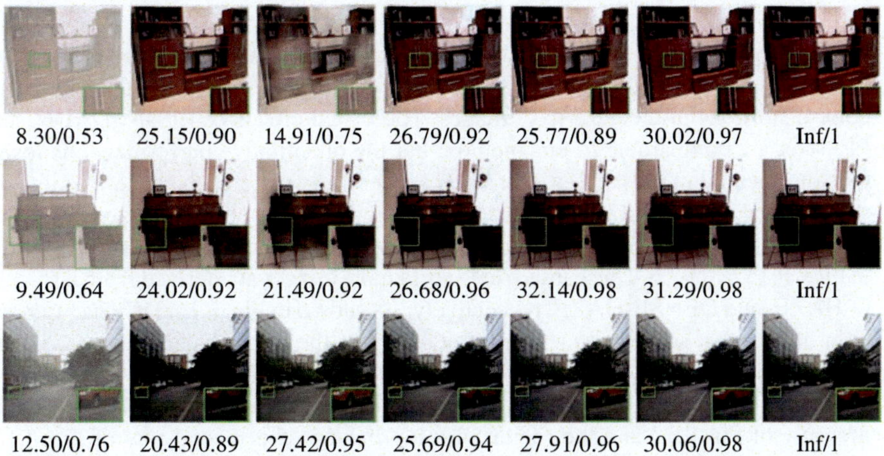

Fig. 8.14 Dehazing results on the synthetic datasets. © [2021] IEEE. Reprinted, with permission, from [Pengyue Li, Jiandong Tian, Yandong Tang, Guolin Wang, and Chengdong Wu, "Deep Retinex Network for Single Image Dehazing," IEEE Transactions on Image Processing, Vol.30, pp. 1100–1115, 2021]

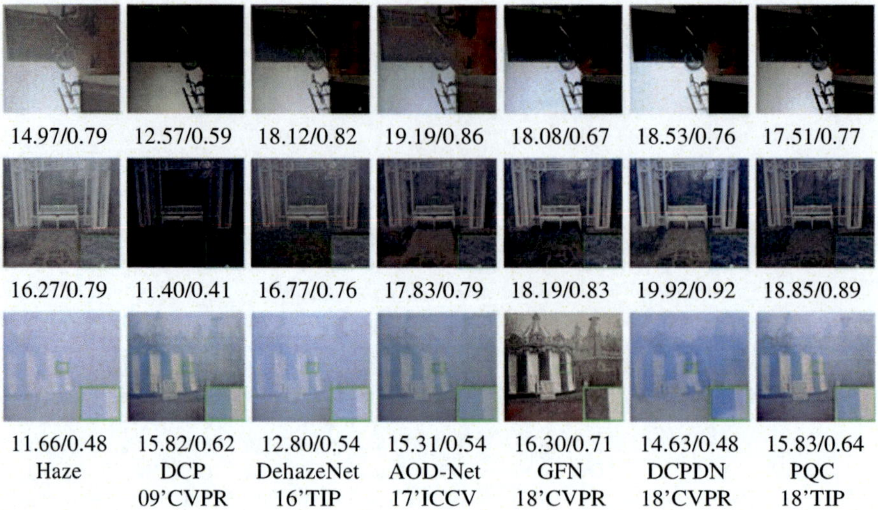

Fig. 8.15 Dehazing results on the real-world hazy datasets. © [2021] IEEE. Reprinted, with permission, from [Pengyue Li, Jiandong Tian, Yandong Tang, Guolin Wang, and Chengdong Wu, "Deep Retinex Network for Single Image Dehazing," IEEE Transactions on Image Processing, Vol. 30, pp. 1100–1115, 2021]

hazy images usually have residual haze. Although DCP and GFN remove haze more thoroughly than the other methods, they tend to produce over-saturated regions or color distortions in the haze-free images. The visual effects of EPDN, DADN, and FFA-Net are comparable to our method, but the quantitative performance is lower than our method in most cases. Our RDN successfully restores the haze-free images with normal colors and less residual haze. Similarly, the visual dehazed results of real-world hazy images sampled from real-world benchmark datasets are presented in Figs. 8.15 and 8.16. These three real-world hazy images are derived from I-HAZE, O-HAZE, and DENSE-HAZE, respectively. As shown in Fig. 8.15, DCP can remove light haze, but it sometimes leaves some haze for the heavy hazy image (see, e.g., the third row of Fig. 8.15). Due to the imbalance between the three confidence maps, GFN tends to produce over-saturated regions (see, e.g., the region of the door in the first row of Fig. 8.15 and the region of the ground in the second row of Fig. 8.15) or color distortions (see, e.g., the third row of Fig. 8.15) in the haze-free images. For DehazeNet, AOD-Net, DCPDN, and PQC the dehazed results for real-world hazy images are similar to those for synthetic hazy images, which generally contain residual haze. As shown in Fig. 8.16, GridNet, DADN, and FFA-Net also tend to have residual haze in the haze-free images. EPDN can remove light haze well, but it tends to produce color cast. Compared with other methods, our method has less residual haze and less color distortion on real-world dehazed images.

We also compared the RDN with state-of-the-art methods on real-world hazy images collected from the Internet. As shown in Figs. 8.17 and 8.18, our method outperforms the other methods for haze removal and the suppression of over-saturation.

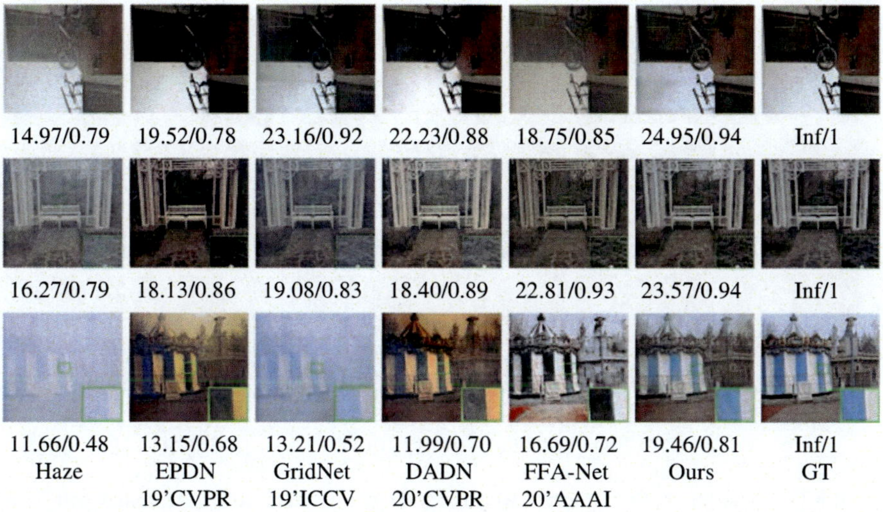

Fig. 8.16 Dehazing results on the real-world hazy datasets. © [2021] IEEE. Reprinted, with permission, from [Pengyue Li, Jiandong Tian, Yandong Tang, Guolin Wang, and Chengdong Wu, "Deep Retinex Network for Single Image Dehazing," IEEE Transactions on Image Processing, Vol. 30, pp. 1100–1115, 2021]

The other methods tend to yield either over-dehazing or under-dehazing. Specifically, DehazeNet usually cannot handle heavy haze well and tends to leave some haze in dehazed images. Similar results are also observed from AOD-Net, DCPDN, PQC, GridNet, DADN, and FFA-Net. DCP, GFN, and EPDN sometimes suffer from color distortions and darken some regions of images as indicated by the red boxes. Moreover, we also evaluated the dehazing results on real-world images published by previous algorithms. As shown in Figs. 8.19 and 8.20, for DehazeNet, AOD-Net, PQC, GridNet, DADN, and FFA-Net, some haze still remains in the dehazed images. DCP and EPDN usually tend to produce over-saturated regions indicated by the red boxes and slight color distortion. From the overall visual effect, the dehazing results of GFN and DCPDN are more natural, but they also produce slight color saturation in some local regions indicated by the red boxes. Compared with other methods, although our method also leaves slight haze in some local regions of haze-free images (e.g., around faces and background areas in stadium), it makes a trade-off between residual haze and color distortion; that is, it leaves less residual haze and less color distortion. Generally, the computer monitor settings and viewing angles maybe affect the visual effect (Table 8.4).

In addition, we also compare the residual illumination map of our method with that of the other methods, including Single-Scale Retinex (SSR), Multi-Scale Retinex (MSR), and Multi-Scale Retinex with Color Restoration (MSRCR) [26]. MSR includes three scales: 15 (SSR1), 80 (SSR2), and 250 (SSR3). The quantitative PSNR and SSIM results are shown in Table 8.6. It is observed that our RDN surpasses other

Haze | DCP 09'CVPR | DehazeNet 16'TIP | AOD-Net 17'ICCV | GFN 18'CVPR | DCPDN 18'CVPR

Fig. 8.17 Dehazing results on the real-world hazy images. © [2021] IEEE. Reprinted, with permission, from [Pengyue Li, Jiandong Tian, Yandong Tang, Guolin Wang, and Chengdong Wu, "Deep Retinex Network for Single Image Dehazing," IEEE Transactions on Image Processing, Vol.30, pp. 1100–1115, 2021]

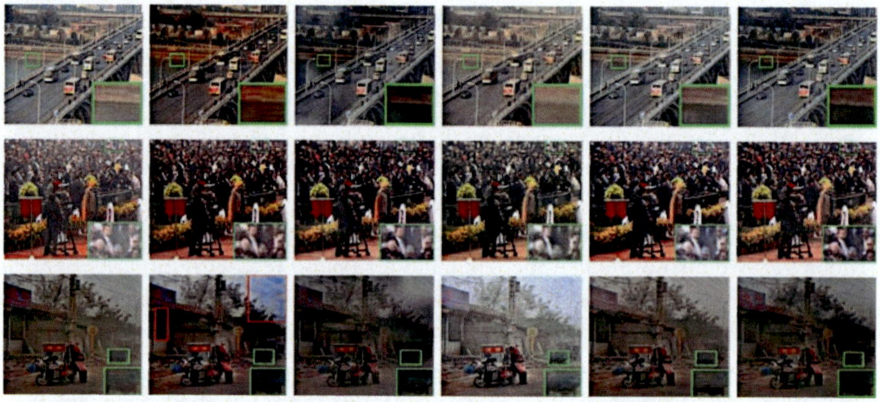

PQC 18'TIP | EPDN 19'CVPR | GridNet 19'ICCV | DADN 20'CVPR | FFA-Net 20'AAAI | Ours

Fig. 8.18 Dehazing results on the real-world hazy images. © [2021] IEEE. Reprinted, with permission, from [Pengyue Li, Jiandong Tian, Yandong Tang, Guolin Wang, and Chengdong Wu, "Deep Retinex Network for Single Image Dehazing," IEEE Transactions on Image Processing, Vol. 30, pp. 1100–1115, 2021]

methods for both PSNR and SSIM on synthetic and real-world datasets, including RESIDE-ITS, RESIDE-ITS, I-HAZE, and O-HAZE. Specifically, the PSNR values of our method outperform those of the suboptimal method with margins of 6.86 dB, 5.39 dB, 3.99 dB, and 4.81 dB on aforementioned four datasets, respectively, and the SSIM values of our method also outperform those of the suboptimal method with

Fig. 8.19 Dehazing results evaluated on real-world images released by the authors of previous methods. © [2021] IEEE. Reprinted, with permission, from [Pengyue Li, Jiandong Tian, Yandong Tang, Guolin Wang, and Chengdong Wu, "Deep Retinex Network for Single Image Dehazing," IEEE Transactions on Image Processing, Vol.30, pp. 1100–1115, 2021]

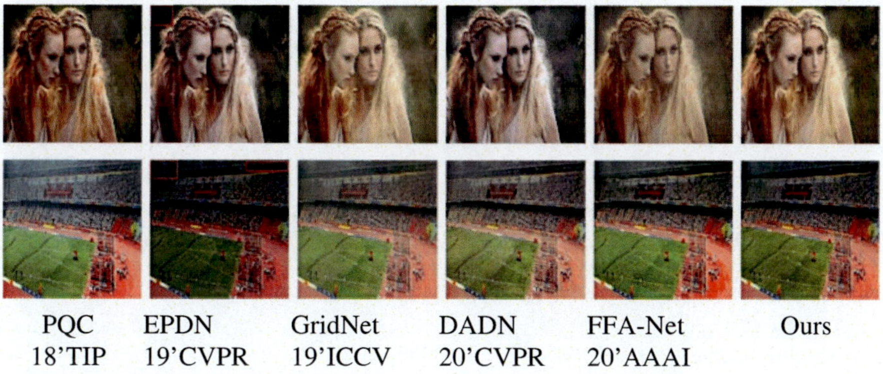

Fig. 8.20 Dehazing results evaluated on real-world images released by the authors of previous methods. © [2021] IEEE. Reprinted, with permission, from [Pengyue Li, Jiandong Tian, Yandong Tang, Guolin Wang, and Chengdong Wu, "Deep Retinex Network for Single Image Dehazing," IEEE Transactions on Image Processing, Vol.30, pp .1100–1115, 2021]

Table 8.4 Quantitative results of the estimated residual illumination map under different scales

Scale	Scale1	Scale2	Scale3	Multi-scale
PSNR	13.62	13.90	15.68	19.54
SSIM	0.62	0.63	0.78	0.82

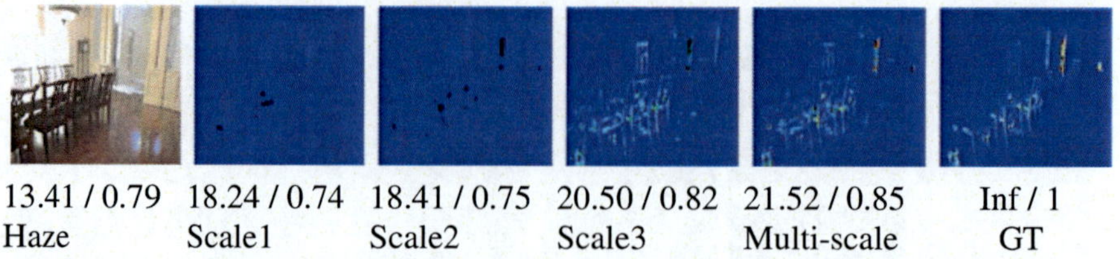

Fig. 8.21 Estimated residual illumination maps using one scale and multiple scales. © [2021] IEEE. Reprinted, with permission, from [Pengyue Li, Jiandong Tian, Yandong Tang, Guolin Wang, and Chengdong Wu, "Deep Retinex Network for Single Image Dehazing," IEEE Transactions on Image Processing, Vol.30, pp. 1100–1115, 2021]

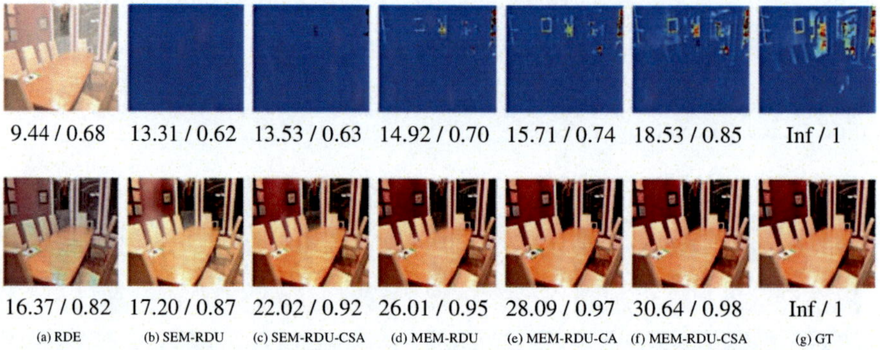

Fig. 8.22 Estimation results of the illumination map and the dehazed image using different models. © [2021] IEEE. Reprinted, with permission, from [Pengyue Li, Jiandong Tian, Yandong Tang, Guolin Wang, and Chengdong Wu, "Deep Retinex Network for Single Image Dehazing," IEEE Transactions on Image Processing, Vol.30, pp. 1100–1115, 2021]

gains of 23%, 10%, 6%, and 6% on four datasets, respectively. Figure 8.23 gives the comparison of visual effect on both synthetic and real-world datasets. It is observed that the residual illumination map estimated by our method is closer to ground truth.

8.3.4.3 Ablation Analysis

To reveal the avail of each module in RDN, we conduct ablation studies, containing multiscale examination and RDN overall structure examination. In addition, we examine the influence of the loss functions on both the estimated residual illumination map and the dehazed image.

Effectiveness of the multiscale network. To test the influence of the multi-scale module on the results, we conduct four comparative experiments using a small-scale image (scale1), a medium-scale image (scale2), a large-scale image (scale3), and multiple scale images to estimate the residual illumination map. In single-scale mode (scale1, scale2, scale3), we conduct the related experiments with only one scale path

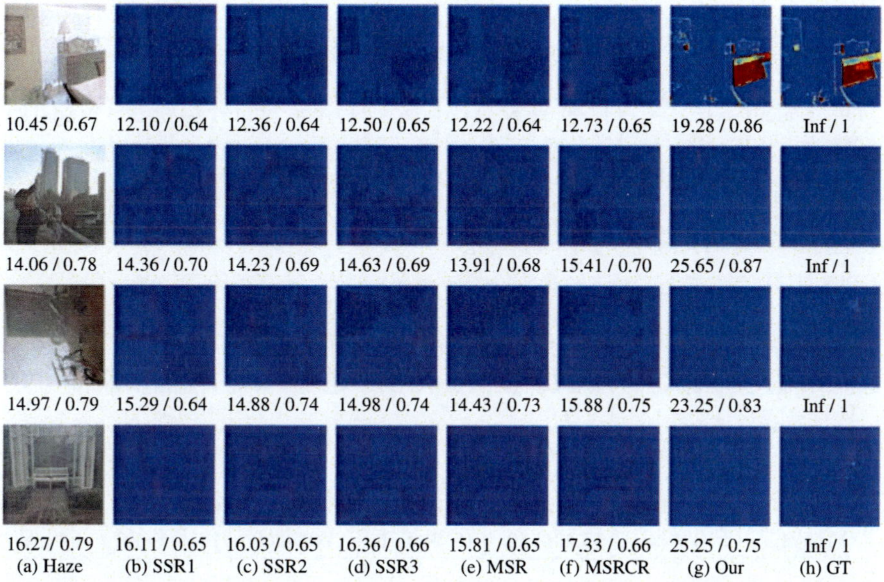

Fig. 8.23 Estimation results of the illumination map using different methods. © [2021] IEEE. Reprinted, with permission, from [Pengyue Li, Jiandong Tian, Yandong Tang, Guolin Wang, and Chengdong Wu, "Deep Retinex Network for Single Image Dehazing," IEEE Transactions on Image Processing, Vol. 30, pp. 1100–1115, 2021]

to estimate residual illumination map. In multi-scale mode, we conduct the related experiment with three scale paths to estimate residual illumination map. The small-scale image, medium-scale image, and large-scale image refer to 1/8, 1/4, 1/2 of the original image size, respectively. The PSNR and SSIM values of the residual illumination map at different scales are presented in Table 8.4. It can be observed that the PSNR and SSIM values gradually increase with the scale enlargement. When three scales are jointly applied, the values of PSNR and SSIM have significant improvement. This result indicates that the multi-scale network can greatly improve the result by fusing the multi-scale information. Figure 8.21 shows visual results of the estimated residual illumination maps. It can be seen that the network can capture global information from the small-scale image and extract local detailed information (e.g., edge and texture information) from the large-scale image. The multi-scale images yield a trade-off between these results and enable the estimated illumination map to be closer to the ground-truth map.

Effectiveness of the overall structure. To assess the necessity and effectiveness of each module in our network (RDN), we conducted six ablation experiments involving (1) a residual dense U-Net (RDU) without an illumination estimation module, (2) a single-scale illumination estimation module and RDU without CSA operations (SEM-RDU), (3) a single-scale illumination estimation module and RDU with CSA operations (SEM-RDU-CSA), (4) a multi-scale illumination estimation module

Table 8.5 Quantitative PSNR and SSIM results of different models

Models	RDU	SEM-RDU	SEM-RDU-CSA	MEM-RDU	MEM-RDU-CA	MEM-RDU-CSA
Illumination map	-/-	13.99/0.63	14.31/0.63	17.47/0.79	17.93/0.80	19.54/0.82
Dehazed image	20.02/0.85	21.01/0.91	24.71/0.94	28.31/0.96	30.24/0.97	32.41/0.98

Table 8.6 Quantitative PSNR and SSIM results of different methods

Methods	SSR1	SSR2	SSR3	MSR	MSRCR	OURS
RESIDE-ITS	12.23/0.57	12.29/0.58	12.39/0.58	12.06/0.57	12.68/0.59	19.54/0.82
RESIDE-OTS	15.00/0.72	15.01/0.71	15.10/0.72	14.58/0.71	15.82/0.72	21.21/0.82
I-HAZE	14.25/0.66	14.20/0.66	14.34/0.66	13.94/0.66	14.79/0.67	18.78/0.73
O-HAZE	15.44/0.60	15.42/0.60	15.49/0.61	15.07/0.60	16.30/0.62	21.11/0.68

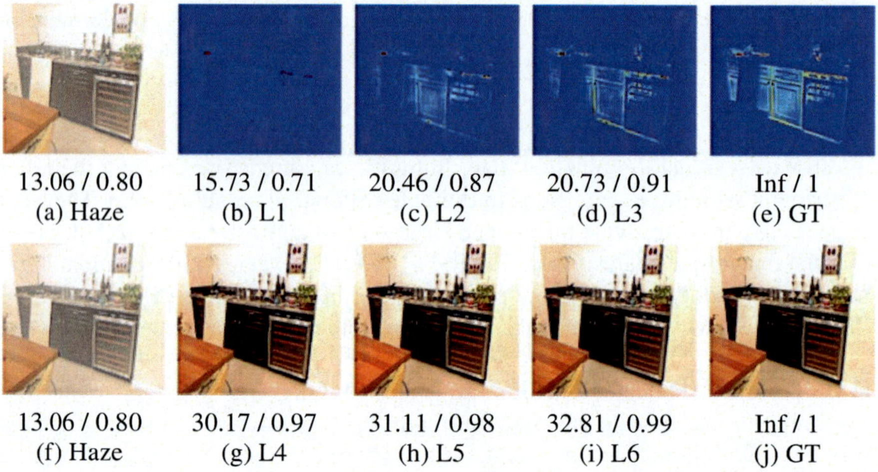

Fig. 8.24 Estimated results using different loss functions. Top: illumination maps. Bottom: dehazed images. © [2021] IEEE. Reprinted, with permission, from [Pengyue Li, Jiandong Tian, Yandong Tang, Guolin Wang, and Chengdong Wu, "Deep Retinex Network for Single Image Dehazing," IEEE Transactions on Image Processing, Vol. 30, pp. 1100–1115, 2021]

Table 8.7 Comparison of estimated residual illumination maps using different loss functions

Loss function	L1	L2	L3
PSNR	14.69	16.57	19.07
SSIM	0.64	0.75	0.82

Table 8.8 Comparison of estimated dehazed images using different loss functions

Loss function	L4	L5	L6
PSNR	30.89	31.15	32.41
SSIM	0.97	0.97	0.98

Table 8.9 Quantitative results of different CSA modules

Module	CSA-MP	CSA-AP	CSA-DWC	CSA-PWC	CSA
PSNR	32.171	32.085	31.733	31.547	32.414
SSIM	0.983	0.982	0.978	0.971	0.985

and RDU without CSA operations (MEM-RDU), (5) a multi-scale illumination estimation module and RDU with channel attention operations (MEM-RDU-CA), and (6) a multi-scale illumination estimation module and RDU with CSA operations (MEM-RDU-CSA).

The ablation experiments were performed on the synthesized ITS dataset. The PSNR and SSIM values of six models on both residual illumination maps and dehazed images are provided in Table 8.5. RDU is a basal network that only uses a residual dense U-Net to learn the dehazed image. It enables the PSNR and SSIM of the dehazed image to reach 20.02 dB and 0.85, respectively. SEM-RDU adds a single-scale illumination estimation module to RDU. This operation not only enables the network to estimate the residual illumination map, but also increases the PSNR by 0.99 dB and the SSIM by 6% of the dehazed image. SEM-RDU-CSA adds the spatial and channel attention block to the RDU component of SEM-RDU. It further boosts 3.70 dB on PSNR and 3% on SSIM for the dehazed image. MEM-RDU replaces the illumination estimation subnetwork of SEM-RDU with a multi-scale module. By comparing SEM-RDU and MEM-RDU, it can be observed that the multi-scale operation contributes 3.48 dB on PSNR and a 16% improvement on SSIM for the residual illumination map. MEM-RDU-CA adds a channel attention block to the RDU component of MEM-RDU. The channel attention operation further increases the PSNR by 1.93 dB and the SSIM by 1% for the dehazed image. MEM-RDU-CSA adds spatial attention to the RDU component of MEM-RDU-CA. It makes the PSNR and SSIM of the dehazed image reach 32.41 dB and 0.98, respectively, and those of the residual illumination map reach 19.54 dB and 0.82, respectively. Visual results are shown in Fig. 8.22. From the first row of Fig. 8.22, it is observed that the multi-scale module is able to effectively estimate the illumination regions with relatively large residual illumination values, which represent large differences of illumination between the haze image and the ground truth. By comparing (d) and (f) in the second row of Fig. 8.22, it can be observed that the CSA module can make the dehazing results smooth and reduce under-dehazing (e.g., the region between the chairs) and over-dehazing (e.g., the red wall) issues. In addition, we also conduct experiments to demonstrate the necessity of each component in the CSA by removing each of them.

Fig. 8.25 Estimation results of the dehazed image using different CSA modules. © [2021] IEEE. Reprinted, with permission, from [Pengyue Li, Jiandong Tian, Yandong Tang, Guolin Wang, and Chengdong Wu, "Deep Retinex Network for Single Image Dehazing," IEEE Transactions on Image Processing, Vol. 30, pp. 1100–1115, 2021]

Specifically, we remove the maximal pooling layer, the average pooling layer, the depth-wise convolution layer and the point-wise convolution layer, respectively, to generate four varietal CSA structures including CSA-Maxpooling (CSA-MP), CSA-Avgpooling (CSA-AP), CSA-Depthwise Conv (CSA-DWC), CSA-Pointwise Conv (CSA-PWC). The quantitative results are shown in Table 8.9. Visual comparisons are shown in Fig. 8.25. It can be observed that each component of the CSA is necessary and removal of any component of the CSA will reduce the values of PSNR and SSIM (Figs. 8.23, 8.24 and 8.25).

Effectiveness of The Loss Function. To examine the influence of different loss functions on the estimated residual illumination map during network optimization, we compare the differences of using three different loss functions to estimate the residual illuminance map. Similarly, for the learning of a dehazed image, we also compare the differences of using three different loss functions. Specifically, for the estimation of the residual illuminance map, we train the residual illumination map estimation subnetwork (RIMEN) using the mean absolute error loss (L1) as the baseline. Further training uses the mean absolute error loss and smooth loss as the loss function (L2). The third refinement loss function (L3) adds SSIM loss to L2. For the learning of the dehazed image, we set the loss function of RIMEN to L3 and change the loss function of the channel and spatial attention dehazing U-Net (CASDUN) to learn the dehazed image. Similarly, we first use the mean absolute error loss (L4) as the baseline. Then, to efficiently preserve edges, we add edge loss to L4 as the fifth refinement loss function (L5). The sixth refinement loss function (L6) adds SSIM loss to L5. The quantitative experimental results are summarized in Table 8.7 and Table 8.8. By comparing the values of the adjacent two columns in Table 8.7 and Table 8.8, we find that adding a certain loss function can improve the values of both the PSNR and SSIM compared to the reference values. The quantitative performance evaluated on the hazy image dataset demonstrates the effectiveness of each loss function. Visual comparisons are shown in Fig. 8.24. It is observed that smooth loss, edge loss, and SSIM loss all improve the matching between the estimated residual illumination map and the ground truth and the dehazed image and the ground truth (Tables 8.6, 8.9).

Table 8.10 Parameter numbers and running time (in s) of different methods. C and G denote CPU and GPU, respectively

Method	DCP	DehazeNet	AOD-Net	GFN	DCPDN	PQC	EPDN	GridNet	DADN	FFA-Net	Ours
Time	39 s	2 s	0.08 s	0.49 s	0.43 s	50 s	0.28 s	0.57 s	0.44 s	1.10 s	0.38 s
GPU/CPU	(C)	(C)	(G)	(G)	(G)	(C)	(G)	(G)	(G)	(G)	(G)
Parameters	–	-	1716	–	66,889,917	–	17,379,249	958,051	54,591,388	4,455,913	845,800

8.3.4.4 Runtime and Parameters Analysis

We also evaluated the complexity and computational efficiency of the aforementioned state-of-the-art methods. We conduct experiments on the RESIDE-ITS dataset which contains five hundred 620×460 images. DCP, DehazeNet, and PQC are implemented on CPU. Other deep learning-based methods are implemented on GPU. The average running time and parameter numbers are shown in Table 8.10. C and G denote CPU and GPU, respectively. Our method takes about 0.38s to deal with an image of size 620×460. It is observed that our method achieves comparable efficiency to other deep learning-based methods except AOD-Net and EPDN and also keeps a lightweighted framework.

In this section, we proposed a deep retinex network for image dehazing. Our retinex-based method not only avoids the traditional model errors associated with the simplified scattering model and the dependence on the prior information of traditional retinex-based dehazing methods but also achieves a trade-off between over-dehazing and under-dehazing. In the method, we first deduce a retinex-based dehazing decomposition model. Then, based on our model, we design a multiscale residual dense network to learn the residual illumination map. Sequentially, we obtain a coarse haze-free image by our model. Finally, the dehazing refinement subnetwork embeds both channel-wise and pixel-wise attention mechanisms into a U-Net to reinforce haze-free image learning. The experimental results on synthetic and real-world datasets validate our method.

References

1. C.O. Ancuti, C. Ancuti, C. Hermans, P. Bekaert, A fast semi-inverse approach to detect and remove the haze from a single image, in *Asian Conference on Computer Vision* (Springer, 2010), pp. 501–514
2. T.O. Aydin, R. Mantiuk, K. Myszkowski, H.-P. Seidel, Dynamic range independent image quality assessment. ACM Trans. Graph. (TOG) **27**(3), 1–10 (2008)
3. B. Cai, X. Xu, K. Jia, C. Qing, D. Tao, Dehazenet: an end-to-end system for single image haze removal. IEEE Trans. Image Process. **25**(11), 5187–5198 (2016)
4. W.-T. Chen, J.-J. Ding, S.-Y. Kuo, Pms-net: robust haze removal based on patch map for single images, in *Proceedings of the IEEE Conference on Computer Vision and Pattern Recognition* (2019), pp. 11681–11689
5. V.W. De Dravo, J.Y. Hardeberg, Stress for dehazing, in *2015 Colour and Visual Computing Symposium (CVCS)* (IEEE, 2015), pp. 1–6
6. Z. Deng, L. Zhu, X. Hu, C.-W. Fu, X. Xu, Q. Zhang, J. Qin, P.-A. Heng, Deep multi-model fusion for single-image dehazing, in *Proceedings of the IEEE International Conference on Computer Vision* (2019), pp. 2453–2462
7. R. Fattal, Single image dehazing. ACM TOG **27**(3), 1–9 (2008)
8. A. Galdran, A. Alvarez-Gila, A. Bria, J. Vazquez-Corral, M. Bertalmío. On the duality between retinex and image dehazing, in *Proceedings of the IEEE Conference on Computer Vision and Pattern Recognition* (2018), pp. 8212–8221
9. N. Hautiere, J.-P. Tarel, D. Aubert, E. Dumont, Blind contrast enhancement assessment by gradient ratioing at visible edges. Image Anal. Stereology **27**(2), 87–95 (2008)

10. K. He, J. Sun, X. Tang, Single image haze removal using dark channel prior. IEEE Trans. Pattern Anal. Mach. Intell. **33**(12), 2341–2353 (2011)
11. K. He, J. Sun, X. Tang, Guided image filtering. IEEE Trans. Pattern Anal. Mach. Intell. **35**(6), 1397–1409 (2012)
12. J. Hu, L. Shen, G. Sun, Squeeze-and-excitation networks, in *Proceedings of IEEE CVPR* (2018), pp. 7132–7141
13. Q. Huynh-Thu, M. Ghanbari, Scope of validity of PSNR in image/video quality assessment. Electronics Lett. **44**(13), 800–801 (2008)
14. J.-H. Kim, W.-D. Jang, J.-Y. Sim, C.-S. Kim, Optimized contrast enhancement for real-time image and video dehazing. J. Vis. Commun. Image Represent. **24**(3), 410–425 (2013)
15. E.H. Land, The retinex theory of color vision. Sci. Am. **237**(6), 108–129 (1977)
16. B. Li, X. Peng, Z. Wang, J. Xu, D. Feng, Aod-net: all-in-one dehazing network, in *Proceedings of the IEEE International Conference on Computer Vision* (2017), pp. 4770–4778
17. B. Li, W. Ren, D. Fu, D. Tao, D. Feng, W. Zeng, Z. Wang, Benchmarking single-image dehazing and beyond. IEEE Trans. Image Process. **28**(1), 492–505 (2018)
18. R. Li, J. Pan, Z. Li, J. Tang, Single image dehazing via conditional generative adversarial network, in *Proceedings of the IEEE Conference on Computer Vision and Pattern Recognition* (2018), pp. 8202–8211
19. Y. Li, Q. Miao, W. Ouyang, Z. Ma, H. Fang, C. Dong, and Y. Quan. Lap-net: Level-aware progressive network for image dehazing, in *Proceedings of the IEEE International Conference on Computer Vision* (2019), pp. 3276–3285
20. X. Liu, Y. Ma, Z. Shi, and J. Chen. Griddehazenet: Attention-based multi-scale network for image dehazing, in *Proceedings of the IEEE International Conference on Computer Vision* (2019), pp. 7314–7323
21. L.T. Maloney, B.A. Wandell, Color constancy: a method for recovering surface spectral reflectance. J. Opt. Soc. Am. A **3**(1), 29–33 (1986)
22. E.J. McCartney, F.F. Hall, *Optics of the Atmosphere: Scattering by Molecules and Particles* (Wiley, New York, 1976)
23. K. Mei, A. Jiang, J. Li, M. Wang, Progressive feature fusion network for realistic image dehazing, in *Asian Conference on Computer Vision* (Springer, Berlin, 2018), pp. 203–215
24. G. Meng, Y. Wang, J. Duan, S. Xiang, C. Pan, Efficient image dehazing with boundary constraint and contextual regularization, in *Proceedings of the IEEE International Conference on Computer Vision* (2013), pp. 617–624
25. S.G. Narasimhan, S.K. Nayar, Contrast restoration of weather degraded images. IEEE Trans. Pattern Anal. Mach. Intell. **25**(6), 713–724 (2003)
26. A.B. Petro, C. Sbert, J.-M. Morel, Multiscale retinex. Image Process. On Line 71–88 (2014)
27. X. Qin, Z. Wang, Y. Bai, X. Xie, H. Jia, Ffa-net: feature fusion attention network for single image dehazing, in *AAAI* (2020), pp. 11908–11915
28. Y. Qu, Y. Chen, J. Huang, Y. Xie, Enhanced pix2pix dehazing network, in *Proceedings of the IEEE Conference on Computer Vision and Pattern Recognition* (2019), pp. 8160–8168
29. W. Ren, L. Ma, J. Zhang, J. Pan, X. Cao, W. Liu, M.-H. Yang. Gated fusion network for single image dehazing, in *Proceedings of the IEEE Conference on Computer Vision and Pattern Recognition* (2018), pp. 3253–3261
30. Z. Rong, W.L. Jun, Improved wavelet transform algorithm for single image dehazing. Optik-Int. J. Light Electron Opt. **125**(13), 3064–3066 (2014)
31. O. Ronneberger, P. Fischer, T. Brox, U-net: convolutional networks for biomedical image segmentation, in *Proceedings o ICMICCAI* (2015), pp. 234–241
32. S. Santra, R. Mondal, B. Chanda, Learning a patch quality comparator for single image dehazing. IEEE Trans. Image Process. **27**(9), 4598–4607 (2018)
33. Y. Shao, L. Li, W. Ren, C. Gao, N. Sang, Domain adaptation for image dehazing, in *Proceedings of the IEEE/CVF Conference on Computer Vision and Pattern Recognition* (2020), pp. 2808–2817
34. R.T. Tan, Visibility in bad weather from a single image, in *2008 IEEE Conference on Computer Vision and Pattern Recognition* (IEEE, 2008), pp. 1–8

35. Y. Wang, J. Yang, W. Yin, Y. Zhang, A new alternating minimization algorithm for total variation image reconstruction. Siam J. Imaging Sci. **1**(3), 248–272 (2008)
36. Z. Wang, A.C. Bovik, H.R. Sheikh, E.P. Simoncelli, Image quality assessment: from error visibility to structural similarity. IEEE TIP **13**(4), 600–612 (2004)
37. S. Woo, J. Park, J.-Y. Lee, I. So Kweon, Cbam: convolutional block attention module, in *Proceedings of the European Conference on Computer Vision (ECCV)* (2018), pp. 3–19
38. B. Xie, F. Guo, Z. Cai, Improved single image dehazing using dark channel prior and multi-scale retinex, in *2010 International Conference on Intelligent System Design and Engineering Application*, vol. 1 (IEEE, 2010), pp. 848–851
39. l. Xu, c. Lu, y. Xu, j. Jia, Image smoothing via l0 gradient minimization. ACM Trans. Graph. (2011)
40. H. Zhang, V.M. Patel, Densely connected pyramid dehazing network, in *Proceedings of the IEEE conference on computer vision and pattern recognition* (2018), pp. 3194–3203
41. Y. Zhang, Y. Tian, Y. Kong, B. Zhong, Y. Fu, Residual dense network for image super-resolution, in *Proceedings of the IEEE Conference on Computer Vision and Pattern Recognition* (2018), pp. 2472–2481

Chapter 9
Underwater Descattering from Light Field

Abstract Traditional imaging methods and computer vision algorithms are often ineffective when images are acquired in scattering media, such as underwater, fog, or biological tissue. Compared with the atmospheric scattering environment, densely underwater scattering modeling is more difficult. In this chapter, we introduce an underwater scattering image restoration and depth estimation algorithm based on light field imaging. First, we present a new single-image restoration algorithm which removes backscatter and attenuation from images better than existing methods, and we apply it to each view in the light field. Second, we combine a novel transmission-based depth cue with existing correspondence and defocus cues to improve light field depth estimation. In densely scattering media, our transmission depth cue is critical for depth estimation since the images have low signal-to-noise ratios which significantly degrades the performance of the correspondence and defocus cues. Finally, we propose shearing and refocusing multiple views of the light field to recover a single image of higher quality than what is possible from a single view. We demonstrate the benefits of our method through extensive experimental results in a water tank.

9.1 Introduction

In this chapter, we propose a novel physically based algorithm to remove backscatter as well as attenuation. Our restoration requires knowing the scene depths, which are a priori unknown. As such, we first perform a rough restoration assuming a constant initialized depth. We apply this restoration to each view in the light field.

After our initial restoration, existing shape from light field algorithms can be used to estimate depth. However, in densely scattering media, the restorations are noisy and have low signal-to-noise ratio because the additive backscatter dominates the camera's dynamic range. As such the common defocus and correspondence cues do not provide reliable depth. To address this, we introduce a new, non-uniform corrected, transmission-based depth cue, which complements the light field cues.

Once the depth is estimated, we can use it to perform a final image restoration. However as mentioned, the restored images are noisy. As such, we also propose

J. Tian, *All Weather Robot Vision*, Research on Intelligent Manufacturing,
https://doi.org/10.1007/978-981-16-6429-8_9

shearing and averaging multiple views of the light field to produce one high-quality image. This produces much higher quality images than simply extracting the central view or simply averaging multiple views.

9.2 Underwater Image Formation Model

As shown in Fig. 9.1, consider a perspective camera placed at the origin, with the image (x, y) coordinates parallel to the world's (X, Y) axes, and the Z-axis aligned with the camera's optical axis. Let the point $\mathbf{X} = (X, Y, Z)$ be the point on the object's surface along the line of sight of pixel $\mathbf{x} = (x, y)$. We have the following relations:

$$\mathbf{x} = \left(f\frac{X}{Z}, f\frac{Y}{Z}\right)^t \quad ; \quad \mathbf{X} = \left(\frac{Z}{f}x, \frac{Z}{f}y, Z\right)^t, \tag{9.1}$$

where f is the focal length. We ignore the (u, v) angular coordinates of the light field in this section.

Let $S = (X_s, Y_s, Z_s)$ be the world coordinates of a near-field point light source, and define $\mathbf{D}(\mathbf{X}) = \mathbf{S} - \mathbf{X}$ as the vector from the object to the source. As in related works [7, 14], we adopt the single scattering model, ignore small angle forward scattering, and thus only consider backscatter from the source. Thus, the radiance arriving at the camera can be expressed as the sum of two terms:

$$I(\mathbf{x}) = I_d(\mathbf{x}) + I_b(\mathbf{x}) \tag{9.2}$$

where I_d is the direct light reflected from the object and I_b is composed of rays of light emitted by the source that are scattered into $\mathbf{X}'s$ line of sight before hitting the surface. This term is known as backscatter.

9.2.1 Direct Radiance Term

As seen in Fig. 9.1, consider an isotropic point source with radiant intensity E. Light from it travels a distance $||\mathbf{D}(\mathbf{X})||$ to the object. Thereafter, the light is reflected by the surface with albedo $\rho(\mathbf{X})$ and travels a further distance $||\mathbf{X}||$ ($||\mathbf{X}|| = \sqrt{X^2 + Y^2 + Z^2}$) to the camera. The direct radiance can be written as follows:

$$I_d(\mathbf{x}) = \frac{E}{||\mathbf{D}(\mathbf{X})||^2} e^{-\sigma||\mathbf{D}(\mathbf{X})||} \rho(\mathbf{X}) e^{-\sigma||\mathbf{X}||} \tag{9.3}$$

where σ denotes extinction coefficient of the medium.

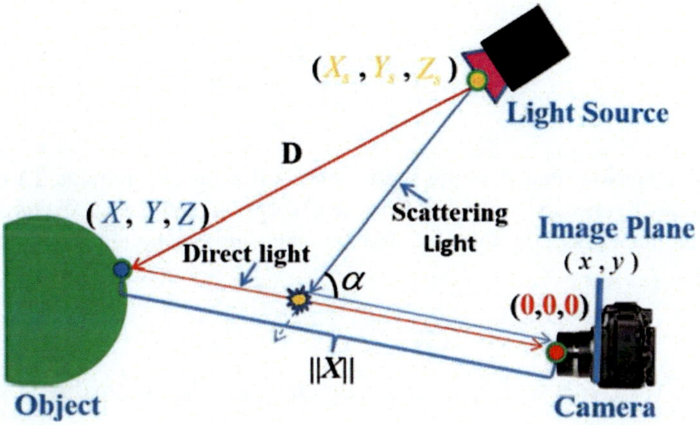

Fig. 9.1 Image formation and light propagation in a scattering medium. The radiance arriving at the camera is the sum of two components: the direct reflected light and the backscatter. The direct light travels distance D from the light to the object and then distance $\|X\|$ to the camera. The backscatter is scattered through angle α directly into the camera, without interacting with the object. ©[2017] IEEE. Reprinted, with permission, from [Tian et al. "Depth and Image Restoration from Light Field in a Scattering Medium", International Conference on Computer Vision, pp. 2401–2410, 2017]

Let $J(\mathbf{x}) = \frac{E}{||\mathbf{D}(\mathbf{X})||^2}\rho(\mathbf{X})$ be the clear image, and we have,

$$I_d(\mathbf{x}) = J(\mathbf{x})\mathrm{e}^{-\sigma(||\mathbf{D}(\mathbf{X})||+||\mathbf{X}||)} \tag{9.4}$$

9.2.2 Backscatter Term

Light which is scattered directly into the camera by the medium without reaching the object is termed backscatter. The fraction of light scattered to each direction is determined by scattering coefficient β and phase function $P(g, \alpha)$. We adopt the common Henyey–Greenstein phase function [4] in this book.

$$P(g, \alpha) = \frac{1}{4\pi} \cdot \frac{1 - g^2}{[1 + g^2 - 2g\cos\alpha]^{3/2}} \tag{9.5}$$

where scattering angle α is given by $\cos(\alpha) = \hat{\mathbf{D}} \cdot \hat{\mathbf{X}}$ ($\hat{\mathbf{D}}$ and $\hat{\mathbf{X}}$ denote normalized vectors) and satisfies $a \in [0, \pi]$. The parameter $g \in (-1, 1)$ controls the relative amounts of forward and backward scattering.

Based on previous work [8, 11], the backscatter is given by

$$I_b(\mathbf{x}) = FL_b \tag{9.6}$$

where

$$L_b(\mathbf{x}) = \int_0^{\|\mathbf{X}\|} \frac{e^{-\sigma||D(r\hat{\mathbf{X}})||}}{||D(r\hat{\mathbf{X}})||^2} P(\alpha) e^{-\sigma r} \mathrm{d}r \tag{9.7}$$

and $F = E\beta$. Note that β is the scattering coefficient and is related to extinction coefficient σ by $\sigma = \beta + \epsilon$, where ϵ is the absorption coefficient. We have absorbed the radiant intensity of the source E and the scattering coefficient β into one effective constant F.

9.2.3 *Extension to Light Sources Outside the Medium*

So far, we have assumed the light source is in the medium; however, this is not always the case. One common application where the light source is near field but outside the medium is indoor autonomous underwater vehicles (indoor AUVs). Our experimental setup also exhibits this configuration. Here we extend the scattering model presented in the previous sections to this case.

Consider the geometry in Fig. 9.2. Light travels from the source and enters the medium at point $\mathbf{X}_w$ without undergoing any scattering. Then the light continues into the medium where it undergoes scattering as before. The direct radiance becomes

$$I_d(\mathbf{x}) = J(\mathbf{x}) e^{-\sigma(||\tau\mathbf{D}(\mathbf{X})||+||\mathbf{X}||)} \tag{9.8}$$

and the backscatter becomes

$$L_b(\mathbf{x}) = \int_0^{\|\mathbf{X}\|} \frac{e^{-\sigma||\tau D(r\hat{\mathbf{X}})||}}{||D(r\hat{\mathbf{X}})||^2} P(\alpha) e^{-\sigma r} \mathrm{d}r \tag{9.9}$$

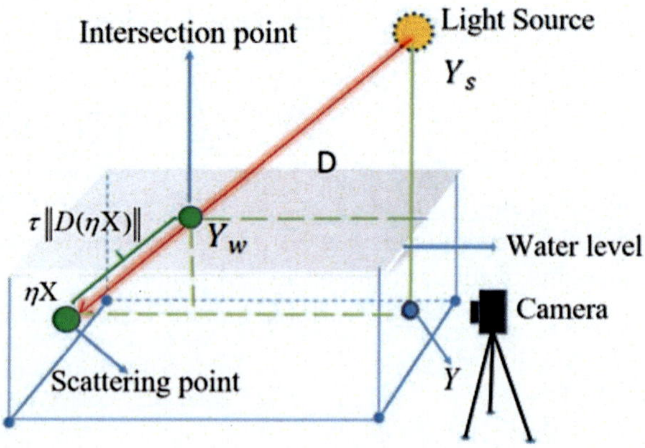

Fig. 9.2 Illustration of the geometry of a near-field source outside the medium.©[2017] IEEE. Reprinted, with permission, from [Tian et al. "Depth and Image Restoration from Light Field in a Scattering Medium", International Conference on Computer Vision, pp. 2401–2410, 2017]

where the only difference from Eq. 9.7 is that the attenuation of the source is scaled by $\tau \in (0, 1]$. Note that when $\tau = 1$ the source is in the water.

τ is given by

$$\tau(\mathbf{X}) = \frac{Y_w - Y}{Y_s - Y} \tag{9.10}$$

where Y_w, Y_s, and Y denote water surface, source location, and scattering point location, respectively, in the vertical direction. It is worth noting that τ is a function of **X**, which means we do not assume all points are equidistant from the source, as in [7].

Note that we have not included terms here to account for refraction nor the Fresnel effect. These terms could easily be added into the model and do not affect the derivations in the rest of the paper. In fact, for the rest of our analysis, we just assume the light is in the medium for notational clarity. In our experimental setup, we found that including the dependence on τ was important, but the refraction and Fresnel effect could safely be neglected (regarding the camera, it was placed close to the thin glass interface so that flat refractive effects were minimal).

9.3 Single-Image Restoration

Substituting Eq. 9.4 into Eq. 9.2, we get

$$I(\mathbf{x}) = J(\mathbf{x})\mathrm{e}^{-\sigma(||\mathbf{D}(\mathbf{X})||+||\mathbf{X}||)} + I_b(\mathbf{x}) \tag{9.11}$$

To recover the restored image, we need to solve Eq. 9.11 for J which requires estimating the unknown medium parameters F, σ, and g, as well as the depth for each pixel. To make the optimization simpler, we first estimate the medium parameters by examining pixels which only contain backscatter. Once the medium parameters are known, an initial restoration is computed by assuming the entire scene is at a constant known depth Z_{ref}. These restored images are used to estimate depth as described in the following section, and then the estimated depth is used to compute a final restoration. Our experiments show that further iteration does not improve the results.

For pixels that only contain backscatter, $J(\mathbf{x}) = 0$ and Eq. 9.11 reduces to

$$I(\mathbf{x}) = I_b(\mathbf{x}) \tag{9.12}$$

Let V be a set of points that only contain backscatter (we will describe how to find such a set of points in the next section). We estimate F, σ, and g by minimizing

$$\min_{g} \min_{\sigma,F} \sum_{\mathbf{x}\in V} \|I(\mathbf{x}) - I_b(\mathbf{x})\| \tag{9.13}$$

while assuming $Z = Z_{\text{ref}}$. The inner optimization is solved using the simplex method [5], while the outer optimization is solved by brute force search with a step size of 0.01 over the limited range $g \in [0.7\text{–}1.0]$ valid for water [8].

We allow light sources with different intensities E^c in each channel $c \in \{R, G, B\}$. We solve the optimization in Eq. 9.13 for the blue channel and then compute F^R and F^G by

$$F^c = \frac{1}{|V|} \sum_{\mathbf{x} \in V} \left(\frac{I^c(\mathbf{x})}{I^B(\mathbf{x})}\right) F^B, \; F^c = E^c \beta, c \in \{R, G\} \tag{9.14}$$

where $|V|$ is the number of pixels in set V. To find a good set V of pixels that only contain backscatter, we first convert the input image into HSV color space and extract its saturation and intensity channels. We then extract regions, whose intensity and saturation are both low, using Otsu's method [10]. In these regions, we further sample points at twenty pixel intervals in both the x and y directions. Finally, we take the minimum over five pixel neighborhoods for these sample points to generate set V.

The method that finds set V is inspired by the dark channel prior (DCP) proposed by He et al. [3]. The slight difference is that we did not extract the minimum of different color channels, because we allow more general light sources, i.e., E^c is different in each channel in our model. Furthermore, to make sure the points in V only contain backscatter contribution, we apply the DCP on low saturation and intensity pixels rather than on the whole image as in [3].

9.4 Depth Estimation

After the initial image restoration, assuming a constant depth Z_{ref}, we estimate the true depth using shape from light field [13]. However, for densely scattering media, where the restored images have a low signal-to-noise ratio, the defocus and correspondence cues are not enough to recover accurate depth. As such, we introduce our new transmission-based depth cue which can be combined with the defocus and correspondence cues to recover better depth.

9.4.1 *Defocus and Correspondence Cues*

We first use the following equation from Ng et al. [9] to shear the LF data to various depths.

$$J_\kappa(\mathbf{x}, \mathbf{u}) = J(\mathbf{x} + \mathbf{u}(1 - \frac{1}{\kappa}), \mathbf{u}) \tag{9.15}$$

where J is the initially haze removed 2D LF input image, $J_\kappa(\mathbf{x}, \mathbf{u})$ is the 4D sheared LF images at relative depth κ, $\mathbf{x} = (x, y)$ denotes the spatial coordinates, and $\mathbf{u} =$

(u, v) denotes the angular coordinates. For each pixel, the refocused image $\overline{J_\kappa}$ for the shear value κ is calculated by

$$\overline{J_\kappa}(\mathbf{x}) = \frac{1}{N}\sum_{\mathbf{u}} J_\kappa(\mathbf{x}, \mathbf{u}) \tag{9.16}$$

where N is the number of angular pixels. In our implementation, we shear κ from 0.2 to 2 with 256 steps. Therefore, we have 256 refocused $\overline{J_\kappa}(\mathbf{x})$.

The defocus DE and correspondence CO cues [13] are given by

$$DE_\kappa(\mathbf{x}) = \frac{1}{|W|}\sum_{\mathbf{x}'\in W} |\overline{J_\kappa}(\mathbf{x}') - J(\mathbf{x}', \mathbf{0})| \tag{9.17}$$

$$CO_\kappa(\mathbf{x}) = \frac{1}{N}\sum_{\mathbf{u}} |J_\kappa(\mathbf{x}, \mathbf{u}) - J(\mathbf{x}, \mathbf{0})| \tag{9.18}$$

When a patch is sheared to its correct depth, it will exhibit small variance and defocus. Therefore, we choose the corresponding depth responses by

$$\begin{aligned} Z_D(\mathbf{x}) &= \underset{\kappa}{\operatorname{argmin}}\, DE_\kappa(\mathbf{x}) \\ Z_C(\mathbf{x}) &= \underset{\kappa}{\operatorname{argmin}}\, CO_\kappa(\mathbf{x}) \end{aligned} \tag{9.19}$$

9.4.2 Transmission Depth Cue

Our transmission depth cue comes from the intensity of the backscatter. Taking the minimum over color channels and spatial neighborhoods $\Omega(\mathbf{x})$ of both sides of Eq. 9.11 gives

$$\begin{aligned} \min_c(\min_{\mathbf{y}\in\Omega(\mathbf{x})}(I^c(\mathbf{x})) = \min_c(\min_{\mathbf{y}\in\Omega(\mathbf{x})}(J^c(\mathbf{x}))\cdot \\ e^{-\sigma(||\mathbf{D(X)}||+||\mathbf{X}||)} + I_b^c(\mathbf{x}) \end{aligned} \tag{9.20}$$

According to the DCP prior [3], $\min_c(\min_{\mathbf{y}\in\Omega(\mathbf{x})}(J(\mathbf{x})) = 0$. Let $I^\dagger(\mathbf{x}) = \min_c(\min_{\mathbf{y}\in\Omega(\mathbf{x})}(I^c(\mathbf{x}))$. Then Eq. 9.20 reduces to (after substituting Eqs. 9.6, 9.7)

$$I^\dagger(\mathbf{x}) = F^c \int_0^{||\mathbf{X}||} \frac{e^{-\sigma||D(r\hat{\mathbf{X}})||}}{||D(r\hat{\mathbf{X}})||^2} P(\alpha)e^{-\sigma r}\mathrm{d}r \tag{9.21}$$

Equation 9.21 is a nonlinear equation for the depth $\|\mathbf{X}\|$, and could be solved using nonlinear optimization. However, we found this to be very time consuming and often did not converge to a good solution. As such, we made the following linear approximation which solved both of these problems.

First we change the variable of integration in Eq. 9.21 to $\mathrm{d}r = \|X\|\mathrm{d}s$ giving

$$I^{\dagger}(\mathbf{x}) = \|\mathbf{X}\| F^{c} \int_{0}^{1} \frac{\mathrm{e}^{-\sigma ||D(\|\mathbf{X}\| s\hat{\mathbf{X}})||}}{||D(\|\mathbf{X}\| s\hat{\mathbf{X}})||^{2}} P(\alpha)\mathrm{e}^{-\sigma\|\mathbf{X}\| s}\mathrm{d}s \tag{9.22}$$

Next we let $I_b^{\dagger}$ be defined as the integral in Eq. 9.22, and remove the global parameter F^c by normalization, giving

$$I^{\dagger}(\mathbf{x}) = \|\mathbf{X}\| I_b^{\dagger}(\mathbf{x}) \tag{9.23}$$

Finally, we solve Eq. 9.22 for $\|\mathbf{X}\|$ by making the approximation that $I_b^{\dagger}(\mathbf{x})$ depends on the constant depth Z_{ref} instead of the unknown depth Z. Let $Z_s(\mathbf{x}) = ||\mathbf{X}||$ denote our transmission depth cue.

The spatially varying but depth-independent term $I_b^{\dagger}(\mathbf{x})$ in the approximation can be seen as a non-uniform correction factor to the standard DCP depth algorithm. Although this correction can be neglected in mild scattering, in dense scattering, the backscatter varies spatially and cannot be ignored (see Fig. 9.3). As such, our transmission depth can achieve better results than both the classical DCP and the improved underwater DCP (UDCP).

9.4.3 Depth Fusion and Propagation

As can be seen in Fig. 9.4, any one depth cue on its own does not produce a reliable depth estimate. However, by combining all three complementary cues, we can recover good depth estimates.

To combine the defocus, correspondence, and transmission depth cues, we need determine the confidence weights $\Gamma(\mathbf{x})$ of each component. $\Gamma_D(\mathbf{x})$ and $\Gamma_V(\mathbf{x})$ are given in [13]. We define our transmission depth cue confidence as follows:

$$\Gamma_S(\mathbf{x}) = \sum_{\substack{c_1=\{R,R,G\} \\ c_2=\{G,B,B\}}} |J^{c_1}(\mathbf{x},\mathbf{0}) - J^{c_2}(\mathbf{x},\mathbf{0})| \tag{9.24}$$

The intuition for this is that properly restored images will be less white than images with backscatter. Thus if the color channels are very similar, then there is likely to still be backscatter.

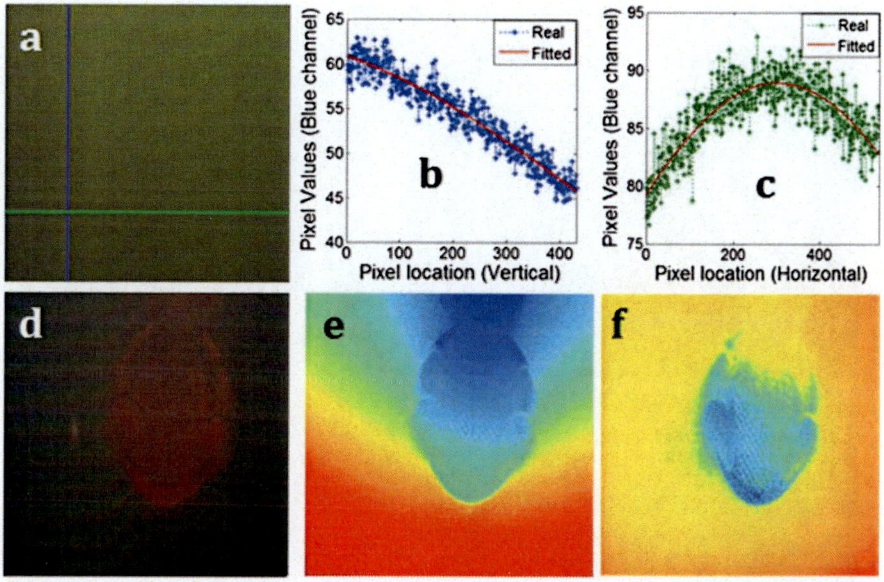

Fig. 9.3 Our transmission depth can account for the non-uniform backscatter. **a** An image of a uniform depth black background with scattering. All the image intensity comes from the backscatter. The blue and green lines show cross sections through the image plotted in (**b**) and (**c**). We can see that pixel values are not spatially uniform. This phenomenon cannot be accounted for by the Narasimhan–Nayar scattering model. As such, depth estimation with methods based on this model, such as DCP fail. **d** An image of a toy lobster in a dense scattering media. **e** Depth from the improved underwater DCP [2] **f** Depth from our new non-uniform corrected transmission depth cue.©[2017] IEEE. Reprinted, with permission, from [Tian et al. "Depth and Image Restoration from Light Field in a Scattering Medium", International Conference on Computer Vision, pp. 2401–2410, 2017]

The goal now is to combine different depth cues and propagate the local depth estimation to regions with low confidence. To weight our data constraint, we want to optimize depth to retain the local depth values with high confidence. The final optimized depth is obtained by

$$\underset{Z}{\text{minimize}}\{\lambda_{\text{source}} \sum_{j=D,V,S} \Gamma_j ||Z - Z_j||^2 + \lambda_{\text{smooth}} ||Z \otimes Q||^2\} \tag{9.25}$$

where $\otimes$ is the convolution operator and Q is the discrete Lapacian filter. We used $\lambda_{\text{source}} = 1$ and $\lambda_{\text{smooth}} = 1$ in our implementation, and solved (Eq. 9.25) using trust region reflective algorithm [1].

Now we can use the estimated depths to obtain an improved image restoration.

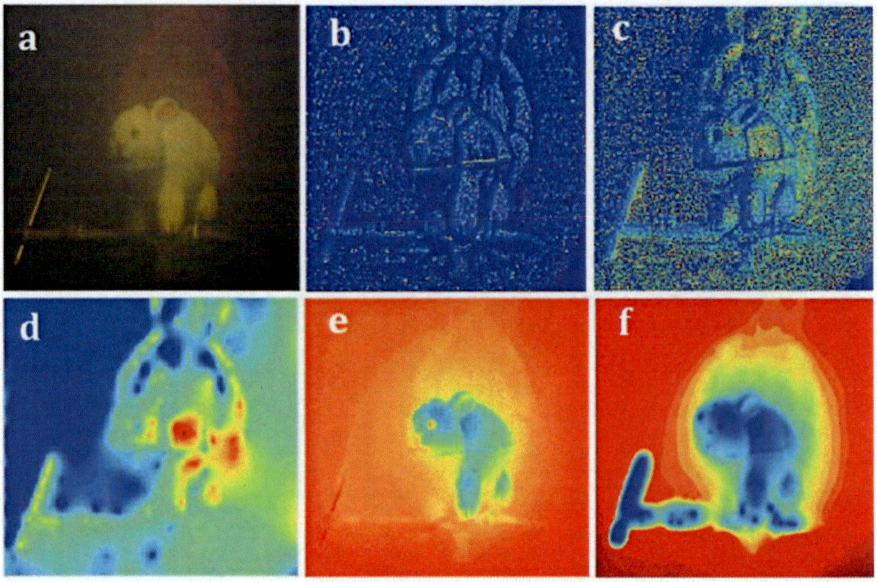

Fig. 9.4 Combining depth cues. We see that any one depth cue is not enough, and only after combing all three we get good results. **a** Input image. **b** Defocus depth. **c** Correspondence depth. **d** Combined LF defocus and correspondence depth. **e** Our new transmission depth. **f** All three depth cues combined. ©[2017] IEEE. Reprinted, with permission, from [Tian et al. "Depth and Image Restoration from Light Field in a Scattering Medium", International Conference on Computer Vision, pp. 2401–2410, 2017]

9.5 High-Quality Images from Shearing and Refocusing LF

Although the backscatter and attenuation are removed by restoration, the resulting images are often noisy (see Figs. 9.5 and 9.6b), especially for densely scattering media. This is because the backscatter takes up most of the dynamic range of the camera, and the attenuated signal from the scene is weak. To recover a single high-quality image, we propose shearing and averaging multiple views of the light field.

Without scattering, a point on a Lambertian surface will appear the same from different views of an in focus LF image. On the other hand, in a scattering medium, these different views will have slightly different scattering paths, and thus the point may appear different. In densely scattering media, after our restoration, the effect becomes very noticeable, as shown in Fig. 9.5. Our restoration algorithm does not explicitly handle this variation of scattering with respect to LF view. However, by combining the different information from multiple views, we can produce a single, all in focus, high-quality image with less noise than is possible from a single view (Fig. 9.6).

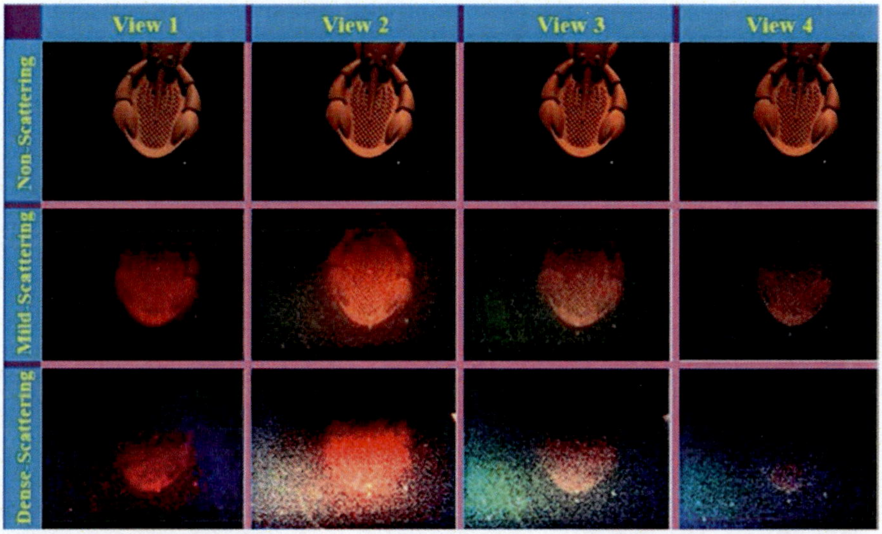

Fig. 9.5 Different views of LF data under different degree scatterings. We can see that without scattering, different views are similar and with good image quality. With scattering, the restored images from different views are different and affected by noise, and this phenomenon becomes more serious with more scattering. ©[2017] IEEE. Reprinted, with permission, from [Tian et al. "Depth and Image Restoration from Light Field in a Scattering Medium", International Conference on Computer Vision, pp. 2401–2410, 2017]

For comparison, we also show that simply averaging the different views into a single 2D image reduces the noise, but adds blur since the rays of the LF have not been properly aligned (Fig. 9.6c). In Sect. 9.6, we will show another comparison that simulates noise reduction from a single large aperture.

9.6 Experiments

First we did a simple simulation experiment to verify our backscatter removal under a wide variety of scattering conditions. Since all the scattering parameters are only contained in the backscatter part, we captured a black background image with known depth in clear water as the ground truth for the simulation. We set the parameter range as $F \in (100, 350), \sigma \in (0.001, 0.1)$, and $g \in (0.5, 1)$. Then we generated 2100 images by randomly selecting the three parameters in their ranges. The pixels have values in the range [0,255]. We then run our restoration method, He'method [3], and Tsiotsios's method [14]. The mean-squared errors (MSE) are 2.45, 36.71, and 79.18, respectively. We can see that our method outperforms existing methods and is robust in a wide variety of media.

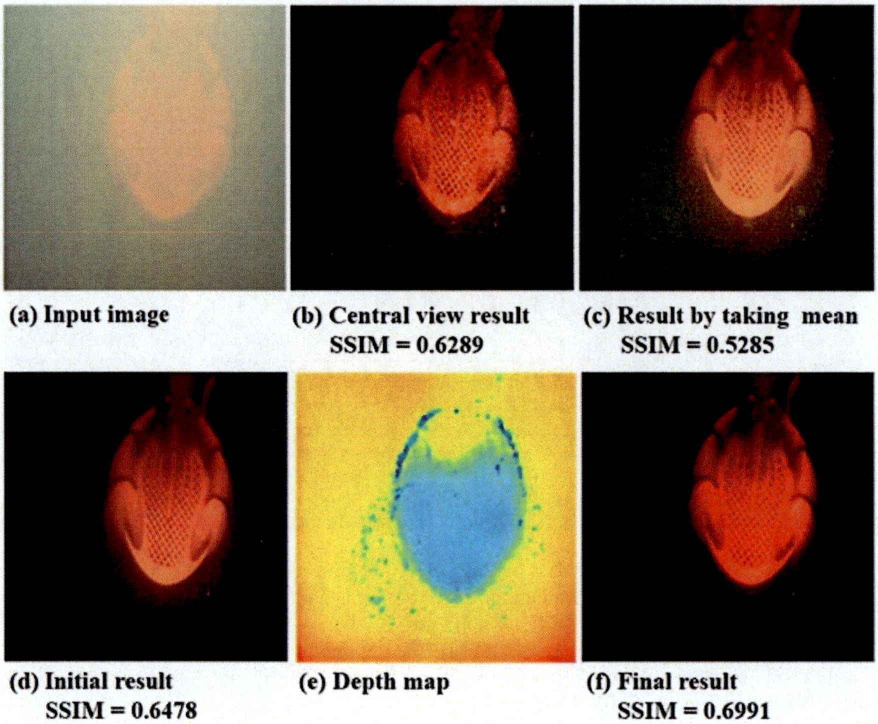

Fig. 9.6 Restoration (**d**) by shearing and refocusing on initially haze removed LF data can produce a better result than restoration only using the center view image (**b**) or simply taking the mean of different view directions (**c**). The final restoration result (**f**) by repeating the same shear and refocus procedure and incorporating a rough estimated depth (**e**) can produce an even clearer result. ©[2017] IEEE. Reprinted, with permission, from [Tian et al. "Depth and Image Restoration from Light Field in a Scattering Medium", International Conference on Computer Vision, pp. 2401–2410, 2017]

Second, we demonstrate our method using real experiments conducted in a glass tank filled with tap water (see Fig. 9.7). A Lytro Illum light field camera with an 18 mm lens was placed 1 cm away from the tank. The light source was an incandescent light bulb placed at $X_s = 5\,\text{cm}$, $Y_s = 65\,\text{cm}$, and $Z_s = 10\,\text{cm}$. The turbidity was increased by adding varying amounts of milk.

We compare our restoration on real captured images with four single-image-based methods: He's method [3], Meng's method [6], UDCP method [2], and Tsiotsios's method [14]. We also compare our depth results with two transmission-based methods [2, 3] and two LF-based methods: Tao's method [12] and Wang's method [15]. Since we are the first to do LF depth estimation in a scattering media, we also compare to a simple "combined" method that first removes the backscatter using [14] and then does depth estimation using [12].

Figure 9.8 shows comparisons of our restoration with existing methods. We can see that the three methods that are based on Narasimhan–Nayar model [2, 3, 6] cannot account for the non-uniform backscatter and thus do not improve the image

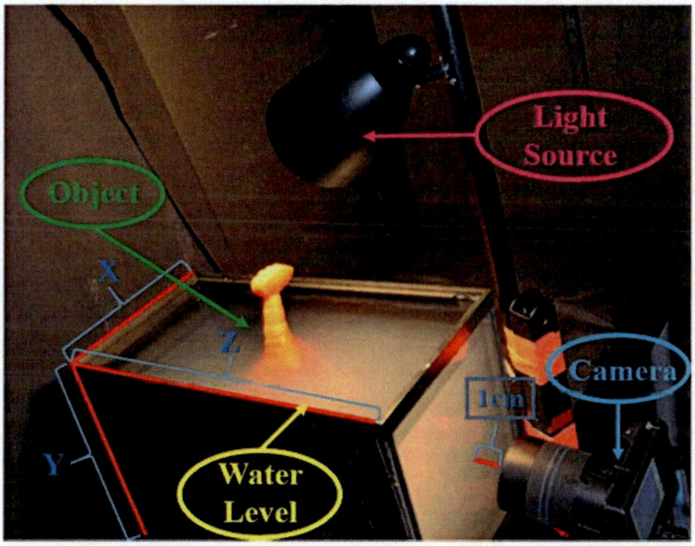

Fig. 9.7 Our experimental setup. ©[2017] IEEE. Reprinted, with permission, from [Tian et al. "Depth and Image Restoration from Light Field in a Scattering Medium", International Conference on Computer Vision, pp. 2401–2410, 2017]

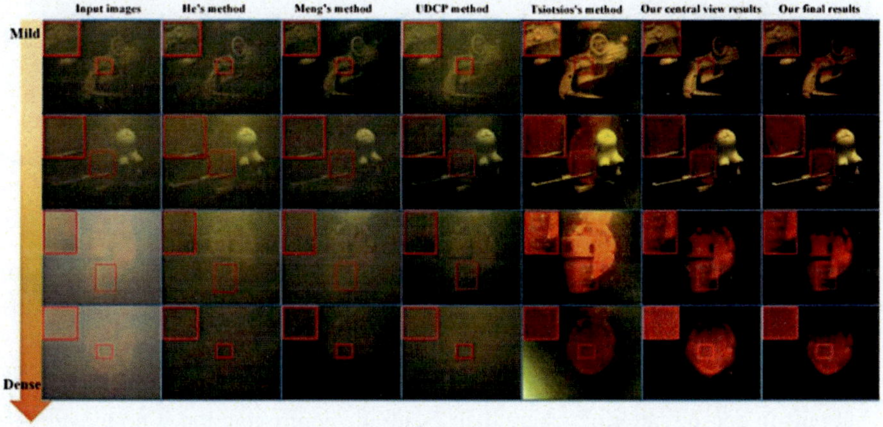

Fig. 9.8 Image restoration comparison. The results for the four other methods compared are derived by running their algorithm on the central view image of LF data. For fairness, we provide both our central view results that only use the central view image and final results that combine information from different views in the LF data. We can see that our central view results are better than other methods. None of the other methods can recover the uniform black background. Our final results are even better than our central view results because they contain less noise. ©[2017] IEEE. Reprinted, with permission, from [Tian et al. "Depth and Image Restoration from Light Field in a Scattering Medium", International Conference on Computer Vision, pp. 2401–2410, 2017]

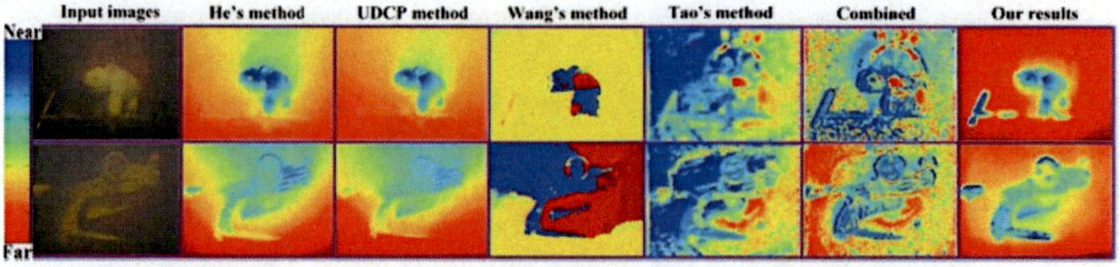

Fig. 9.9 Depth estimation comparison. We can see that our method produces better results than recent related methods as well as the simple "combined" method both in details and uniformity of the background. ©[2017] IEEE. Reprinted, with permission, from [Tian et al. "Depth and Image Restoration from Light Field in a Scattering Medium", International Conference on Computer Vision, pp. 2401–2410, 2017]

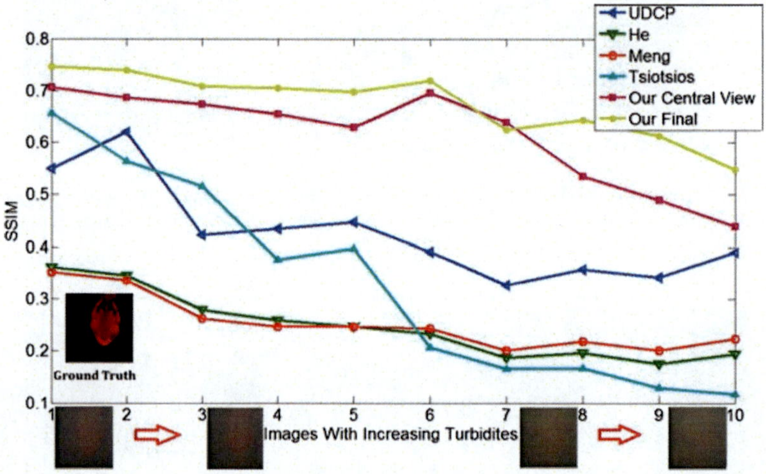

Fig. 9.10 Plot of image quality (SSIM) versus increasing turbidity for various restoration methods. The thumbnails give sample input images for ground truth and turbidity levels 1,3,7,and 9. It is clear that our restorations are better than previous methods across a broad range of turbidities. ©[2017] IEEE. Reprinted, with permission, from [Tian et al. "Depth and Image Restoration from Light Field in a Scattering Medium", International Conference on Computer Vision, pp. 2401–2410, 2017]

quality that much. Tsiotsios's method [14] is better able to remove the scattering, but fails in dense scattering where their quadratic approximation is no longer valid. Our method consistently produces the best results. And after shearing and refocusing multiple views, our results look even better.

Figure 9.9 shows comparisons of our depth with existing methods. Similar to the restoration shown in Fig. 9.8, [2, 3] cannot handle the non-uniform scattering and thus do not give a uniform depth to the background. The LF methods [12, 15] do not take scattering into account and thus produce bad depth. We also did a simple combined experiment for depth estimation (shown as "combined" in Fig. 9.9), which is haze removed by Tsiotsios et al. [14] followed by depth estimation by Tao et al. [12]. From the comparison, we can see that our method produces by far the best depth estimates.

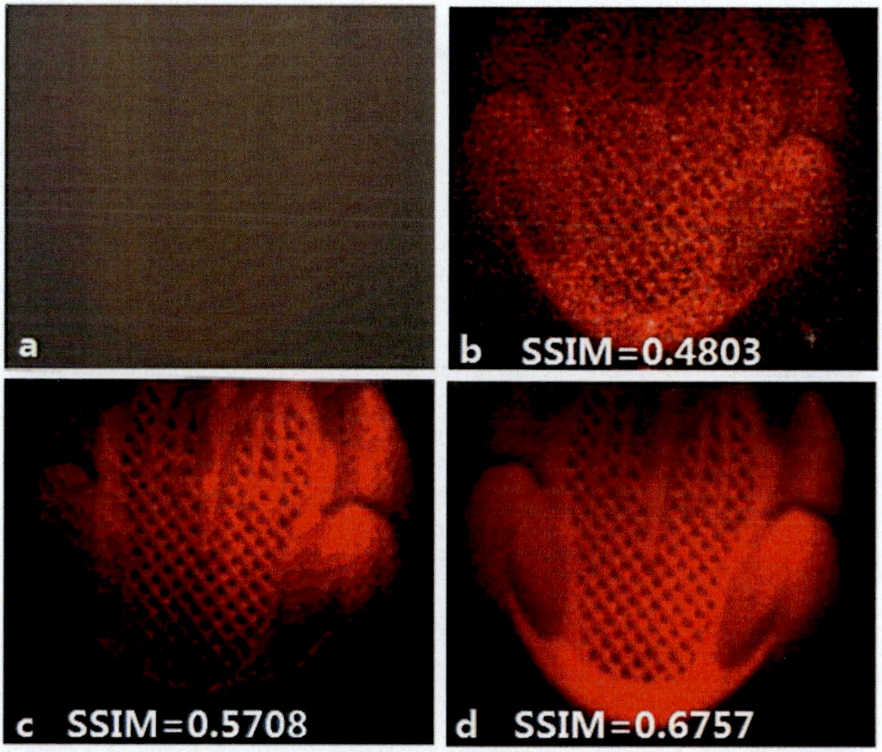

Fig. 9.11 Comparison of LF dehazing with effective large aperture single-image restoration. **a** Central view of input LF, **b** our descattering applied to the central view, **c** our descattering applied to the sheared and refocused effective large aperture 2D single image, **d** our proposed method of descattering the LF followed by shearing and refocusing. We can see that our method produces the best results. ©[2017] IEEE. Reprinted, with permission, from [Tian et al. "Depth and Image Restoration from Light Field in a Scattering Medium", International Conference on Computer Vision, pp. 2401–2410, 2017]

Figure 9.10 shows a plot of turbidity level vs image quality (measured using SSIM) for each of the restoration algorithms. SSIM [16] is a widely used metric that is more consistent with human perception than other metrics like mean-squared error (MSE) and peak signal-to-noise ratio (PSNR). We use an image captured in clear water as ground truth. We can see that our restored central view is better than all previous restoration methods and degrades more gracefully with increasing turbidity. By shearing and refocusing multiple views, we get even better results across all turbidity levels.

Shearing and refocusing reduce noise by averaging over many small apertures. For fair comparison, we compare our method with a baseline that simulates a single large aperture with less noise (Fig. 9.11). We shear and refocus the light field to a single 2D image, which we then remove the backscatter from using our backscatter subtraction. Our method clearly outperforms the large aperture baseline showing that combining information from many subapertures of a descattered LF is better than using a single large aperture.

In summary, our transmission depth cue can effectively be combined with the standard LF depth cues to produce superior depth estimates than existing methods. These better depth estimates combined with our near-field illumination scattering model allow for better haze removal than existing methods. Finally, by shearing and refocusing our dehazed LFs, we can achieve better restorations than is possible from a single large aperture image as well as each single view.

We have proposed the novel use of light field imaging in a scattering medium. Toward this end, we have made three main contributions: an image restoration algorithm for near-field illumination, a novel transmission depth cue which we combine with existing light field cues, and a multi-view image fusion for improved signal-to-noise ratio. Our image restoration algorithm outperforms existing methods, particularly for densely scattering media where prior methods often fail. Our combined depth estimation produces better depth estimates than previous methods. And our sheared and averaged final image has much less noise and looks nicer than other methods of restoring single view images.

References

1. T.F. Coleman, Y. Li, A reflective newton method for minimizing a quadratic function subject to bounds on some of the variables. SIAM J. Optim. **6**(4), 1040–1058 (1996)
2. P. Drews, E. Nascimento, F. Moraes, S. Botelho, M. Campos, Transmission estimation in underwater single images, in *Proceedings of the IEEE International Conference on Computer Vision Workshops* (2013), pages 825–830
3. K. He, J. Sun, X. Tang, Single image haze removal using dark channel prior. IEEE Trans. Pattern Anal. Mach. Intell. **33**(12), 2341–2353 (2011)
4. L.G. Henyey, J.L. Greenstein, Diffuse radiation in the galaxy. Astrophys. J. **93**, 70–83 (1941)
5. J.C. Lagarias, J.A. Reeds, M.H. Wright, P.E. Wright, Convergence properties of the nelder-mead simplex method in low dimensions. SIAM J. Optim. **9**(1), 112–147 (1998)
6. G. Meng, Y. Wang, J. Duan, S. Xiang, and C. Pan. Efficient image dehazing with boundary constraint and contextual regularization, in *Proceedings of the IEEE International Conference on Computer Vision* (2013), pages 617–624
7. S.G. Narasimhan, S.K. Nayar, B. Sun, S.J. Koppal, Structured light in scattering media, in *Tenth IEEE International Conference on Computer Vision (ICCV'05) Volume 1*, volume 1 (IEEE, 2005), pages 420–427
8. S.G. Narasimhan, M. Gupta, C. Donner, R. Ramamoorthi, S.K. Nayar, H.W. Jensen, Acquiring scattering properties of participating media by dilution, in *ACM Transactions on Graphics (TOG)*, volume 25 (ACM, 2006), pages 1003–1012
9. R. Ng, M. Levoy, M. Brédif, G. Duval, M. Horowitz, P. Hanrahan, *Light field photography with a hand-held plenoptic camera*. Ph.D. thesis, Stanford University (2005)
10. N. Otsu, A threshold selection method from gray-level histograms. Automatica **11**(285–296), 23–27 (1975)
11. B. Sun, R. Ramamoorthi, S.G. Narasimhan, S.K. Nayar, A practical analytic single scattering model for real time rendering. ACM Trans. Graph. (TOG) **24**(3), 1040–1049 (2005)

12. M.W. Tao, S. Hadap, J. Malik, R. Ramamoorthi, Depth from combining defocus and correspondence using light-field cameras, in *Proceedings of the IEEE International Conference on Computer Vision* (2013), pages 673–680
13. M.W. Tao, P.P. Srinivasan, J. Malik, S. Rusinkiewicz, R. Ramamoorthi, Depth from shading, defocus, and correspondence using light-field angular coherence, in *Proceedings of the IEEE Conference on Computer Vision and Pattern Recognition* (2015), pages 1940–1948
14. C. Tsiotsios, M.E. Angelopoulou, T.-K. Kim, A.J. Davison, Backscatter compensated photometric stereo with 3 sources, in *Proceedings of the IEEE Conference on Computer Vision and Pattern Recognition* (2014), pages 2251–2258
15. T.-C. Wang, A.A. Efros, R. Ramamoorthi, Occlusion-aware depth estimation using light-field cameras, in *Proceedings of the IEEE International Conference on Computer Vision* (2015), pages 3487–3495
16. Z. Wang, A.C. Bovik, H.R. Sheikh, E.P. Simoncelli, Image quality assessment: from error visibility to structural similarity. IEEE TIP **13**(4), 600–612 (2004)

Chapter 10
Applications and Future Work

Abstract In this chapter, we will show the applications of some illumination processing methods introduced in the previous chapters in robot vision, including UAV target tracking under complex illumination conditions, road and navigation line extraction in rural scenes, passable area recognition, tracking in rainy scenes, and object detection in highlight. The results show the effectiveness and value of our models and preprocessing algorithms. Finally, the future research in this field is prospected.

10.1 Application Examples of All-Weather Robot Vision

10.1.1 Unmanned Aerial Vehicle (UAV) Target Tracking Under Complex Illumination Conditions

The essence of target tracking method is to establish the relationship between targets in sequence images according to their characteristics. In the outdoor environment, if the target enters or leaves the complex illumination area such as shadows during the movement, the characteristics such as brightness, texture, and edges of the target will change dramatically, which will easily lead to tracking failure. To solve this problem, this section presents an outdoor target tracking method based on the intrinsic images in Chap. 5. Firstly, the regions where the target passes through the shadows are transformed into intrinsic images. Afterward the mean shift tracking algorithm is used to accurately locate the target in intrinsic images.

From Formulas (5–12), we have

$$\begin{aligned}&\log(F_R + 14) + \log(F_G + 14) - \beta \cdot \log(F_B + 14)\\ &= \log(f_R + 14) + \log(f_G + 14) - \beta \cdot \log(f_B + 14)\\ &= I_r\end{aligned} \tag{10.1}$$

$$\beta = \frac{\log K_R + \log K_G}{\log K_B} \tag{10.2}$$

J. Tian, *All Weather Robot Vision*, Research on Intelligent Manufacturing,
https://doi.org/10.1007/978-981-16-6429-8_10

In other words, define operator $\psi = [11\text{-}\beta]$ and add 14 to each of the three channels of RGB and then do ψ operations, which can make the pixels inside and outside the shadow share the same value. In this way, an intrinsic image insensitive to the shadow can be obtained. The pixel value in the intrinsic image is independent of whether the point is in the shadow or not, which can maintain the consistency of the same object in the shadow area. Therefore, the intrinsic-image deriving method shown above is applied to the tracking of outdoor targets for UAV in this book.

In order to determine the transformation parameters in the current sequence image, we detect the shadow in the initial frame of the sequence image and then estimate the scale coefficient under the current illumination condition through the detection results. Firstly, the shadow detection method proposed in literature [7] is used to obtain the shadow area of the image and calculate the mean brightness of R, G, and B channels of the shadow region and non-shadow region, respectively. They are recorded as follows: $\overline{R_s}$, $\overline{G_s}$, $\overline{B_s}$ and $\overline{R_{ns}}$, $\overline{G_{ns}}$, $\overline{B_{ns}}$. The proportional parameters of each channel are calculated according to the following formula:

$$K_R = \left(\frac{\overline{R_{ns}} + 14}{\overline{R_s} + 14}\right)^{2.4}, K_G = \left(\frac{\overline{G_{ns}} + 14}{\overline{G_s} + 14}\right)^{2.4}, K_B = \left(\frac{\overline{B_{ns}} + 14}{\overline{B_s} + 14}\right)^{2.4}$$

Since the shooting time of images in the same sequence is similar, we can consider that the illumination conditions are consistent in the tracking process, and the same set of parameters can be used in the sequence images. After the parameters K_R, K_G, and K_B under the current illumination are obtained, the transformation parameters are calculated according to formula 10.2. According to formula 10.1, the image can be transformed into a shadow-invariant space, and the transformed result is an intrinsic image I_{r0}.

In addition, it should be pointed out that I_{r0} is a grayscale image. In the practical application, it is found that the pixel values in I_{r0} after the original transformation are usually a decimal between [−6 and 3]. In order to facilitate the tracking in the next step, we carry out the following linear transformation of the image:

$$I_r = (I_{r0} + b) \cdot k \tag{10.3}$$

The parameter is set to $b = 6$ and $k = 30$. In this way, the contrast of the transformed image can be guaranteed, and the range of most pixels can be between 0 and 255. Finally, we make values less than 0 equal to 0 and values larger than 255 equal to 255.

Using the intrinsic-image deriving method proposed in this book, the images of the target in the sequence before and after entering the shadow are transformed under the same parameters. Then, the histogram distribution of the same target in the shadow and outside the shadow is compared in the original gray image and the intrinsic image (as shown in Figs. 10.1 and 10.2).

Figure 10.1 shows the contrast of a car before and after entering the shadow, which was a video taken in the morning. The calculation parameters of the intrinsic images

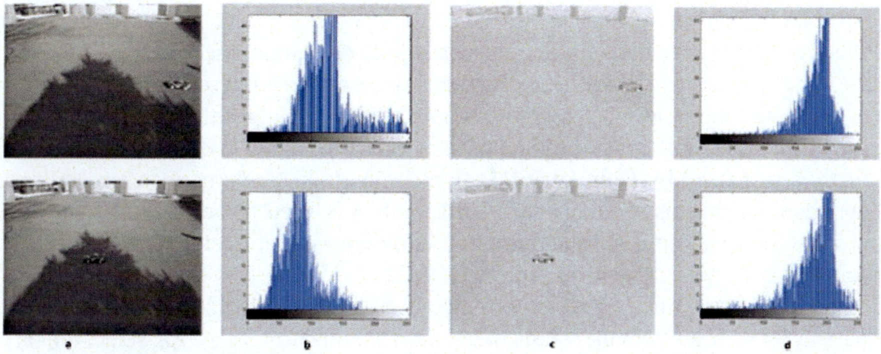

Fig. 10.1 Histogram comparison of car target. **a** grayscale images **b** histograms of the grayscale images **c** intrinsic images **d** histograms of the intrinsic images

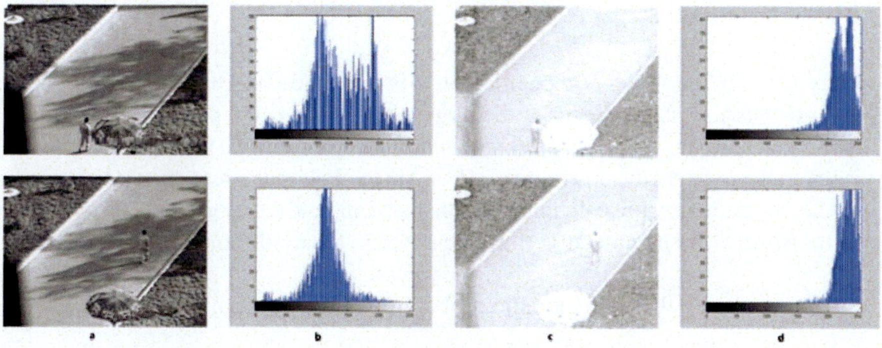

Fig. 10.2 Histogram comparison of pedestrian target. **a** grayscale images **b** histograms of the grayscale images **c** intrinsic images **d** histogram of the intrinsic images

are as follows: $K_R = 5.56$, $K_G = 5.26$, $K_B = 3.59$, $\beta = 2.64$. From the figure, we can see that there is a big difference between the shadow and the non-shadow area in the gray image. When the target enters the shadow area from outside, its brightness value decreases, and the change of histogram of its gray value also shows this point. By contrast, there is little difference between the shadow and non-shadow areas in the intrinsic image. The difference in the histogram of the target before and after entering the shadow region in the intrinsic image is also little.

Figure 10.2 shows a comparison of a pedestrian before and after entering the shadow area. The image was taken in the afternoon. The calculation parameters of the intrinsic images are as follows:$K_R = 3.52, K_G = 3.15, K_B = 2.27, \beta = 2.91$. It can be seen from the figure that the target keeps the appearance consistent in the intrinsic images, and the shadow region is transformed without introducing difference from the surrounding area. The target feature cannot maintain consistency inside and outside the shadow in the grayscale image. Target tracking is based on the consistency of a certain feature of the target to establish the relationship in the sequence images.

The pixel values in the intrinsic image are independent of whether the pixel is in the shadow, which can maintain the consistency of the same object inside and outside the shadow area, and it is verified by the above experimental results.

After the intrinsic images are obtained, mean shift algorithm is used for target tracking. Mean shift tracking method is a classical kernel-based efficient tracking method proposed by Comaniciu [1], which has high tracking accuracy and wide applicability. In this section, mean shift method is used to verify the effect of the proposed intrinsic-image transformation in outdoor target tracking. The steps are as follows:

(a) The first frame is selected to detect the shadow, calculate the transformation parameter ψ, transform the whole image into the intrinsic image I_{r0}, select the target to be tracked, and calculate the model probability distribution of the target:

$$\widehat{q}_u = C\sum_{i=1}^{n} k\left(\|\frac{x_i - x_0}{h}\|^2\right)\delta[b(x_i) - u] \tag{10.4}$$

where C is the regularization coefficient, k is the Gaussian kernel function, h is the window width, x_0 is the center of the target position, x_i is the pixel position, and δ is the Kronecke function, u=1,2...m, here m=52, $b(x_i)$=$ceil(I_r(x_i)/5)$.

(b) Assuming that the position x_{k-1} ($k \geq 2$) in the k-1 image sequence is known, then for the kth frame image, calculate the intrinsic image $I_r(x)$ around x_{k-1} according to the transformation parameter ψ that calculated in the first frame image, $\|x - x_{k-1} \leq 2h\|$.

(c) Calculating the characteristic probability distribution of candidate targets:

$$\widehat{p}_u(y) = C_h\sum_{i=1}^{n_k} k\left(\|\frac{x_i - y}{h}\|^2\right)\delta[b(x_i) - u] \tag{10.5}$$

where y is the center of the candidate target, its initial value y_0 is set as x_{k-1}, and the location of the candidate target center next time is calculated:

$$\widehat{y}_{j+1} = \frac{\sum_{i=1}^{n} x_i w_i g\left(\frac{\|\widehat{y}_j - x_i\|^2}{h^2}\right)}{\sum_{i=1}^{n} w_i g\left(\frac{\|\widehat{y}_j - x_i\|^2}{h^2}\right)} \tag{10.6}$$

where $w_i = \sum_{u=1}^{m} \delta[b(x_i) - u]\sqrt{\frac{q_u}{p_u(y_i)}}, g(x) = -k'(x)$. We perform iterative calculation until $\|\widehat{y}_{j+1} - \widehat{y}_j\| \leq \varepsilon$ and then determine $\widehat{y}_{j+1}$ as the current target center x_i.

In order to verify the effectiveness of the target tracking algorithm proposed in this book, some target tracking experiments in shadow scenes are compared, as shown in Figs. 10.3, 10.4, and 10.5. We present not only the tracking results of the proposed method based on intrinsic images, but also the tracking results of the mean shift methods [1] in the original shaded images and the L1 tracking method in the literature [4].

Fig. 10.3 Target tracking results based on intrinsic images

Fig. 10.4 Target tracking results based on original shadow images

The experimental results in Figs. 10.3, 10.4, and 10.5 show that in three groups of sequential images, the tracking method proposed in this book can achieve correct tracking both the target entering and leaving the shadow regions. If the intrinsic images are not used, the mean shift tracking method directly tracks the original shadow images, and the targets can be correctly tracked before the targets enter the shadows. However, once the targets enter the shadow areas, the tracking method will lose the targets, and the positioning results will stop at the shadow boundaries.

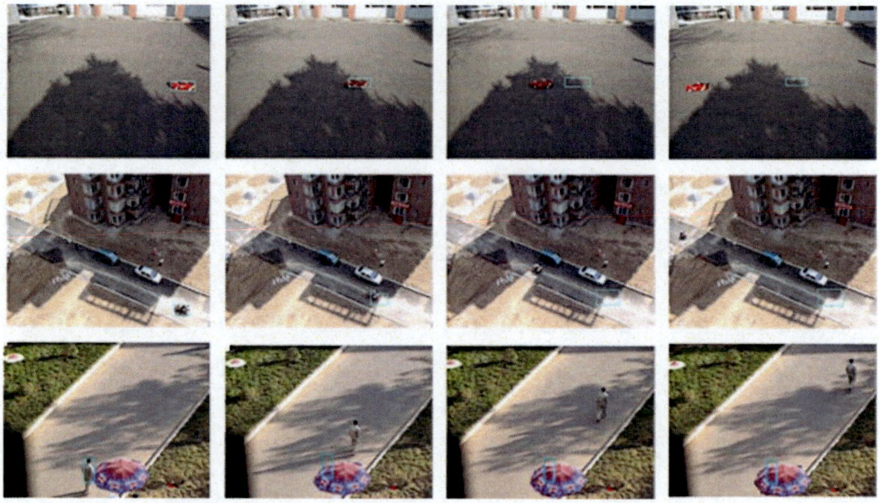

Fig. 10.5 Target tracking results based on L_1 method

In the three groups of sequence images, L_1 method can correctly track the target in the pedestrian sequence, because the brightness attenuation caused by shadow in the image sequence is not drastic compared with other sequences. But this method loses the targets after the targets enter the shadow area when tracking the target of motorcycle and car in drastic illumination changes.

Through three groups of comparative experiments, it can be seen that in the case of outdoor shadow, the intrinsic images are not sensitive to shadows, so the mean shift tracking method on intrinsic images can make the tracking process run correctly. At the same time, if mean shift tracking method is used solely, the targets will be lost; L_1 tracking method has a certain shadow resistance, but it will fail when the illumination change caused by shadow is relatively severe.

10.1.2 Road and Navigation Line Extraction in Rural Scenes

The automatic navigation system of robots based on vision is a hot research topic, and road detection is the key step of a visual navigation system. There are usually shadow and illumination changes in road scenes. In the actual environment, plants, vehicles, and trees often produce changes of shadow and illumination which will cause a great disturbance to road extraction. This section describes a road extraction algorithm for shaded rural and suburban road images. The algorithm firstly segments the intrinsic image obtained based on the illumination orthogonal decomposition. Then, based on the method of voting function, the region of interest was extracted from the results

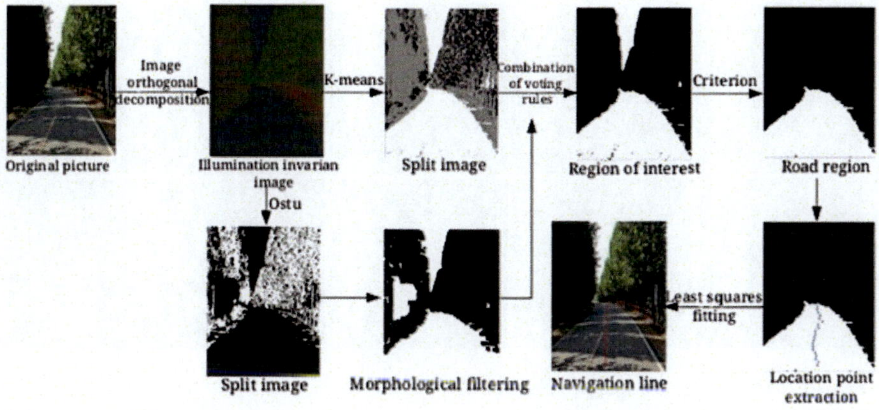

Fig. 10.6 Algorithm flowchart

of Otsu method and K-means clustering segmentation, afterward the road area was extracted by the method of subregion analysis.

As shown by the algorithm flowchart in Fig. 10.6, the road and navigation line extraction algorithm of shadow road image described in this section mainly includes the following steps:

(1) The method based on orthogonal decomposition in Chap. 5 is used to extract color intrinsic image from shadow road image;

(2) Color intrinsic-image segmentation and morphological processing based on Otsu method and K-means algorithm;

(3) Using the voting function to obtain the region of interest by combining the segmentation results of K-means and Otsu method;

(4) Extract the road area by analyzing the subregions of interest;

(5) The navigation line positioning points are obtained by scanning;

(6) The visual navigation line is extracted from the location point by least square fitting.

After orthogonal decomposition, the special solution of the image u_p can be called as color intrinsic or intrinsic image, mark it as I_p, as shown in Fig. 10.7. The first line presents several different shadow road images, and the color intrinsic image shown in the second line is obtained after orthogonal decomposition. It can be seen from the figure that the color intrinsic image has filtered out most of the shadow regions and has the consistency of illumination, which is more conducive to the subsequent road segmentation.

According to the color intrinsic images in Fig. 10.7, there are obvious distinctions between the road area and other regions. Therefore, we first use the Otsu method to segment the intrinsic images. The specific steps are as follows: firstly, transform the color intrinsic image I_p (Fig. 10.8a) into a gray image I_{pg} and then use the Otsu method to segment I_{pg} to get the segmented binary image I_{pb} (Fig. 10.8b). Otsu method is based on the segmentation of pixels, which will produce some outliers

Fig. 10.7 Color intrinsic-image extraction

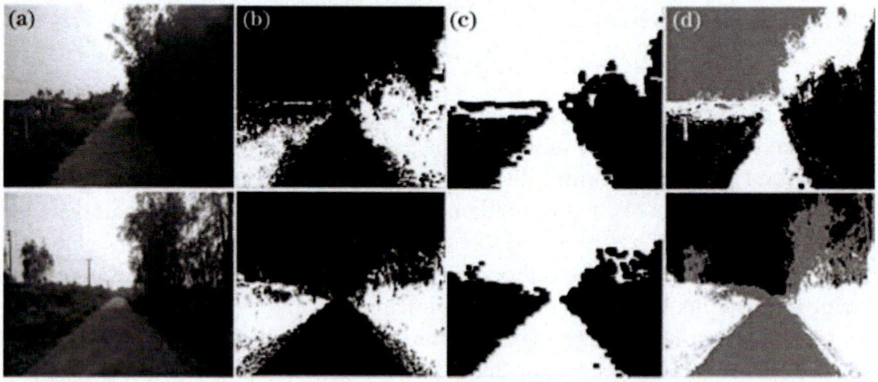

Fig. 10.8 Segmentation results based on color intrinsic images. **a** Color intrinsic images; **b** binary images; **c** morphological operation results; **d** K-means clustering results

after segmentation. To make the segmentation of the road area more complete, the isolated points are filtered out by morphological method. 3×3 rectangular template is used for erosion operation to obtain the segmentation result I_{ps} (Fig. 10.8c).

Sometimes the road area cannot be separated from the image after segmentation by the Otsu method, the situation happens such as the road area shown in Fig. 10.8c which is not an independent closed area. Therefore, the K-means-based clustering method is used to segment the intrinsic image and combined with the segmentation results of the Otsu method to assist the extraction of road areas. Based on K-means clustering [3] method, the intrinsic image I_p is clustered into three categories, and the clustering segmentation results as shown in Fig. 10.8d are obtained. Since the regions after K-means clustering segmentation are not easy to distinguish and there are a large number of non-road areas in the road clustering area, combining with the results of Otsu segmentation is beneficial to extract the road region more accurately.

According to the results of K-means clustering and the result of Otsu segmentation, the voting function of road region estimation is proposed. For each region of clustering $R_i(i = 1, 2, 3)$, the voting function is set as follows:

(1) Maximum area voting rule: The segmentation result of Otsu method coincides with the three regions of clustering segmentation, and the larger area is more likely a road area. According to the rule of maximum area, the voting function is set as follows:

$$V_{si} = sum(sum(I_{pg} \cap R_i)), (i = 1, 2, 3) \quad (10.7)$$

where $sum(\cdot)$ denotes the sum function.

(2) Bottom sampling voting rules: The road area is generally at the bottom of the image. Therefore, by sampling the bottom of the three areas after clustering, the voting function is set as follows:

$$V_{bi} = sum(sum(I_{pb} \cap R_i \cap S)), (i = 1, 2, 3) \quad (10.8)$$

Among them, S is the sampling area at the bottom, and its sampling area is set as the area of image height T_1 to T_2. In order to make the two voting functions have the same weight, the voting function is normalized in this book. In summary, the estimated voting function of the road area is as follows:

$$V_i = V_{si}/\max(V_{si}) + V_{bi}/\max(V_{bi}), (i = 1, 2, 3) \quad (10.9)$$

where $\max(\cdot)$ is the maximum function. According to the voting function, the ROI of road estimation is selected as follows:

$$I_{roi} = R_n \cap I_{pb}, n = argmax(V_i), (i = 1, 2, 3) \quad (10.10)$$

According to the method of voting, the area of interest shown in Fig. 10.6 can be obtained. However, there are some small interference areas in the region, which are not all road areas. Therefore, further selection is needed. For each closed region L_i of the region of interest, its area S_{Li} and center of gravity C_{Li} are calculated, where $I_{roi} = \sum_{i=1}^{m} L_1$, m is the number of closed regions. The rules for road selection are as follows:

$$S_{Li} \geq \max(S_{Li}) * T_s \cap C_{Li} * T_c \quad (10.11)$$

Among them, T_s and T_c are the empirical thresholds of area and center of gravity, respectively. When each closed area in the region of interest T_{roi} satisfies the formula 10.11, it is defined as the road region. By discriminating the first five regions with large areas, the final road region I_{road} is obtained, as shown in Fig. 10.6.

In general, the front of the camera's field of view is the direction in which the vehicle is moving. In intelligent navigation, vehicles realize autonomous navigation by adjusting the driving direction in real time according to the navigation line of the road. We extract the centerline from the acquired road region and define it as the navigation line.

Firstly, the road navigation starting point is selected. In order to improve the efficiency of the algorithm, the road region was scanned with the fixed step length S_T to obtain the left and right boundary point coordinates, the midpoint is calculated according to the left and right boundary point coordinates, and the navigation line location point set $Q(x, y)$ of the whole road area is obtained, as shown in the location point set in Fig. 10.6.

Then, the least square fitting is performed on the locating point set to obtain the red navigation line as shown in Fig. 10.6. The equation of the line fitting is as follows:

$$y = ax + b \tag{10.12}$$

where a and b are fitting coefficients obtained by least square fitting, respectively.

In order to verify the performance of the proposed road extraction algorithm, it is compared with the algorithm in literature [2] and literature [8] through the road image dataset. Figure 10.9 shows the road extraction results of the three algorithms in different road scenarios. The first line in Fig. 10.9a–d shows the road extraction results of our algorithm; the second and third lines are the results of Kong's algorithm and Wang's algorithm, respectively. Through comparison, it can be seen that our algorithm is better than the other two for roads extraction with shadows.

In order to show the effect of this algorithm, the visual navigation lines of road images in test dataset are extracted. Figure 10.10 shows the results of the navigation line extraction in different cases. From Fig. 10.10, we can seen that the algorithm in this book can accurately extract the navigation lines from images with complicated illumination.

10.1.3 Finding Passable Region for Robots from Single Image

Vision-guided obstacle avoidance and navigation, especially in unstructured environments, are a challenging issue in robotics. According to the characteristics of scene imaging, we analyze the relationship between the relative depth of the scene and the image features and then identify the passable regions from a single image. In this method, shadows have a great influence on the performance of the algorithm, and they are often mistakenly recognized as obstacles. After preprocessing the test image with the intrinsic-image deriving algorithm, the robustness of the passable region detection algorithm to shadows is obviously improved.

When the parameter values of road and navigation line extraction are within a certain range, the change of parameter values has little influence on the accuracy of the algorithm. The parameters used for road area discrimination are all limited

Fig. 10.9 Comparison of road extraction results with different roads under different conditions using three kinds of algorithm. **a** Curve road with shadow; **b** straight road with complicate shadow; **c** road without complicate shadow; **d** special shadow road

by the threshold range based on prior knowledge, in which case they have universal applicability and good stability for most road images.

Depth information is very important for obstacle avoidance. There are two types of depth: absolute depth and relative depth. Absolute depth is the accurate distance from objects to observer, whereas relative depth is just the comparative distance from objects to observer. Absolute depth is hard to obtain in monocular vision because it is an ill-posed problem in mathematics. In fact, though human cannot tell the absolute depth by their eyes, they can do well in obstacle avoidance and path planning. It implies that robots may find passable regions only depending on the relative depth information. The geometric characteristics of camera imaging enable the relative depth information to be retained although the absolute depth disappears during imaging. In a single image, the common image features that can infer the relative depth information include object size (Near objects look larger than farther ones), texture clarity, occlusion, edge direction, pixel position, etc. The edge information of some images especially the oblique edge can show the direction of depth growth, as illustrated in Fig. 10.11. As shown in Fig. 10.12, near objects have better texture resolution than distant objects.The nature of lens perspective projection makes the depth and pixel position have a great relationship, as illustrated in Fig. 10.13, if the ground is flat and there are no obstacles, the farther away the image is, the closer the image will be to the top of the image.

Fig. 10.10 Visual navigation line extraction results. **a** Curve road with shadow; **b** curve road with complicate shadow; **c** straight road with complicate shadow; **d** country road without shadow; **e** country road with pedestrian; **f** country road with different backgrounds

Fig. 10.11 Edges labeled with red lines indicate the depth growing direction. Reprinted, with permission, from [Jiandong Tian, Wei Dong, and Yandong Tang, "Finding Passable Regions for Robots From a Single Still Image", International Journal of Robotics and Automation, Vol. 24, Issue:3, pp.228–234, 2009.]. Copyright ©2009 ACTA Press

We will use the following information to obtain the depth growing direction on a single image and then obtain the passable region. Firstly, Canny edge detector is utilized to extract the edge features. The gotten edge image is marked as image I_1. Then given a pixel (x, y) of image f, the clarity and detail information are extracted by Eq. 10.13:

$$D(x, y) = f(x+1, y) + f(x-1, y) + f(x, y+1) + f(x, y-1) - 4^{*} f(x, y) \tag{10.13}$$

Fig. 10.12 Details of near scenes are clearer than those of distant scenes. Reprinted, with permission, from [Jiandong Tian, Wei Dong, and Yandong Tang, "Finding Passable Regions for Robots From a Single Still Image", International Journal of Robotics and Automation, Vol. 24, Issue:3, pp.228–234, 2009.]. Copyright © 2009 ACTA Press

Fig. 10.13 Relationship between pixel position and depth: **a** a real image **b** pin-hole model. Reprinted, with permission, from [Jiandong Tian, Wei Dong, and Yandong Tang, "Finding Passable Regions for Robots From a Single Still Image", International Journal of Robotics and Automation, Vol. 24, Issue:3, pp.228–234, 2009.]. Copyright © 2009 ACTA Press

The image I_2 is obtained by binarizing D, and the binarization threshold is calculated by the following formula:

$$T = \frac{1}{M * N} \iint D(x, y)dxdy \tag{10.14}$$

where $M * N$ is the size of image D.

Pixels with similar color may have the similar depth and passable property in an image. To measure the similarity, we select an initial region both in image I_1 and image I_2 as a reference. In the edge image I_1, we get line segment L_j for each column as follows: The line starts from a point at the bottom of the image and ends at the point where it meets the first edge. As shown in Fig. 10.13, for flat passable region, the farther away the image is, the closer the image will be to the top of the image. In this way, we get the following initial regions in the edge image:

$$R_1 = \{(x, y) \mid (x, y) \in \cup L_j (1 \le j \le c)\} \tag{10.15}$$

where c equals to the column number of image . In image I_2, we obtain the initial region R_2 using the following formula:

$$R_2 = \{(x, y) \mid D(x, y) < \delta T\} \tag{10.16}$$

Here, δ is empirically set to 0.8; T equals to the threshold in 10.14. The initial region in the original RGB image is $R_1 \cap R_2$. The mean value of the initial region is calculated in each color channel as follows:

$$M_k = \frac{1}{\text{Area}(R_1 \cap R_2)} \iint_{R_1 \cap R_2} f(x, y, k)dxdy \tag{10.17}$$

where f is the original image and Area$(R_1 \cap R_2)$ denotes the area of $R_1 \cap R_2$. Then the similarity image can be obtained by the following:

$$\begin{cases} 1 & if \;\; f(x, y, k) \in [T_1 M(k) \quad T_2 M(k)] \\ 0 & \text{others} \end{cases} \tag{10.18}$$

T_1 and T_2 are the thresholds to measure the degrees of similarity. We set T_1 and T_2 to be 0.9 and 1.1, respectively, here. Now we obtain a similarity image I_3. In image I_3, all the pixels in the initial region Area$(R_1 \cap R_2)$ are set to 1, whereas others are determined by 10.18. These three feature images will be used as the basis to identify the passable area.

The above three feature images are sensitive to illumination changes. We will use two illumination processing methods as a comparison. The first one is the normalized RGB method that is often used in the field of computer vision, as shown in formula 10.19, and the second one is the intrinsic-image deriving method based on the trichromatic attenuation model in Chap. 5.

$$\begin{bmatrix} r \\ g \\ b \end{bmatrix} = \begin{bmatrix} R/(R+G+B) \\ G/(R+G+B) \\ B/(R+G+B) \end{bmatrix} \tag{10.19}$$

Figure 10.14 shows the passable region detection results with the normalized RGB illumination preprocessing method, which can obtain good results when the illumination is uniform or the illumination changes are weak (as shown in Fig. 10.14(c)). However, in the image with strong shadows shown in Fig. 10.15, the detection algorithm cannot effectively identify the passable regions by using the normalized RGB illumination processing method, and the shadow areas are recognized as obstacles. By contrast, depending on our intrinsic-image illumination processing method, the algorithm can correctly identify the passable regions without being affected by the shadows. That means we can obtain better region detection results by using our illumination processing method than RGB normalized method in strong shadows.

10.1.4 Tracking in Rainy Scenes

For tracking in rainy scenes, we first restore clear images using our matrix decomposition rain removal method in Chap. 7 and then use the fusion tracker [6] for object tracking.

Currently, there is no tracking dataset in rain scenes. Thus, we first synthesize two rain datasets on *UCSD* and *MOT17* to evaluate single person tracking and multi-people tracking, respectively. Figure 10.16 shows two rain examples on *UCSD*. As observed, the rain image has a low-quality density map, since the people count is much lower than the ground truth. After rain removal using our method in Chap. 7, the image quality and the estimated density map can both be improved. The same also applies to the *MOT17* dataset (see Fig. 10.17).

In Table 10.1, we report the results of S-KCF and our fusion tracker in different scenes (denoted as Tracker + Scene). Here, "R", "D", and "C" represent the rain scene, derained scene, and clear scene, respectively. For example, "Fusion + R" means tracking using rain images and their corresponding low-quality density maps. We can find that tracking performance can be improved when the rain is removed. For example, P@10 of S-KCF is improved from 0.3326 to 0.3970. Also, tracking can be improved when the fusion tracker is applied even in rains scenes (Fusion + R performs better than S-KCF + R). Overall, we can conclude that both rain removal method and fusion tracker can contribute to tracking in rain scenes.

In Table 10.2, we also report the MOT results on synthetic *MOT17* rain scenes. When scenes are degraded by rain, performance of FusionCNN-MT is significantly dropped (e.g., MOTA dropped from 53.1 to 43.7), and thus it is necessary to do preprocessing when tracking in rain scenes. When rain scenes are first processed by our rain removal method, FusionCNN-MT can be improved (see FusionCNN-MT + D), and it can achieve comparable results with FusionCNN-MT + C (MOTA of 50.5 vs. 53.1).

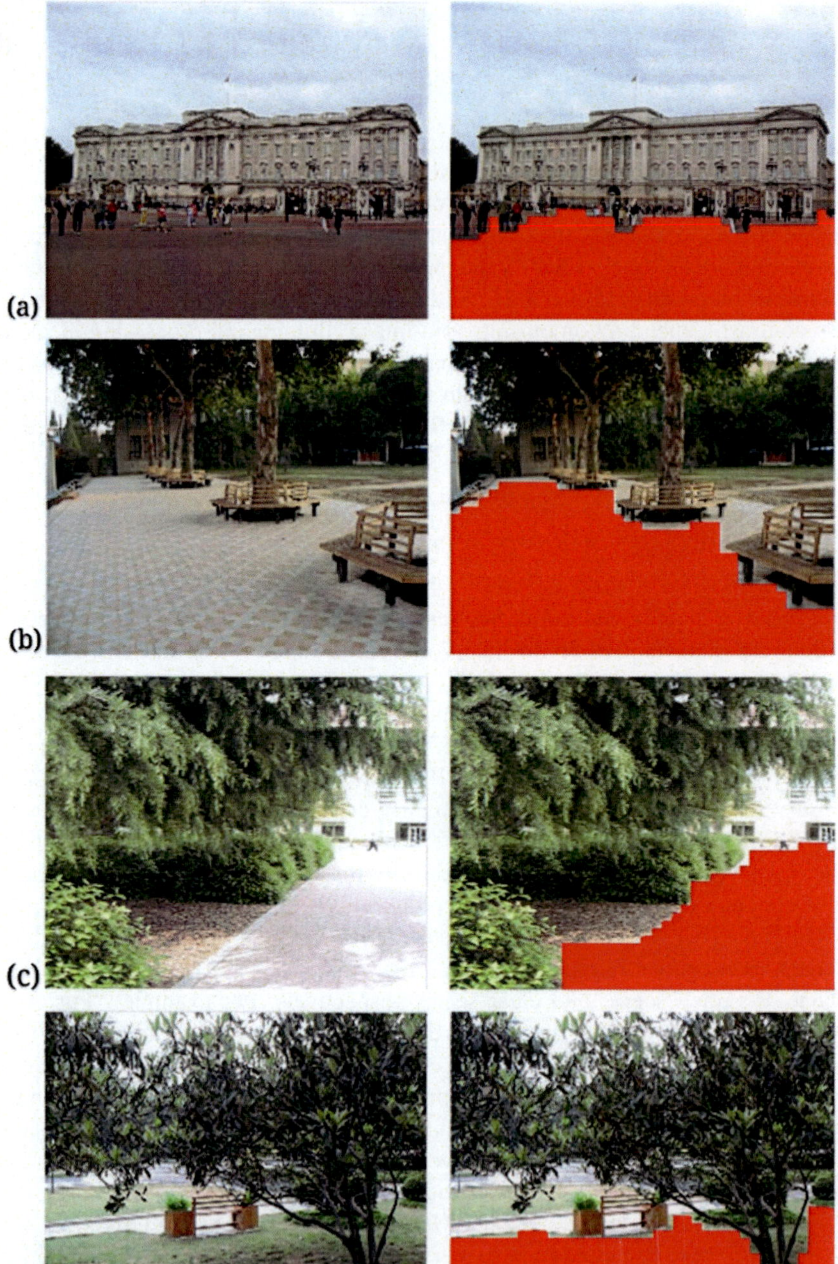

Fig. 10.14 Results of passable region detection without obvious shadow. Reprinted, with permission, from [Jiandong Tian, Wei Dong, and Yandong Tang, "Finding Passable Regions for Robots From a Single Still Image", International Journal of Robotics and Automation, Vol. 24, Issue:3, pp.228–234, 2009.]. Copyright © 2009 ACTA Press

Fig. 10.15 Results of passable region detection on images with obvious shadows. Left: original images; middle: results of normalized RGB illumination processing method; right: results of our illumination processing method

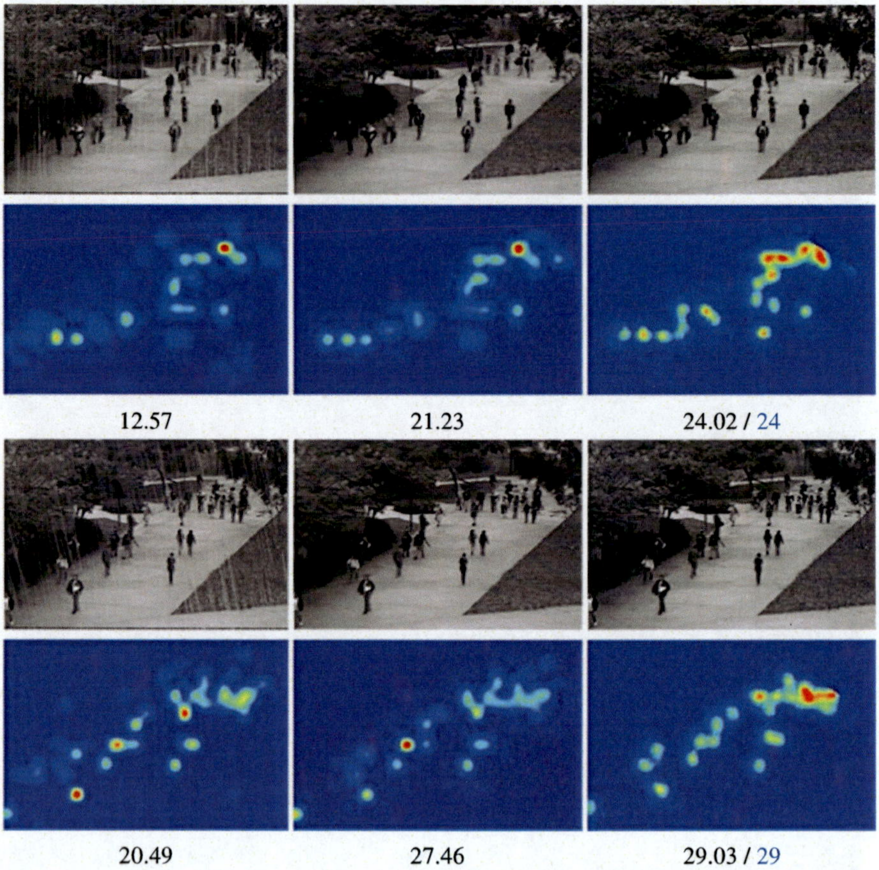

Fig. 10.16 Two synthetic rain examples on *UCSD*. The first (middle, last) column is the rain (recovered, clear) image and its corresponding estimated density map. Estimated people count is also reported under the density map (blue one is the ground truth)

We also apply our rain removal method to a video where is no rain (denoted as "DC"), and the tracking performance will become slightly worse, since video backgrounds are blurred after image processing. For single person tracking (Table 10.1), though S-KCF+ DC performs worse than S-KCF+ C, the result of Fusion + DC is almost the same with Fusion + C (P@10 of 0.5863 vs 0.5999). Thus, we can conclude that rain removal method can be applied all the time without considering efficiency. For multi-object tracking, the same conclusion can also be drawn from Table 10.2.

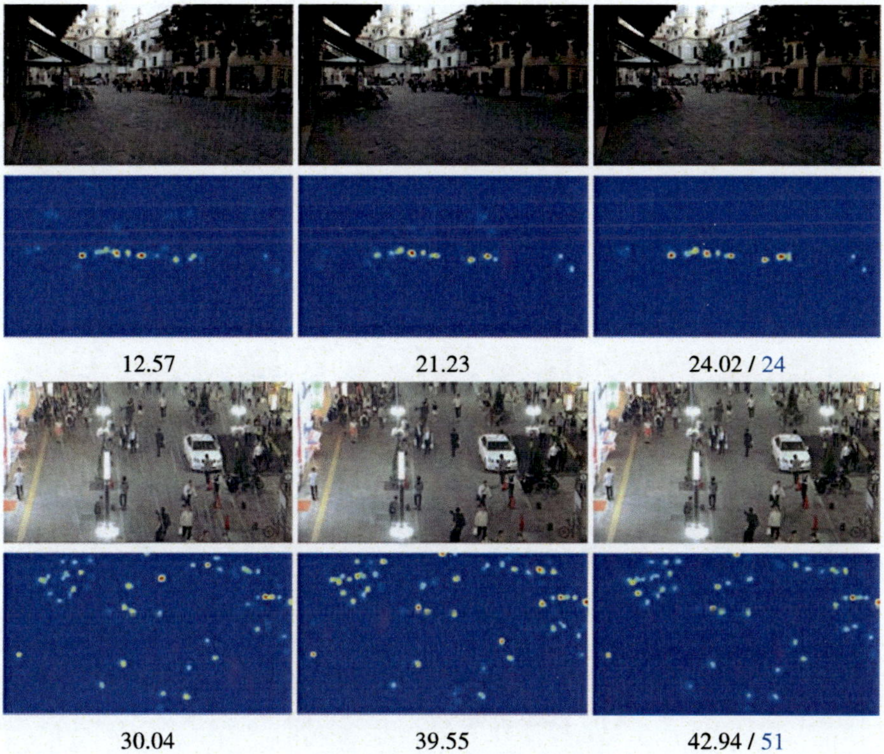

Fig. 10.17 Two synthetic rain examples on *MOT7*. The first (medium, last) column is the rain (recovered, clear) image and its corresponding estimated density map. Estimated people count is also reported under the density map (blue one is the ground truth

Table 10.1 Tracking results on synthetic rain dataset *UCSD*

Model	S-KCF + R	S-KCF + D	S-KCF + C	S-KCF + DC	Fusion + R	Fusion + D	Fusion + C	Fusion + DC
P@10	0.3326	0.3970	0.4356	0.4128	0.4406	0.5301	0.5999	0.5863
IoU@0.5	0.3866	0.4897	0.5148	0.5120	0.5179	0.6193	0.6378	0.6219

10.1.5 Effect of Highlight Removal on Detection

Currently, all the highlight removal methods work in constrained environment where the light source is single. For outdoor environment, illumination is very complicated due to the presence of reflections and refractions. Thus, most of the highlight removal methods do not handle well for natural highlight images. In Fig. 6.19 of Chap. 6, we show that our proposed highlight removal method can work well on two natural images. However, the two images are local areas of a scene, which means

Table 10.2 Multi-object tracking results on synthetic rain dataset *MOT17*

Tracker	MOTA↑	IDF1↑	MOTP↑	FAF↓	MT↑	ML↓	FP↓	FN↓	IDS↓	FM↓
FusionCNN-MT + R	43.7	46.5	76.0	3.3	18.0%	39.6%	57,977	255,719	4,063	**5,230**
FusionCNN-MT + D	50.5	45.1	**76.2**	3.2	**21.8%**	**38.0%**	57,513	**216,875**	4,990	5,554
FusionCNN-MT + C	**53.1**	**50.1**	75.8	**2.0**	20.4%	39.4%	**35,295**	225,606	**3681**	6,271
FusionCNN-MT + DC	52.3	49.7	75.9	2.4	21.1%	39.6%	36,451	231,104	3923	6,148

Fig. 10.18 Detection results for highlight (first row) and highlight-free (second row) images

that reflection and refraction are simple. For complex scenes, we need to jointly use polarizer and our proposed method to remove highlight. To synthesize highlight for a video sequence is difficult, since both optical and physical properties need to be met simultaneously. We thus try to verify that detection can be improved after highlight removal, which implicitly proves the effect on tracking.

In Fig. 10.18, we take *three* pairs of outdoor highlight images by using or not using polarizer. Polarizer can remove highlight in most of the areas, but the image taken with it still has some global specular reflection. To achieve more satisfactory result, we use our proposed method to further refine the image. We then use faster-RCNN [5] to produce detections for these images. As observed, both location precision and classification score can be improved when highlight is removed. For example, for the middle image, more objects can be detected after highlight removal. Note that classification may be inaccurate since we only use 20 classes for training. In conclusion, object detection can be improved after highlight removal, which will also improve object tracking that depends heavily on detection.

10.2 Research Prospect

Although great progress has been made in the research of robot vision, the reliability of vision tasks under the conditions of complex illumination or bad weather is still a key technical problem to be solved. So far, the academia has realized the great impact of weather variations on the stability of visual algorithms and has carried out extensive research. However, there is still a long way to go before the research results can be actually and extensively applied. Nowadays, there are two main issues in image preprocessing algorithms for all-weather robot vision. The first is that many algorithms are lack of robustness. Taking the defogging task as an example, the current algorithms are only effective for mild fog and haze. As to the thick fog, the algorithms are almost ineffective. Therefore, it is generally necessary to configure expensive fog penetrating cameras to solve the problem of heavy fog in engineering. The second is that many algorithms are lack of real-time performance. The main reasons can be summarized from three aspects. Firstly, the natural environment is too complex, and it is very difficult to accurately model; secondly, the models have many parameters, and some image information has been lost. The task of image restoration under this condition is a sick problem; finally, the image database for shadow, highlight, rain, snow, and fog removal are lacking. A visual algorithm often needs a database that contains a large number of original images and corresponding ground truth for training, test, detection, restoration, or evaluation. In some research fields, such as face recognition, there already have been many mature and recognized databases, but in the field of rain, snow,, and fog removal, there is still lack of such databases. To build such a database is a huge and very difficult work. We need to collect images with the same illumination for different scenes. Especially, the desnow ground-truth image acquisition is the most difficult task. We need to avoid the impact of snow on the ground, branches, vehicles, or other objects.

At the practical application level, the robot is also required to be able to automatically determine different illumination and weather conditions. Previous algorithms are manually determined with different illumination and weather conditions, such as whether the scene has highlight or shadow, whether any fog, rain, or snow happens. However, with the rapid development of technology, robots need to understand the environment more intelligently, that is to say, robots need to automatically determine the illumination and weather conditions and automatically select the corresponding processing algorithm by itself.

When researchers of robot vision systems develop specific vision tasks, they often do not consider the impact of illumination and weather changes. It is not until the test or actual task had been executed that they find the necessity of illumination and weather processing. If they load the illumination processing algorithm at this time, they may need to reconsider the design of the data interface and resource allocation. That is inconvenient for vision tasks. We believe that it is difficult to completely solve these problems by just relying on algorithm improvements. Robot vision is not the same as computer vision. The mobility and embeddedness of robot vision determine that it is difficult to use big data and large energy consumption methods, just like computer vision did. Robots also need to respond quickly to their surrounding dynamic environment. We incline to the view that, in the future, taking the stability and real-time performance of the algorithms as the starting point, using the combination of software and hardware and the combination of multi-mode signals to solve vision tasks will be a very promising solution. In view of this, we plan to adopt the combination of software and hardware to develop a new all-weather camera to fundamentally solve these problems. The brand-new camera can collect more information and establish a relationship between the robot and the external environment. Moreover, it can give the robot the ability to actively detect the environment in which it is located and the ability to automatically determine the illumination and weather conditions and then select the corresponding processing algorithm. Finally, the camera can shield the complex illumination and bad weather and output images that are not affected by illumination or weather conditions. That means, this camera will bring great convenience to the robot vision, and there is no need to consider the complex problems of illumination and weather processing during vision algorithm development and robot working.

At present, deep learning technology can greatly improve the performance of most visual tasks. However, the existing research has shown that deep learning technology cannot significantly improve the visual system's ability under complex illumination and bad weather conditions. This perhaps can also explain that relying merely on image data (even big data) is not enough to solve these difficult problems. We believe that robot vision is not necessarily limited to the image data provided by existing cameras. Therefore, we will adopt the combination of software and hardware and make full use of the interaction between the robot body and its outside surroundings to obtain as much information as possible and effectively associate the physical illumination environment with the image data. We will use the content of the previous nine chapters of this book as theoretical guidance, combining advanced imaging technology such as RGB-NIR CCD, new optical components (array polarizer), ISP circuit board, lens, and image processing algorithms to develop an "all-weather camera" that would automatically eliminate highlight and shadows. It can also have the ability to resist variations in global illumination and automatically remove rain, snow, and fog in images. The solution we proposed makes full use of external environmental information and hardware architecture and is more targeted for robot vision. It may provide an effective method to break through the bottleneck problem of stability and real-time performance of the vision system under dynamic illumination and meteorological conditions. As a new sensing method in robot vision, our camera could greatly improve the adaptability of robot vision and other outdoor vision systems (such as surveillance, autonomous driving, military applications) to illumination and weather conditions, enabling the robot has the all-weather working ability.

In the aspect of theoretical research, we will try to establish a unified processing model of rain, snow, and fog. This is because they have similarities in the physical imaging mechanism: Scene radiance added a part of stray light, and in mathematics they are all in the form of the summation of the two parts. In terms of visual appearance, we may consider it as rain or snow when we are in close viewing, while we may think it as fog when we are at a distance. So, it is necessary to deal with rain, snow, and fog uniformly. In terms of complex illumination processing, we will also try to establish a unified illumination processing model to unify the detection and removal of shadows and highlights under one theoretical framework.

References

1. D. Comaniciu, V. Ramesh, P. Meer, Kernel-based object tracking. IEEE Trans. Pattern Anal. Mach. Intell. **25**(5), 564–577 (2003)
2. H. Kong, J.-Y. Audibert, J. Ponce, General road detection from a single image. IEEE Trans. Image Process. **19**(8), 2211–2220 (2010)
3. J. MacQueen et al., Some methods for classification and analysis of multivariate observations, in *Proceedings of the fifth Berkeley symposium on mathematical statistics and probability*, vol. 1 (Oakland, CA, USA, 1967), pp. 281–297
4. X. Mei, H. Ling, Robust visual tracking and vehicle classification via sparse representation. IEEE Trans. Pattern Anal. Mach. Intell. **33**(11), 2259–2272 (2011)
5. S. Ren, K. He, R. Girshick, J. Sun, Faster r-cnn: towards real-time object detection with region proposal networks, in *Proceedings of NIPS* (2015), pp. 91–99
6. W. Ren, D. Kang, Y. Tang, A. Chan, Fusing crowd density maps and visual object trackers for people tracking in crowd scenes, in *Proceedings of the IEEE Conference on Computer Vision and Pattern Recognition* (2018)
7. J. Tian, L. Zhu, Y. Tang, Outdoor shadow detection by combining tricolor attenuation and intensity. EURASIP J. Adv. Signal Process. **2012**(1), 1–8 (2012)
8. W. Wenfeng, D. Weili, L. Yong et al., An efficient road detection algorithm based on parallel edges. Acta Opt. Sinica **35**(7), 0715001 (2015)